I0789937

Advanced Semiconductor-on-Insulator Technology and Related Physics 15

Editors:

Y. Omura
Kansai University
Osaka, Japan

H. Ishii
Toyohashi University of Technology
Aichi, Japan

B.-Y. Nguyen
SOITEC
Peabody, Massachusetts, USA

S. Selberherr
Technische Universität Wien
Wien, Austria

F. Gámiz
University of Granada
Granada, Spain

J. A. Martino
University of São Paulo
São Paulo, Brazil

J.-P. Raskin
Université Catholique de Louvain
Louvain-la-Neuve, Belgium

Sponsoring Division:

 Electronics and Photonics

Published by
The Electrochemical Society
65 South Main Street, Building D
Pennington, NJ 08534-2839, USA
tel   609 737 1902
fax 609 737 2743
www.electrochem.org

ecstransactions ™

Vol. 35, No. 5

Copyright 2011 by The Electrochemical Society.
All rights reserved.

This book has been registered with Copyright Clearance Center.
For further information, please contact the Copyright Clearance Center,
Salem, Massachusetts.

Published by:

The Electrochemical Society
65 South Main Street
Pennington, New Jersey 08534-2839, USA

Telephone 609.737.1902
Fax 609.737.2743
e-mail: ecs@electrochem.org
Web: www.electrochem.org

ISSN 1938-6737 (online)
ISSN 1938-5862 (print)
ISSN 2151-2051 (cd-rom)

ISBN 978-1-56677-866-4 (Hardcover)
ISBN 978-1-60768-216-5 (PDF)

Printed in the United States of America.

Preface

The Fifteenth International Symposium on Advanced Semiconductor-On-Insulator Technology and Related Physics was part of the 219th Meeting of the Electrochemical Society, held on May 1 - May 6, 2011, in Montreal, Canada. The Electronics and Photonics Division of the Electrochemical Society sponsored the symposium.

In the 21st century, SOI-based integrated circuits are now well commercialized and many new device technologies based on advanced materials including SiGe, SiC and diamond have been proposed. SOI activists had been hoping for the emergence of many new applications.

The symposium do not disappoint with its 13 impressive invited reviews, including 2 plenary talks, and 33 contributed papers, including 6 poster papers. The symposium has preserved its international nature, with 5 papers from North America, 23 from Europe, 10 from Asia, and 8 from South America. Many of the papers include collaborations between researchers from institutions on different continents.

This issue of *ECS Transactions* contains 44 papers presented at the conference. I would like to thank all the authors for submitting excellent manuscripts and submitting them on time despite the tight schedule. Most of the papers in these proceedings follow the same sequence as the symposium presentations in order to make it convenient to follow the text at the conference site.

It is my great pleasure to acknowledge the symposium co-organizers – Francisco Gamiz, Bich-Yen Nguyen, Joao Antonio Martino, Siegfried Selberherr, Jean-Pierre Raskin, and Hiromu Ishii – for their active involvement and efforts in spite of their heavy workload. Many thanks to the ECS staffs: Paul Urso, John Lewis, and Timothy Fest for their support and advice. On behalf of the organizers, I like to express our great appreciation of the invited speakers who set the tone for the meeting and I thank all symposium speakers and co-authors for coming to Montreal to share their recent exciting data and their technical expertise. Their contributions made the conference a success. The organizers thank the Electronics and Photonics Division for their technical support.

Yasuhisa Omura
Osaka, Japan
February 2011

ECS Transactions, Volume 35, Issue 5
Advanced Semiconductor-on-Insulator Technology and Related Physics 15

Table of Contents

Preface   *iii*

Chapter 1
Opening and Plenary-1

Silicon Spintronics: Challenges and Perspectives   3
  J. Fabian

Chapter 2
Plenary-2

Trends and Challenges in Si and Hetero-Junction Tunnel Field Effect   15
Transistors
  C. Claeys, D. Leonelli, R. Rooyackers, A. Vandooren, A. Verhulst,
  M. Heyns, G. Groeseneken, and S. De Gendt

Chapter 3
Materials-1

SiGe and Ge on Insulator Wafers   29
  N. Daval, C. Figuet, C. Aulnette, D. Landru, C. Drazek, K. K. Bourdelle,
  E. Guiot, F. Letertre, B. Nguyen, and C. Mazure

Au-Catalyst Induced Low Temperature (~250 °C) Layer Exchange   39
Crystallization for SiGe On Insulator
  J. Park, M. Kurosawa, N. Kawabata, M. Miyao, and T. Sadoh

Strain Nano-Engineering: SSOI as a Playground   43
  O. Moutanabbir, A. Hähnel, M. Reiche, W. Erfurth, A. Tarun,
  N. Hayazawa, S. Kawata, F. Naumann, and M. Petzold

Lateral-Liquid Phase Epitaxy of (101) Ge-On-Insulator from Si Template   51
by Metal-Induced Crystallization
  M. Kurosawa, N. Kawabata, R. Kato, T. Sadoh, and M. Miyao

v

Growth-Direction Dependent Rapid-Melting-Growth of Ge-On-Insulator (GOI) and its Application to Ge Mesh-Growth
H. Yokoyama, Y. Ohta, K. Toko, T. Sadoh, and M. Miyao
                                                                                    55

Chapter 4
Electron Device Physics-1

A Simulation Comparison between Junctionless and Inversion-Mode MuGFETs
J. Colinge, A. Kranti, R. Yan, I. Ferain, N. Dehdashti Akhavan, P. Razavi, C. Lee, R. Yu, and C. Colinge
                                                                                    63

Comparative Study of Random Telegraph Noise in Junctionless and Inversion-Mode MuGFETs
A. Nazarov, C. Lee, A. Kranti, I. Ferain, R. Yan, N. Dehdashti Akhavan, P. Razavi, R. Yu, and J. Colinge
                                                                                    73

Hysteresis Effects in FinFETs with ONO Buried Insulator
S. Chang, M. Bawedin, W. Xiong, J. Lee, and S. Cristoloveanu
                                                                                    79

Scaling Scheme and Performance Perspective of Cross-Current Tetrode (XCT) SOI MOSFET for Future Ultra-Low Power Applications
Y. Omura, K. Fukuchi, D. Ino, and O. Hayashi
                                                                                    85

Chapter 5
Characterization-1

Novel SOI Structures and Characterization Strategy
S. Cristoloveanu
                                                                                    93

Evaluation of Interface Trap Density in Advanced SOI MOSFETs
M. Bawedin, S. Cristoloveanu, S. Chang, M. Valenza, F. Martinez, and J. Lee
                                                                                    103

Chapter 6
Poster Session

Humidity Effects on Substrate Bonding for Silicon-On-Glass
A. Usenko
                                                                                    111

Subband Structure Engineering in Silicon-On-Insulator FinFETs Using Confinement
Z. Stanojevic, V. Sverdlov, and S. Selberherr
                                                                                    117

Single Crystal Silicon Thin Film on Polymer Substrate by Double Layer Transfer Method 123
J. Senawiratne and A. Usenko

Zero Temperature Coefficient of Current Gain Cutoff Frequency and Maximum Oscillation Frequency for Various SOI and Si Bulk MOSFETs 129
M. Emam, D. Vahoenacker-Janvier, and J. Raskin

Research of SOI Microelectromechanical Sensors with a Monolithic Tensoframe for High-Temperature Pressure Transducers 135
L. V. Sokolov

Chapter 7
Electron Device Physics-2

Global and/or Local Strain Influence on p- and n MuGFET Analog Performance 145
P. G. Agopian, J. A. Martino, E. Simoen, and C. Claeys

Fin Pitch Impact on Biaxial/Uniaxial Strain Engineering of Triple-Gate Devices 151
M. Rodrigues, V. Sonnenberg, J. A. Martino, N. Collaert, E. Simoen, and C. Claeys

Transport Properties of 3D Vertically Stacked SiGe and SiGeC Nanowires 157
A. El Hajj Diab, E. Saracco, I. Ionica, C. Bonafos, J. Damlencourt, and S. Cristoloveanu

FISH SOI MOSFET: An Evolution of the Diamond SOI Transistor for Digital ICs Applications 163
S. P. Gimenez and D. M. Alati

Nonlinear Properties of Si Based Substrates for Wireless Systems and SoC Integration 169
C. Roda Neve and J. Raskin

Chapter 8
Devices and Circuits

FDSOI Process Technology for Subthreshold-Operation Ultra-Low-Power Electronics 179
S. A. Vitale, P. W. Wyatt, N. Checka, J. Kedzierski, and C. L. Keast

vii

An Analytical Model for the Non-Linearity of Triple Gate SOI MOSFETs 189
R. T. Doria, J. A. Martino, E. Simoen, C. Claeys, and M. A. Pavanello

New Capacitorless Dynamic Memory Compatible with SOI and Bulk CMOS 195
N. Rodriguez, F. Gamiz, and S. Cristoloveanu

Radiation-Induced Pulse Noise in SOI CMOS Logic 201
D. Kobayashi, K. Hirose, H. Ikeda, and H. Saito

Chapter 9
MEMS and Photonics-1

A Tunable Color Filter Using Sub-Micron Grating Integrated 213
with Electrostatic Actuator Mechanism
H. Miyao, K. Takahashi, M. Ishida, and K. Sawada

Chapter 10
MEMS and Photonics-2

Nanomechanical Testing of Free-Standing Monocrystalline Silicon Beams 221
U. K. Bhaskar, S. Houri, V. Passi, T. Pardoen, and J. Raskin

Silicon Photonics Devices Based on SOI Structures 227
S. Itabashi, K. Yamada, H. Fukuda, T. Tsuchizawa, T. Watanabe,
H. Shinojima, H. Nishi, R. Takahashi, Y. Ishikawa, and K. Wada

Chapter 11
Materials and Characterization-2

Ultra-Thin Film SOI/BOX Substrate Development, Its Application 239
and Readiness
W. Schwarzenbach, X. Cauchy, O. Bonnin, N. Daval, C. Aulnette,
C. Girard, B. Nguyen, and C. Maleville

Performance of SOI MOSFETs with Ultra-Thin Body and Buried-Oxide 247
A. Ohata, Y. Bae, S. Cristoloveanu, C. Fenouillet-Beranger, P. Perreau,
and O. Faynot

TiN/HfSiON for Analog Applications of nMuGFETs 253
M. Rodrigues, M. Galeti, J. A. Martino, N. Collaert, E. Simoen,
and C. Claeys

X-ray Radiation Effects in Circular-Gate Transistors   259
   K. H. Cirne, M. Silveira, J. A. De Lima, L. E. Seixas Jr., and S. P. Gimenez

Chapter 12
Nanoscale Simulations

Thermoelectric Properties of Silicon-On-Insulator Nanostructures   267
   Z. Aksamija and I. Knezevic

Properties of Silicon Ballistic Spin Fin-Based Field-Effect Transistors   277
   D. Osintsev, V. Sverdlov, Z. Stanojevic, A. Makarov, J. Weinbub,
   and S. Selberherr

The Roles of the Electric Field and the Density of Carriers in the Improved   283
Output Conductance of Junctionless Nanowire Transistors
   R. T. Doria, M. A. Pavanello, R. D. Trevisoli, M. de Souza, C. Lee,
   I. Ferain, N. Dehdashti Akhavan, R. Yan, P. Razavi, R. Yu, A. Kranti,
   and J. Colinge

Stress Relaxation Empirical Model for Biaxially Strained Triple-Gate Devices   289
   R. D. Trevisoli, J. A. Martino, E. Simoen, C. Claeys, and M. A. Pavanello

Transport-Confined Multi-Barrier FETs: A New Paradigm for Low-Leakage   295
High On-Current Transistors
   A. Afzalian and D. Flandre

Chapter 13
Device Physics and Technology

Numerical Modeling of Noise and Transport in SOI Devices   303
   B. Meinerzhagen, A. Pham, S. Hong, and C. Jungemann

Functionalization of Silicon Nanowires for Specific Sensing   313
   V. Passi, E. Dubois, C. Celle, S. Clavaguera, J. Simonato, and J. Raskin

Ultra Low Power 3-D Flow Meter in Monolithic SOI Technology   319
   N. Andre, B. Rue, G. Scheen, L. A. Francis, D. Flandre, and J. Raskin

Performance of Ultra-Low-Power SOI CMOS Diodes Operating at Low   325
Temperatures
   M. de Souza, B. Rue, D. Flandre, and M. A. Pavanello

Author Index   331

ix

Facts about ECS

The Electrochemical Society (ECS) is an international, nonprofit, scientific, educational organization founded for the advancement of the theory and practice of electrochemistry, electrothermics, electronics, and allied subjects. The Society was founded in Philadelphia in 1902 and incorporated in 1930. There are currently over 7,000 scientists and engineers from more than 70 countries who hold individual membership; the Society is also supported by more than 100 corporations through Corporate Memberships.

The technical activities of the Society are carried on by Divisions. Sections of the Society have been organized in a number of cities and regions. Major international meetings of the Society are held in the spring and fall of each year. At these meetings, the Divisions and Groups hold general sessions and sponsor symposia on specialized subjects.

The Society has an active publications program that includes the following.

Journal of The Electrochemical Society — JES is the peer-reviewed leader in the field of electrochemical and solid-state science and technology. Articles are posted online as soon as they become available for publication. This archival journal is also available in a paper edition, published monthly following electronic publication.

Electrochemical and Solid-State Letters — ESL is the first and only rapid-publication electronic journal covering the same technical areas as JES. Articles are posted online as soon as they become available for publication. This peer-reviewed, archival journal is also available in a paper edition, published monthly following electronic publication. It is a joint publication of ECS and the IEEE Electron Devices Society.

Interface — *Interface* is ECS's quarterly news magazine. It provides a forum for the lively exchange of ideas and news among members of ECS and the international scientific community at large. Published online (with free access to all) and in paper, issues highlight special features on the state of electrochemical and solid-state science and technology. The paper edition is automatically sent to all ECS members.

Meeting Abstracts (formerly Extended Abstracts) — Abstracts of the technical papers presented at the spring and fall meetings of the Society are published on CD-ROM.

ECS Transactions — This online database provides access to full-text articles presented at ECS and ECS-sponsored meetings. Content is available through individual articles, or as collections of articles representing entire symposia.

Monograph Volumes — The Society sponsors the publication of hardbound monograph volumes, which provide authoritative accounts of specific topics in electrochemistry, solid-state science, and related disciplines.

For more information on these and other Society activities, visit the ECS website:

www.electrochem.org

CHAPTER 1

OPENING AND PLENARY-1

2

ECS Transactions, 35 (5) 3-12 (2011)
10.1149/1.3570771 ©The Electrochemical Society

Silicon Spintronics: Challenges and Perspectives

Jaroslav Fabian

Institute for Theoretical Physics, University of Regensburg, 93040 Regensburg, Germany

> Spintronics has significantly advanced our understanding of the electronic spin phenomena in metals and semiconductors. It has also given practical magnetoelectronic devices, as symbolized by the giant magnetoresistance (GMR) and tunneling magnetoresistance (TMR) read heads working in the hard drives. These practical devices comprise ferromagnetic metal layers separated by other metals or insulators. The next advance should involve semiconductors. Special attention is paid to silicon, due to its dominance in the information technology and favorable spin properties.

Spin transistors

Electron spins in semiconductors behave pretty much like in metals. One advantage of semiconductors is their ability to support space charges, which makes them useful as devices whose operation can be controlled by electric fields. Spins in semiconductors can thus also be influenced by the fields, which should make for interesting spintronic devices (1, 2).

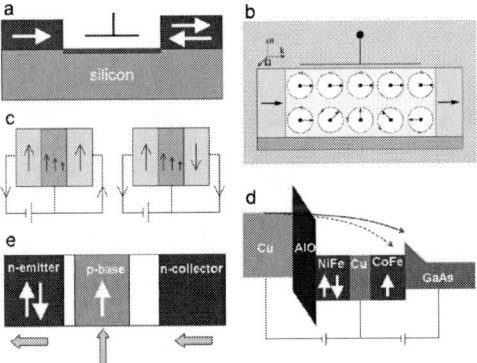

Figure 1. Spin transistors. (a) Silicon spin field-effect transistor, (b) Datta-Das transistor, (c) Johnson all-metal spin transistor, (d) hot electron spin transistor, and (e) magnetic bipolar junction transistor. From Ref. (2).

A silicon spin transistor is a holy grail of the field. It is yet to be demonstrated. There are several schemes proposed for spin transistors, shown in Fig. 1. The most straightforward is a silicon field-effect transistor, see Fig. 1.a., analyzed by Sugahara and Tanaka (3,4). Here the source and drain are ferromagnetic, while the channel is in silicon.

3

The functionality of the transistor is affected by both the gate and the orientation of the magnetizations. This extended functionality should be useful for multi-bit and reconfigurable logic. Spintronic transistors are not intended to compete with conventional transistors for charge-based operations. They should bring more functionality and design flexibility, and perhaps less power dissipation when manipulating the spin-based information.

The most notorious spin transistor is that of Datta-Das (5), in Fig. 1.b. This transistor is based on the channel field-effect transistor, but the gate is supposed to change the spin rather than charge state of the channel. In particular, the electrons in the semiconductor channel feel the spin-orbit interaction due to the semiconductor lattice as well as the confining structure and the external gate field. The gate field can modulate the strength of the spin-orbit interaction to the effect of rotating the electron spins in the channel. In Fig. 1.b. the electrons are injected from a ferromagnetic source with a spin pointing to the right. The upper path shows the electrons with their spins preserved, illustrating a turned off spin-orbit interaction. These electrons readily arrive in the drain with parallel magnetization (spin orientation). If the gate voltage changes the strength of the spin-orbit interaction so that the electron spin manages to rotate by 180 degrees as it arrives at the drain, these electrons are reflected off the drain and stay in the channel (until they flip their spin), increasing the resistances. This spin-based ON and OFF switching is the basis of the spin transistor action.

There are also metal-based spin transistors proposed. One example is the Johnson spin transistor (6), shown in Fig. 1.c., which is a three terminal (this is why the name transistor) spin valve, with ferromagnetic emitter and collector, and nonmagnetic base. While the collector current can change its direction by changing the relative magnetizations of the ferromagnetic layers from parallel to antiparallel, such a metal-based transistor cannot amplify currents. Similarly, the hot electron spin transistor, Fig. 1.d., is based largely on a metallic base (7). The emitter is also metallic, while the collector can be a semiconductor, such as GaAs or Si. The action of this transistor happens in the base, in which a metallic spin valve is placed. Depending on the relative magnetizations of the ferromagnets of the valve, the collector current is either large or small, demonstrating also the spin injection into the collector, as the transport through the base is spin selective. Hot electrons are flown from the emitter over a tunnel barrier. These electrons lose their energy rather fast, so only those continue to the collector over a Schottky barrier whose energy stays above that barrier. The energy loss of electrons in ferromagnet depends on the electrons' spin; this is why the electrons are spin polarized as they enter the collector.

Finally, the magnetic bipolar (junction) transistor is the spin-polarized version of the pnp or npn bipolar junction transistor. The scheme is shown in Fig. 1.e. It is an all-semiconductor transistor, with depletion layers between the active regions. One or more regions are assumed ferromagnetic, based on ferromagnetic semiconductors such as GaMnAs or InMnAs. In addition, the active regions can support nonequilibrium spin generated by electric or optical means, for example by optical orientation (1,2). The intended functionality of the magnetic bipolar transistor is a magnetic control of the amplification—magnetoamplification (8). Such an action has recently been experimentally demonstrated in a InMnAs-based magnetic bipolar transistor (9).

Spintronics requirements: spin injection, spin detection, and spin manipulation

A spintronic device requires spin injection, spin detection, and spin manipulation (1,2), illustrated in Fig. 2. The spin needs to survive for a sufficiently long time before it relaxes. It turns out that the lifetime of spin in silicon is by far greater than in GaAs, making silicon ideal for spintronic applications (2). Spin has already been electrically injected into silicon, and there is a possibility of creating it optically, with circularly polarized light as well. Other important advances are being made with silicon nanostructures, especially quantum dots, for spin qubit quantum information processing. This technology has its own specifics and its successful realization puts certain physical requirements on the electronic structure and spin coherence in these nanostructures.

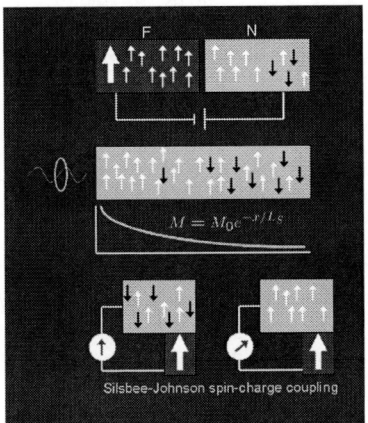

Figure 2. Spintronics requirements. The top figure shows electrical spin injection. Driving current from a normal (N) metal to a metallic ferromagnet (F) leads to spin accumulation in the normal metal as spin-polarized electrons from the ferromagnet flow to the normal metal. The middle figure sketches spin relaxation. Once the spin is injected (the figure illustrates an optical spin injection from a circularly polarized light), they relax due to the combined action of the spin-orbit interaction and momentum scattering. From the source of injection the spin decays exponentially into the normal metal. The decay is given by the spin relaxation or spin diffusion length. The bottom figure shows spin detection using the classic Silsbee-Johnson scheme. If a ferromagnetic electrode is placed on top of a normal metal in which there is a nonequilibrium spin, a voltage between the ferromagnetic electrode and the normal metal develops, in proportion to the amount of the spin accumulation. This coupling between the nonequilibrium spin and voltage goes under the name of spin-charge coupling. From Ref. (2).

Below I review selected properties of silicon relevant for spintronic applications, based on our research.

Why silicon spintronics?

The potential of silicon for spintronics is typically underscored by the facts that silicon has a long spin relaxation time (compared to the heavier zinc-blende semiconductors) of up to microseconds, its main isotopes carry no nuclear so the hyperfine interaction is inefficient, and the possibility of further isotope purification can reduce the spin dephasing hyperfine interaction even more marginal. On top of that comes the obvious integration with the existing charge-based information technology. Indeed, the holy grail of spintronics, the spin transistor which could offer new functionalities to the existing technologies, is still to be demonstrated. Even the simplest and perhaps the most useful form of it, the silicon MOSFET with ferromagnetic source and drain has not yet been reported in a useful form. There appears no obvious (fundamental) obstacle to building such a transistor, so perhaps it is only a matter of time that a silicon spin transistor would be reported. This absence of fundamental obstacles is the reason why, despite waiting to be demonstrated, the silicon spin field-effect transistor is on the International Roadmap for Semiconductors as an emerging logic device.

Electron spin relaxation in bulk silicon

The spin relaxation of mobile carriers in metals and semiconductors is due to the interplay of the spin-orbit interaction and momentum scattering. There are three main mechanisms of spin relaxation of conduction electrons (and holes) in semiconductors (1, 2). The *Elliott-Yafet* (10, 11) mechanism is essentially spin-flip scattering off impurities and phonons. Electrons in a lattice experience spin-orbit interaction making the spin not a good quantum number. Since spin-orbit coupling is weak compared to the typical Fermi level, electron states can still be labeled spin up and down according to the magnetization orientation. However, these states always contain a small admixture of the other spin, so a momentum scattering due to impurities or phonons has a chance to flip the spin. This mechanism is especially important for elemental metals and semiconductors such as silicon.

The *Dyakonov-Perel* (12) mechanism acts in semiconductors without space inversion symmetry such as GaAs. In fact, it is the most important mechanism in the widely studied GaAs. The mechanism is based on motional narrowing. In the absence of space inversion symmetry the spin-orbit interaction removes the spin degeneracy at a given electron momentum. This looks like as if the electron experienced an effective magnetic field dependent on the momentum. The electron spin precesses in this field. The precession axis and frequency depend on the momentum. As the electron changes the momentum upon scattering, the axis and the frequency of the precession changes. These are random changes. From the perspective of the moving electrons the effective magnetic field changes randomly. This model of a spin in randomly fluctuating magnetic field shows spin relaxation and decoherence (2), somewhat as a spin experiencing a random walk.

The third important mechanism is that of *Bir, Aronov, and Pikus* (13). This mechanism is efficient in heavily p-doped semiconductors. Holes lose their spin rather quickly, due to their strong spin-orbit coupled states. The hole spins relax as fast as their momenta, via the Elliot-Yafet mechanism. As the electron spins can be transferred to holes via exchange coupling, the electron spin is dumped by the holes into the lattice which absorbs the spin angular momentum.

In semiconductor nanostructures such as semiconductor quantum dots electron spins can also relax and decohere due to the hyperfine interaction with the nuclei of the lattice. The hyperfine interaction is also important for relaxing the spins of electrons on the donor states. In silicon this interaction is rather inhibited due to the absent nuclear spin of the main isotope.

Since silicon has space inversion symmetry, the only relevant mechanism for the spin relaxation is the Elliott-Yafet one. The most information we have about the electron spins in silicon comes from the electron spin resonance (2, 14) and spin injection (15, 16, 17, 18, 19) experiments. While the former is sensitive to both itinerant and bound (on donor sites) spins, the spin injection probes the conduction electrons only. This is why the electrical spin injection data at temperatures below about 150 K, at which the number of electrons localized on the donor sites is significant, are very important.

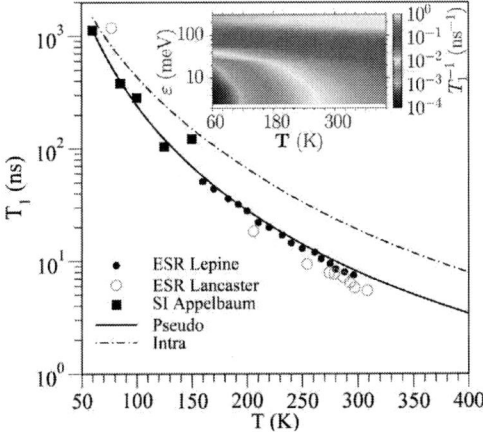

Figure 3. Spin relaxation time in silicon. The symbols are the experimental data from electron spin resonance (ESR) and spin injection (SI). The solid line is the calculation using a pseudopotential model for the electronic structure and electron-phonon coupling, and an adiabatic bond-charge model for phonons. The dashed line is an approximation, limiting the phonon-induced transitions to a single valley. The inset shows the spin relaxation rate as a function of temperature and Fermi energy. From (20).

Recent theoretical investigations has confirmed the Elliott-Yafet as the main mechanism of the spin relaxation of conduction electrons in silicon (20). The calculation is shown in Fig. 3. The agreement with experiment is very good. The momentum relaxation is due to phonons. At low temperatures the spin relaxation time is microseconds; at room temperature it is about 10 nanosecond, still very long.

Electron spin relaxation in silicon quantum dots

Forming quantum dots defined electrostatically on a 2d electron gas in silicon heterostructures, one can realize the Loss and DiVincenzo (21) proposal for spin-based quantum computing with exchange fields as the interactions for the qubit gating; see Fig. 4. The Loss-DiVincenzo proposal can be realized with silicon quantum dots, for example as a spin-resonance transistor based on electrons bound on donor states, that utilizes differences in the g factor of Si and Ge to address individual spins that move in the structure with a Ge concentration gradient. A simulation as well as careful examination of the prerequisites for spin-based quantum computing with silicon quantum dots has been reported in (22).

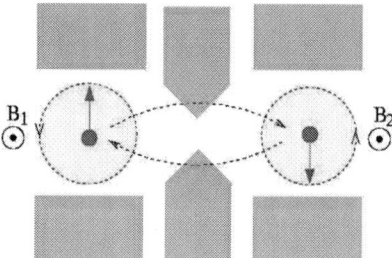

Figure 4. Loss-DiVincenzo model of spin-qubit quantum computing in coupled quantum dots. The dots are defined in a two-dimensional electron gas formed at a semiconductor interface by top gates. A negative potential on the gates depletes the region underneath. The dots can be occupied by one (extra) electron and the coupling between the dots is controlled by top gates between the dots. The tunneling between the dots leads to exchange coupling. This electrically controllable exchange coupling can induce quantum gates (unitary operations acting on the electron spins). The prerequisite of such an operation is the lifetime of the electron spins much longer than the interaction time.

The important parameter in characterizing silicon quantum dots is the valley splitting of the underlying electron gas. Silicon has six valleys, whose energies split into two levels by the (001) confinement. The two lower energy valleys further split by asymmetric confinement or strain. Whether or not the quantum dot states form by a single (superposition of valleys) state or by two states depends on this latter splitting.

Literature provides values from micro to milivolts. It has recently been argued that due to the valley and orbital couplings in strained Si quantum wells, valley splitting from the perspective of further quantum dot confinement is not even a valid concept (23). Recent calculations of the valley splittings can be found in (24, 25). It is argued that the valley splitting is vital to achieve efficient quantum computing operations in silicon quantum dots (26, 27).

We have recently performed numerical calculations of the phonon induced spin relaxation times in silicon coupled dots (28). The silicon heterostructures give rise to the spin-orbit fields of the Bychkov-Rashba and Dresselhaus types. The latter are typically not considered for silicon, as the conventional wisdom shows that it comes from the bulk inversion asymmetry, which is absent in the inversion symmetric bulk silicon. However, silicon interfaces (as well as plain silicon non-reconstructed surfaces) have the C_{2v} symmetry that decomposes into the D_{2d} and C_{4v}, historically called Dresselhaus and Bychkov-Rashba, respectively. In magnitude the two spin-orbit fields are about an order of magnitude below those found in GaAs.

In Fig. 5 I give an example of the calculated spin relaxation rate in a silicon double dot of a variable interdot distance (tunneling energy) for an in-plane magnetic field of 4 Tesla (28). The spikes in the relaxation rate appear due to spin-orbit coupling induced anticrossings in the energy spectrum, called spin hot spots (29, 30, 31, 32). The knowledge about the occurrence and the positions of the spin hot spots seems indispensable for the realization of qubits based on lateral quantum dots as the spin lifetime is drastically decreased at these points.

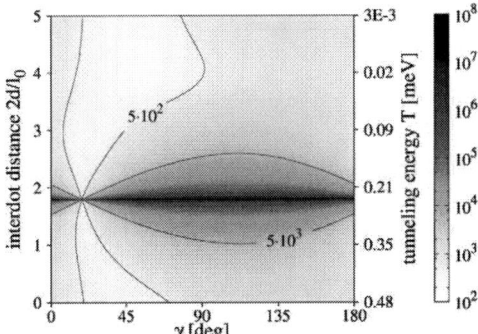

Figure 5. Calculated spin relaxation rate in a laterally coupled silicon quantum dot. The spin relaxation rate is plotted as a function of the magnetic field orientation gamma and the interdot distance/tunneling energy. The color map for the spin relaxation rate is in inverse seconds. Upon increasing the tunneling energy one always (with the exception of one specific orientation of the magnetic field) crosses the region of the spin hot spots in which the spin relaxation is as fast as the orbital relaxation. From (28).

Optical spin orientation

Optical spin orientation is the historically first method for generating nonequilibrium spins in semiconductors with a direct band gap such as GaAs. Circularly polarized light carries angular momentum which is given to the electrons upon excitation from the valence to the conduction band. The transfer of spin is facilitated by the spin-orbit interaction. In zinc-blende semiconductors like GaAs one can achieve 50% of spin polarization in the conduction band upon illumination with 100% circularly polarized light.

Silicon is an indirect gap semiconductor with less efficient optical excitations as phonons need to supply or carry away momentum. In addition, the spin-orbit interaction in silicon is rather weak, making the transfer of spin from photons to electrons more limited. However, there appears to be no fundamental reason why spin cannot be generated in bulk silicon optically. Indeed, recent realistic calculations (33) of indirect optical transitions due to circularly polarized light in silicon show that the optically injected spin polarization can be rather large, reaching about 25% al low temperatures and remaining at about 15% at room temperature. The degree of spin polarization (DSP) is shown in Fig. 6 below for temperatures 4 and 300 K, as a function of the excitation light frequency. The conversion mapping from the spin to optical polarization was given by Li and Dery (34) who developed a theory of optical spin detection in silicon.

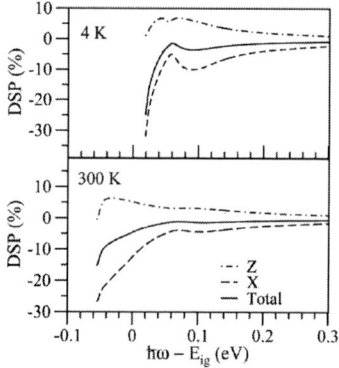

Figure 6. Calculated optical spin orientation efficiency in silicon at 4 and 300 K. Even at room temperature the spin polarization of the optically excited electrons should be 15% at the absorption edge. From (33).

This result is significant as it shows that silicon can be a useful material for studying spin by optical means.

Summary

Silicon is an ideal material for spintronics. It has the longest spin relaxation times (measured to date), of microseconds, due to its weak spin-orbit coupling. Electrical spin injection into silicon has been demonstrated by several groups, and optical spin injection appears a vital possibility. Most exciting is the possibility of integration with conventional electronic technology which is based on silicon. The grand goal is to demonstrate a silicon-based spin transistor.

Acknowledgments

This work is supported by the DFG SFB 689 and SPP 1285.

References

1. I. Žutič , J. Fabian, and S. Das Sarma, *Rev. Mod. Phys.* **76**, 323 (2004).
2. J. Fabian, A. Matos-Abiague, C. Ertler, P. Stano, and I. Žutič, *Acta Phys. Slov.*, **57**, 565 (2007).
3. S. Sugahara and M. Tanaka, *Appl. Phys. Lett.* **84**, 2307 (2004).
4. S. Sugahara and M. Tanaka, *J. Appl. Phys.* **97** 10D503 (2005).
5. S. Datta and B. Das, *Appl. Phys. Lett.* **56**, 665 (1990).
6. M. Johnson, *Science* **260**, 320 (1993).
7. S. van Dyjken, X. Jiang, and S. S. P. Parkin, *Appl. Phys. Lett.* **83**, 951 (2003).
8. J. Fabian, I. Žutič, and S. Das Sarma, *Appl. Phys. Lett.* **84**, 85 (2004).
9. N. Rangaraju, J. A. Peters, and B. W. Wessels, *Phys. Rev. Lett.* **105**, 117202 (2010).
10. R. J. Elliott, *Phys. Rev. B* 96, 266 (1954).
11. Y. Yafet, *Solid State Physics*, Vol. 14, eds. F. Seitz and D. Turnbull, (Academic, New York ,1963), p.2.
12. M. I. Dyakonov and V. I. Perel, *Sov. Phys. Solid State* 13, 3023 (1971).
13. G. L. Bir, A. G. Aronov, and G. E. Pikus, *Sov. Phys. JETP* **42**, 705 (1976).
14. D. J. Lepine, *Phys. Rev. B* **2**, 2429 (1970).
15. I. Žutič, J. Fabian, and S. Erwin, *Phys. Rev. Lett.* **96**, 026602 (2007).
16. I. Appelbaum, B. Huang, and D. J. Monsma, *Nature* **447**, 295 (2007)
17. B. Huang, D. J. Monsma and I. Appelbaum, *Phys. Rev. Lett.* **99**, 177209, (2007).
18. B. T. Jonker, G. Kioseoglou, A. T. Hanbicki, C. H. Li, and P. E. Thompson, *Nature Physics* **3**, 542 (2007).
19. L. Grenet, M. Jamet, P. Noe, V. Calvo, J.-M. Hartmann, L. E. Nistor, B. Rodmacq, S. Auffret, P. Warin, and Y. Samson, *Appl. Phys. Lett.* **94**, 032502 (2009).
20. J. L. Cheng, J. Fabian, and M. W. Wu, *Phys. Rev. Lett.* **104**, 016601 (2010).
21. D. Loss and D. DiVincenzo, *Phys. Rev. A* 57, 120 (1998).
22. M. Friesen, P. Rugheimer, D. E. Savage, M. G. Lagally, D. W. van der Weide, R. Joynt, and M. A. Eriksson, *Phys. Rev. B* **67**, 121301(R) (2003).
23. M. Friesen and S. N. Coppersmith, *Phys. Rev. B* **81**, 115324 (2010).

24. T. Boykin, G. Klimeck, M. A. Eriksson, M. Friesen, S. N. Coppersmith, P. von Allmen, F. Oyafuso, and S. Lee, *Appl. Phys. Lett.* **84**, 115 (2004).
25. V. Sverdlov and S. Selberherr, *Solid-State Electronics* **52**, 1861 (2008).
26. D. Culcer, L. Cywinski, Q. Li, X. Hu, and S. Das Sarma, *Phys. Rev. B* **82**, 155312 (2010).
27. Q. Li, L. Cywinski, D. Culcer, X. Hu, and S. Das Sarma, *Phys. Rev. B* **81**, 085313 (2010).
28. M. Raith, P. Stano, and J. Fabian, arXiv:1101.3858.
29. J. Fabian and S. Das Sarma, *Phys. Rev. Lett.* **81**, 5624 (1998).
30. J. Fabian and S. Das Sarma, *Phys. Rev. Lett.* **83**, 1211 (1999).
31. J. Fabian and S. Das Sarma, *J. Appl. Phys.* **85**, 5075 (1999).
32. J. Fabian and S. Das Sarma, *J. Vac. Sci. Technol. B* **17**, 1708 (1999).
33. J. L. Cheng, J. Rioux, J. Fabian, and J. Sipe, arXiv:1011.2259
34. P. Li, and H. Dery, *Phys. Rev. Lett.* **105**, 037204 (2010).

CHAPTER 2

PLENARY-2

14

Trends and Challenges in Si and Hetero-Junction Tunnel Field Effect Transistors

C. Claeys[1,2], D. Leonelli[1,2], R. Rooyackers[1], A. Vandooren[1], A.S. Verhulst[1],
M.M. Heyns[1,3], G. Groeseneken[1,2] and S. De Gendt[1,4]

[1]imec, Kapeldreef 75, B-3001 Leuven, Belgium
[2]E.E. Dept., KU Leuven, Belgium
[3]MME Dept., KU Leuven, Belgium
[4]Chemistry Dept., KU Leuven, Belgium

This paper gives an overview of the different trends and challenges of fully Si-based and hetero-junction tunnel field effect transistors (TFETs). The different horizontal and vertical approaches are discussed in view of processing aspects and device performance. A benchmarking is given of the state-of-the-art experimental data reported in the literature, enabling to highlight the potentials of these emerging devices.

Introduction

The stringent requirements imposed by the ITRS not only necessitate the implementation of advanced processing modules but also rely on the introduction of alternative and/or new gate concepts. Multi-gate devices such as FinFETs, showing better scaling performance than single-gate devices due to their better short channel behavior [1], have paved the way to the introduction of Si nanowires. To further enhance the electrical performance one can go a step further and switch over to another operating principle of the devices. Quantum flux and spintronics are examples of possible future devices no longer based on the transport of electrical charges.

The device miniaturization for performance improvement has direct consequences for the power consumption which has become a major roadblock. Although the introduction of new materials enabled much reduction in gate leakage current [2], a real breakthrough can only be achieved by a further scaling of the supply voltage. For conventional MOSFETs, based on thermionic emission of carriers over an energy barrier, the subthreshold swing has a fundamental limit of $2.3k_BT/q$. Therefore, alternative device structures based on non-thermal carrier injection mechanisms have been proposed in the literature. Examples are the concept using a three-terminal tunnel device operating with avalanche injection as suggested by Banerjee et al. [3] in 1987 already, and the tunnel field effect transistor using a p-i-n AlGaAs-GaAs structure as proposed by Reddick and Amaratunga [4]. A direct extension of the present CMOS research is making use of tunnelFETs (TFETs), based on Si-based gated p-i-n diodes whereby band-to-band tunneling is used. These devices have a low subthreshold swing (< 60 mV/dec), reduced short-channel effects and enable the fabrication of 3D structures based on vertical nanowires.

The excellent subthreshold behavior of TFETs was first demonstrated by Appenzeller et al. [5] for carbon nanotubes (CNTs) and by Choi et al. [6] for Si devices. TFETs can be realized based on a horizontal or vertical technology using either a planar, double gate or MuGFET approach. Each of these structures has advantages and drawbacks from a viewpoint of process complexity, achievable packing density, and a possible implementation of hetero-structures to tune the bandgap and increase the tunneling efficiency. This overview will discuss different trends and challenges in the field. Technological issues for both vertical and horizontal devices are addressed. Device optimization depends on the used approach and may have to take into account both a large variety of technological parameters such as e.g. implantation profiles, anneal conditions (RTA, spike, laser, SPER etc), gate stack and spacer engineering, and design aspects. Critical issues will be demonstrated by TFET work on going at imec. Attention is also given to aspects related to device characterization. Both fully Si-based and hetero-junction structures are addressed. A benchmarking of some state-of-the-art TFETs reported in the literature is given.

1. Horizontal FinFET-based TFETs

A typical process flow for a horizontal FinFET-based TFET approach is illustrated in Fig. 1, giving the main processing modules and a schematic of the device structure. The TFET devices were fabricated on a (100) SOI substrate with 65 nm thick Si film on top of 145 nm buried oxide. Fin widths down to 10 nm were patterned using 193 nm optical lithography and aggressive resist and hardmask (HM) trimming. The channel of the device is undoped. The gate stack consists of a 100 nm poly-Si layer on top of a 5 nm TiN layer and 2 nm HfO_2 on a 1 nm interfacial SiO_2 layer. After gate patterning, source/drain doping was achieved by an extension implantation of As or BF_2 at 45°. The 30 nm wide nitride spacers are formed on a 5 nm oxide liner. The source/drain regions outside the spacers are highly doped by an extra As/B implantation. After annealing and nickel silicide formation the process is completed with a standard Cu back-end processing. More details are given in [7]

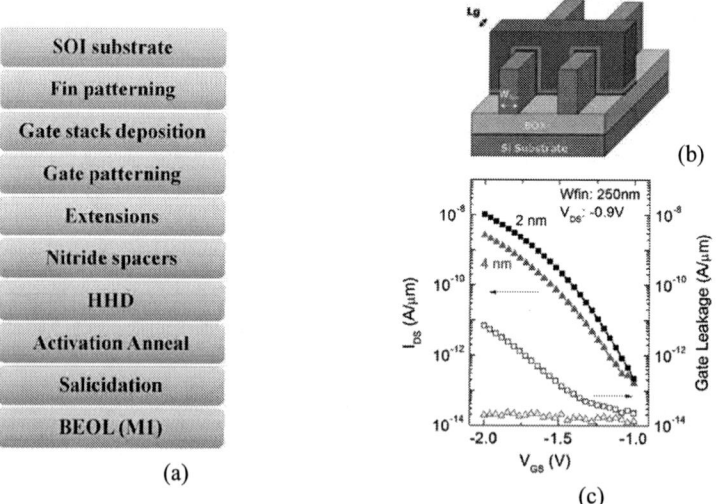

Figure 1. Process flow (a) and schematic cross-section (b) of a horizontal FinFET-based TFET technology. The impact of HfO_2 thickness on I_{on} and gate leakage is shown in (c).

Typical electrical performance data are shown in Fig. 2 [7]. The input characteristics in Fig. 2a indicate an I_{on}/I_{off} ratio of 10^6 at V_{DD}=1.2 V and a minimum point slope of 46 mV/dec (V_{GS}=0.2 V, V_{DS}=-1 V). The I_{on} is close to 10 µA/µm. Some benchmarking with data in the literature will be given in a later section. The output characteristic in Fig. 2b clearly illustrates the potential of these devices in analog applications. The impact of the gate length and fin width is shown in Fig. 2c and Fig. 2d, respectively. Down to 160 nm no gate length dependence is observed, indicating that the band-to-band tunnel transport is not carrier injection limited. For larger lengths the influence of the resistance of the undoped channel becomes dominant. For small fin widths the electrostatic coupling between the gates at the fin sidewalls increases.

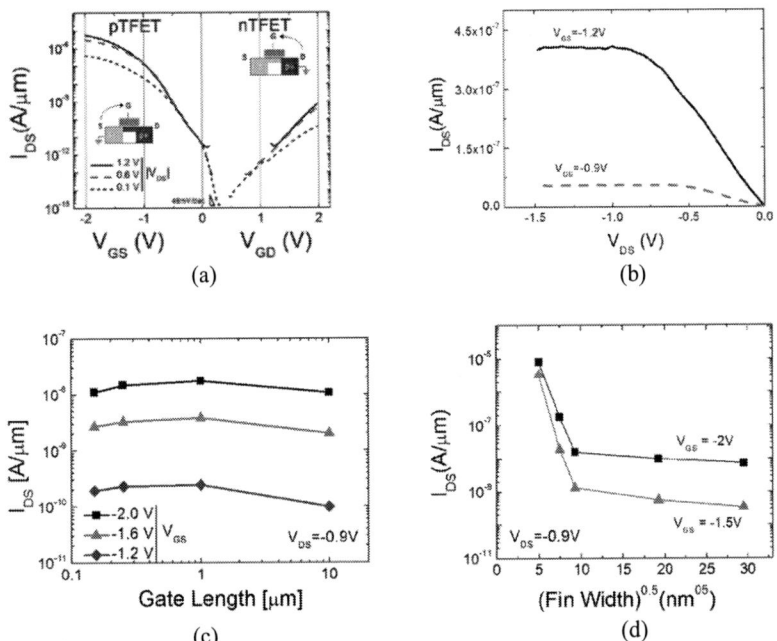

Figure 2. p- and nTFET input (a) and output (b) characteristics for a MuGTFET structure with a 25 nm wide fin (Lg=160 nm), while the gate length (for W_{fin}=85 nm) and fin width dependence are given in (c) and (d), respectively [7].

For a complete modeling of the electrical behavior also the thickness of the silicon film has to be taken into account [5, 8]. A physics-based model using the exact TFET potential profile has been developed, resulting in the following expression for the tunnel current in a double-gate configuration [9]

$$I_{tunnel} = \exp\left[-B_K q\sqrt{E_g}\frac{2t_{gd,eff}+t_{ch}}{\pi}a\sinh[\tan(\frac{E_g}{qV'_{gs}}\frac{\pi}{2})\sin(\frac{t_{gd,eff}}{2t_{gd,eff}+t_{ch}}\pi)]\right] \quad (1)$$

with B_K a material-dependent constant, E_g the bandgap, $t_{gd,eff}$ the effective gate dielectric thickness, V'_{gs} the gate-source voltage minus the flat band voltage and t_{ch} the channel thickness, respectively.

For optimal device performance, i.e., a high drive current combined with a low off current, the gate stack processing, the implant conditions of the extension and the activation anneal of the dopants have to be engineered [7]. Figure 1c clearly illustrates that reducing the thickness of the high-k gate dielectric improves the drive current (I_{on}) but degrades the off current (I_{off}) due to an increased gate tunneling current.

Figure 3 illustrates that using Solid Phase Epitaxial Regrowth (SPER) improves the TFET performance compared to standard thermal or laser annealing [10]. This is related to the impact of the silicide encroachment under the spacer near the gate, impacting the electrical field at the edge affecting the tunnel probability from the silicide [11]. The low temperature SPER step activates the dopants but not fully anneals out the defects enhancing the Ni piping during the salicidation process. The silicide piping only occurs at the n^+ source side due to the interaction of the Ni with the As implantation defects (Fig. 3b). Whereas for standard MOSFETs the high electric field associated with the encroachment would increase the leakage current due to GIDL, in the case of TFETs this has a beneficial effect on the tunneling current. Recently, nickel silicide was used to create a special field-enhancing geometry so that in combination with a high-dopant concentration by segregation TFETs with SS=46 mV/dec and an I_{on}/I_{off} ratio of 10^7 were achieved [12]. This clearly demonstrates that silicide engineering is very promising to boost the TFET device performance.

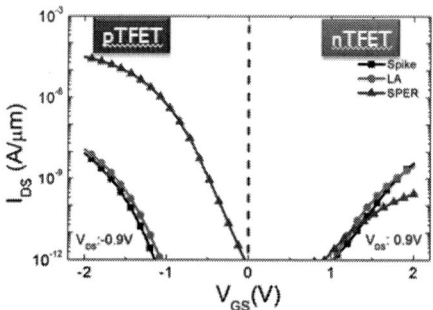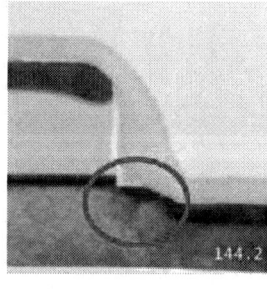

Figure 3. Improvement of the pTFET drive current by using SPER (left) and TEM image of the silicide layer at the n^+ side after SPER (right).

A further improvement can be achieved by using for the source region an etch-back of the Si and subsequently a SiGe epitaxial regrowth in order to lower not only the bandgap but also the series resistance. The basic principle of the approach is illustrated in Fig. 4a, showing a schematic drawing and a TEM image. For standard pMuGFETs the increased mobility leads to an about 25% higher on-state current for the same leakage current, as shown in Fig. 4b [13]. In case of TFETs the improved performance is related to the lower bandgap. Published data for 7 and 25% Ge, pointed out that at constant SS a 15x increase in I_{on} is found for the 25% compared to the 7% Ge case [14].

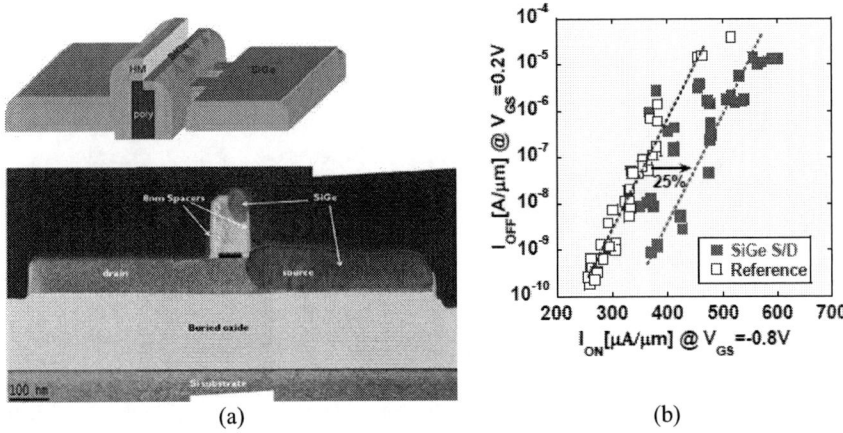

(a)   (b)

Figure 4. Illustration of the source etch-back and SiGe regrowth approach to improve the TFET device performance (a). For standard pMugFETS a 25% higher on-current for the same off-current has been reported [13] (b).

Vertical Si-based TFETs

Instead of a horizontal approach, there is also strong interest in TFETs based on silicon nanowires as the better electrostatic control of the gate over the channel has a beneficial impact on the tunneling efficiency [9, 15]. Besides the already mentioned 3D stacking potential and higher integration density, the vertical TFET also facilitates the extension towards hetero-junction TFETs enabling the use of materials with a lower bandgap. The latter is beneficial for increasing the I_{on}/I_{off} ratio by combining a low off-state current with a higher drive current due to the enhanced tunneling efficiency so that on-state tunneling currents are becoming close to the values reached in CMOS technologies [16]. A schematic process flow for a nanowire based process is given in Fig. 5, starting with a high-doped n$^+$ substrate on which the intrinsic region is epitaxially grown. A top-down approach with an oxide hard mask is used to pattern the nanowires. To isolate the bottom junction from the gate   a high density plasma oxide is deposited followed by chemical-mechanical polishing (CMP) and an oxide etch-back. The gate stack consists of 2nm HfO$_2$, 10 nm TiN and 30 nm amorphous Si. The hardmask is formed by a second HDP oxide, CMP and oxide etch-back. After B implantation for the top junction, nitride spacers are used to isolate the gate from the source contact layer. A doped amorphous Si capping layer allows to contact the NWs and to connect the NWs together. The process is completed with a nickel silicide. More details of the process flow can be found in [17]. A TEM cross-section of a 35 nm NW TFET is shown in Fig. 6, together with a higher magnification of the gate stack area.

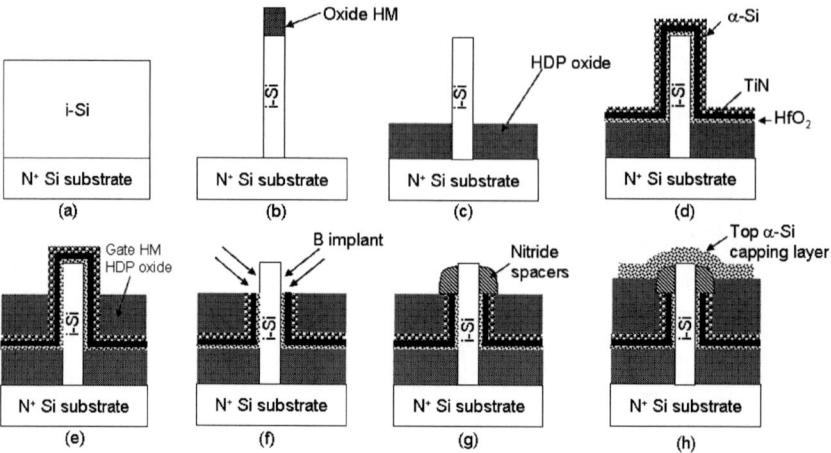

Figure 5: Process flow of a vertical nanowire silicon TFET technology: (a) blanket epitaxy on a highly-doped substrate, (b) NW patterning using e-beam lithography, (c) bottom isolation, (d) metal gate/high-k stack deposition, (e) gate HM formation, (f) gate etch and top junction implantation, (g) top nitride spacer formation, (h) capping layer formation for NW array [17].

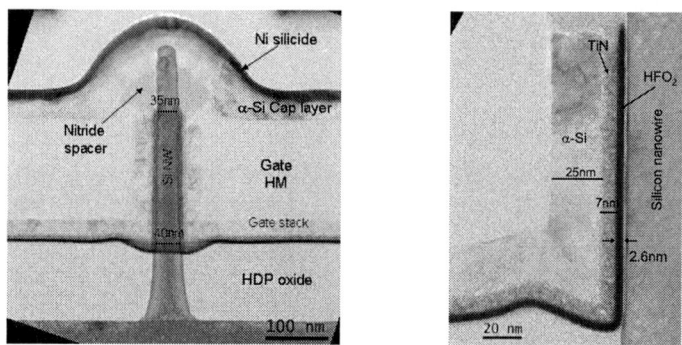

Figure 6: TEM cross-section of a 35 nm NW TFET (left) and details of the gate stack (right).

For vertical TFETs the amount of gate overlap on the drain side is very important to control the ambipolar effect in the p-i-n structure. To avoid that either the n- or the p-TFET is on depending on the gate voltage, one can work with a so-called short-gate approach, whereby not only the ambipolar effect is suppressed but also an increased switching speed and a lower processing complexity are obtained [18].

The abruptness of the source junction is determined by the doping technique used, i.e., ion implanted or an *in-situ* doped epitaxial layer. However, the analysis of the electrical performance of different process splits pointed out that the junction abruptness is not the dominant factor but that the active doping level under the gate and the amount of gate overlap are the most important parameters to control [19].

Simulations have pointed that scaling the gate dielectric has a beneficial impact on the drive current, and is more important than scaling the film thickness [9, 20]. The devices configuration, i.e., single, double or gate all around (GAA) has also an impact on the device performance. The double-gate and GAA structures improve the TFET performance, especially in case of a small channel thickness (< 10 nm) [9].

Hetero-structure TFETs

The band-to-band tunneling depends on the bandgap E_g as indicated in the following equation

$$I_D = AE_g \exp\left[-\frac{\pi m^{*1/2} E_g^{3/2}}{2q\hbar E_s}\right] = AE_s(-B/E_s) \qquad (2)$$

where E_s is the electrical field at the gate-source overlap region. Therefore, to boost up the on-state current for the n-channel TFETs it is possible to use a hetero-structure with a lower bandgap material, such as e.g. Ge (direct bandgap 0.66 eV), for the source in combination with a Si channel and a Si compatible gate stack [16, 21]. To realize CMOS circuits also a hetero-structure has to be used for the n-TFETs, e.g. based on an $In_{0.6}Ga_{0.4}As$ (indirect bandgap 068 eV) source [22]. A schematic CMOS approach together with a bandgap diagram is illustrated in Fig. 7a and 7b. Such a structure can be realized with a vertical NW based configuration using epitaxial deposition techniques. The selection of the source material has not only to take into account the bandgap energy and the tunnel probability, but also the lattice mismatch with Si in order to minimize defect generation such as misfit and threading dislocations. For Ge the mismatch with Si is about 4%, while for e.g. $In_{0.6}Ga_{0.4}As$ the mismatch increases to 8.5%. The strain in the layer has a beneficial impact on the bandgap lowering so that strain engineering is required to achieve a compromise between bandgap energy, tunnel probability, drive current and defect generation.

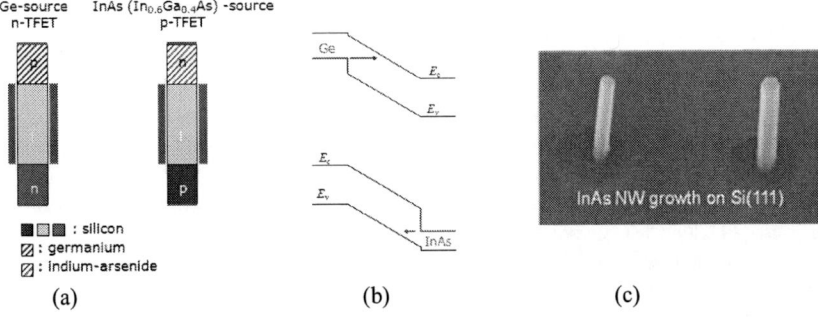

Figure 7. Schematic of a vertical TFET approach for CMOS circuits, based on Ge for nFETs and InGaAs for pFETs source (a), the associated bandgap scheme (b), and illustration of the growth of InAs NWs on Si (c).

Benchmarking

To obtain a better feeling concerning the potential of the different horizontal and vertical TFET approaches, this section discusses some interesting TFET experimental data that has been reported in the literature. Figure 8 illustrates some of the device structures that are further discussed.

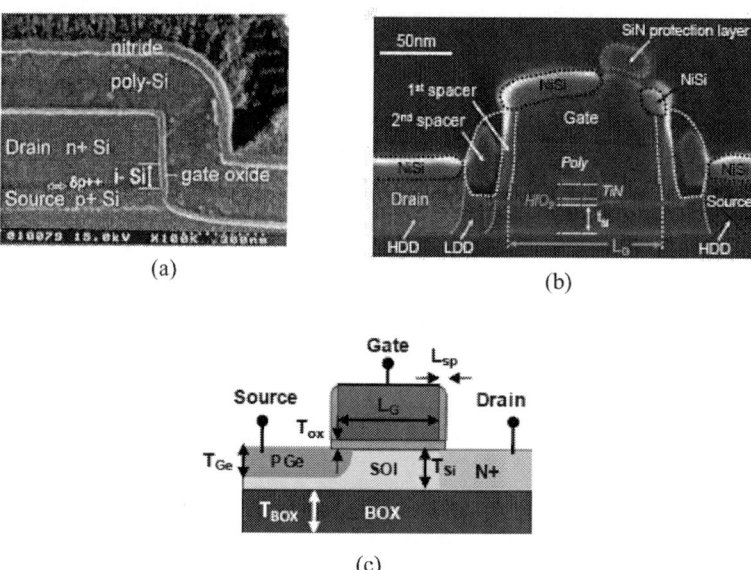

(a)   (b)

(c)

Figure 8: Overview of some horizontal and vertical TFET device concepts that have experimentally been studied: (a) horizontal TFET with a δ-layer at the source [23], (b) horizontal SOI TFET [28], Ge-source hetero-structure [30].

Early reported TFETs were vertical devices with at the source end a highly doped Si δ-layer (Fig. 8a), achieving an on-state current of 10^{-4} A/μm and a subthreshold swing (SS) of about 300 [23]. Further optimization can be done by using a SiGe δ-layer for modulating the tunnel bandgap and engineering the work function [24, 25].

For horizontal devices several groups have reported good results. Horizontal TFETs have been processed in both 130 and 90 nm CMOS technologies [26] and were in 2005 used to demonstrate in a 65 nm technology the fabrication of a 0.68 μm^2 6T-SRAM cell [27]. Mayer $et\ al.$ [28] performed a comparative experimental study of TFETs processed on SOI, Si$_{1-x}$Ge$_x$OI and GeOI. The used FD SOI CMOS process, illustrated in Fig. 8b leads to an off-state current of 30 fA/μm, SS ~42 mV/dec and I_{on}/I_{off}~10^4. The devices show no gate length dependence and the on-current are optimized by the source-drain engineering. The used substrate material has a strong impact on the on-current as illustrated in Fig. 9a. A pronounced current improvement is achieved in case of GeOI, at the expense of a large degradation of the off-state current.

Horizontal TFETs can also be fabricated based on a FinFET approach (Fig. 1b). Fulde *et al.* [29] reported devices with on-currents of 38×10^{-9} A/μm and a SS=210 mV/dec. No gate length dependence is observed, in agreement with the recent data shown in Fig. 2a for the imec process. As reported in section 2, by further optimization imec presently achieved an I_{on}/I_{off} of 10^6 with a SS-46 mV/dec [11].

In case of TFETs with hetero-structures, record performance data have been reported for Ge-source devices, giving at V_{DD}=0.5 V an I_{on}=0.42 μA/μm and I_{off}=012 pA/μm [30]. The thickness of the Si film is 70 nm, the Ge source has a dopant concentration of about 10^{18} cm^{-3} and a 3 nm SiO_2 is used as gate oxide.

A summary of the device performance of some of the above discussed devices is given in Table 1. Although for the moment better results are reported for horizontal TFETS, the vertical approach is fully under development and has a great potential for hetero-structure devices. Much research is ongoing to the introduction of new materials. There is also growing interest in carbon-based materials [e.g. 34, 35], but this is out of scope of the present review.

Table 1: Benchmarking of state-of-the-art TFETs reported in the literature.

Reference (experimental)	Minimum Point SS (mV/dec)	I_{on}/I_{off}	Type	Remarks
Horizontal Integration				
Leonelli *et al.* [7] SSDM 2009	46	10^6	pFTET	MuGFET
Mayer *et al.* [28] IEDM 2008	42	$\sim10^4$	pTFET	Planar SOI
	NA	10^2	pTFET/ nTFET	Planar SOI
Choi *et al.* [31] EDL 2007	52.8	$\sim3\times10^3$	nTFET	Planar SOI
Fulde *et al.* [29] Nanoelectronics Conf. 2008	210	NA	pTET	MugFET
Krishnamohan *et al.* [32] IEDM 2008	50	3×10^3	nTFET	Planar SOI SiGe channel DG
Jeon *et al.* [12] VLSI 2010	46	7×10^7	pTFET	Planar SOI
Kim *et al.* [30] VLSI 2009	~40	3×10^6	pTFET	Planar SOI Ge
Vertical Integration				
Bhuwalka *et al.* [33] Jpn JAP 2006	>300	NA	pTFET	Vertical TFET Si
Vandooren *et al.* [17] Nanoelectronics Workshop 2009	240	$\sim10^5$	nFTET	Vertical TFET Si

A special approach for horizontal TFETs is based on the broken-gap hetero-junction (BG-TFET) [36, 37]. The concept of the band diagram, compared to a staggered-gap approach is shown in Fig. 9b. Recent 1-D simulations point out that a SS<60 mV/dec should be achievable [38].

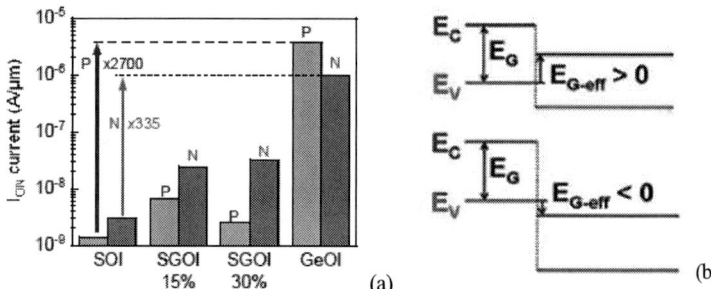

(a)

(b)

Figure 9. (a) Impact of the substrate material on the on-state current of horizontal TFETs processed in a FD SOI CMOS technology [28]. (b) Schematic illustrated of the staggered (top) and broken-gap (bottom) hetero-junction band structure [37].

Conclusions

The main challenges for TFETs remain device performance with a high on-current, a low-off current and a very low subthreshold swing, while keeping the process complexity limited. Low substhreshold swings have been reported but mostly over a rather restricted bias range. There are strong indications that hetero-structure TFETs have a great potential for future low-power/high-performance applications. Therefore, TFETs are considered as the building blocks for coming to a green technology.

Acknowledgments

The authors would to acknowledge the partners of imec's Industrial Affiliation program on Emerging Devices.

References

1. S.-H. Lo, D. A. Buchanan, Y. Taur and W. Wang, *IEEE Electron Dev. Lett.*, **18**, 209 (1997).
2. M.T. Bohr, R.S. Chau, T. Gnani and K. Misty, *IEEE Spectrum*, **44**, no. 10, pp. 29-35 (2007).
3. S. Banerjee, W. Richardson, J. Coleman and A. Chatterjee, *IEEE Electron Dev. Lett.*, **8**, 347 (1987).
4. W.M. Reddick and G.A.M. Amaratunga, *Appl. Phys. Lett.*, **67**, 494 (1995).
5. J. Appenzeller, Y.-M. Lin, J. Knoch, Z. Chen and P. Avouris, *IEEE Trans. Electron Dev.*, **52**, 2568 (2005).

6. W.Y. Choi, B.-G. Park, J.D. Lee and T.-J. K. Liu, *IEEE Electron Dev. Lett.*, **28**, 743 (2007).
7. D. Leonelli, A. Vandooren, R. Rooyackers, S. De Gendt, M.M. Heyns and G. Groeseneken, *Jpn. J. Appl. Phys.*, **49**, 04EDC10 (2010).
8. K. Boucart and A.M. Ionescu, *IEEE Trans. Electron Dev.*, **54**, 1725 (2007).
9. A.S. Verhulst, B. Sorée, D. Leonelli, W.G. Vandenberghe and G. Groeseneken, *J. Appl. Phys.*, **107**, 024518 (2010).
10. D. Leonelli, A. Vandooren, R. Rooyackers, A.S. Verhulst, S. De Gendt, M.M. Heyns and G. Groeseneken, in *Proc. ESSDERC*, p. 170 (2010).
11. D. Leonelli, A. Vandooren, R. Rooyackers, A.S. Verhulst, S. De Gendt, M.M. Heyns and G. Groeseneken, presented at SSDM, Tokyo, Japan (2010).
12. K. Jeon, W.-Y. Loh, P. Patel, C.Y. Kang, J. Oh, A. Bowonder, C. Park, C. S. Park, C. Smith, P. Majhi, H.-H. Tseng, R. Jammy, T.-J. King Liu and C. Hu, In *VLSI Techn. Dig.*, 121 (2010).
13. P. Verheyen, N. Collaert, R. Rooyackers, R. Loo, D. Shamiryan, A. De Keersgieter, G. Eneman, F. Leys, A. Dixit, M. Goodwin, Y.S. Yim, M. Caymax, K. De Meyer, P. Absil, M. Jurczak and S. Biesemans, In *VLSI Techn. Dig.*, 194 (2005).
14. S.J. Koester et al., Electrochem, *Trans. Electrochem. Soc.*, **33(6)**, 357 (2010).
15. Z.X. Chen, H.Y. Yu, N. Singh, N.S. Shen, R.D. Sayanthan, G.Q. Lo and D.L. Kwong, *IEEE Electron Device Lett.*, **30**, 754 (2009).
16. A.S. Verhulst, W.G. Vandenberghe, K. Maex and G. Groeseneken, *J. Appl. Phys.*, **104**, 064514 1 (2008).
17. A. Vandooren, R. Rooyackers, D. Leonelli, F. Iacopi, E. Kunnen, D. Nguyen, M. Demand, P. Ong, L. Willie, J. Moonens, O. Richard, A.S. Verhulst, W.G. Vandenberghe, G. Groeseneken, S. De Gendt and M. Heyns, *Proc. Si Nanoelectronics Workshop*, Kyoto, 21 (2009).
18. A. Verhulst, W.G. Vandenberghe, K. Maex and G. Groeseneken, *Appl. Phys. Lett.*, 91, 053102 (2007).
19. D. Leonelli *et al.* (unpublished)
20. A.S. Verhulst, W.G. Vandenberghe, D. Leonelli, R. Rooyackers, A. Vandooren, G. Pourtois, S. De Gendt, M.M. Heyns and G. Groeseneken, *Trans. Electrochem. Soc.*, **33(6)**, 363 (2010).
21. O.M. Nayfeh, C.N. Chleirigh, J. Hennessy, L. Gomez, J.L. Hoyt and D.A. Antoniadis, *IEEE Electron Dev. Lett.*, **29**, 1074 (2008).
22. A.S. Verhulst, W.G. Vandenberghe, K. Maex, S. De Gendt, M.M. Heyns and G. Groeseneken, *IEEE Electron Dev. Lett.*, **29**, 1398 (2008).
23. K.K. Bhuwalka, S. Sedlmaier, A.K. Ludsteck, C. Tolksdorf, J. Schulze and I. Eisele, *IEEE Trans. Electron Dev.*, **51**, 279 (2004).
24. K.K. Bhuwalka, J. Schulze and I. Eisele, In *Proc. ESSDERC 2004*, p. 241 (2004).
25. K.K. Bhuwalka, J. Schulze and I. Eisele, *IEEE Trans. Electron Dev.*, **52**, 909 (2005).
26. Th. Nirschl, P.-F. Wang, C. Weber, J. Sedlmeii, R. Heinrich, R. Kakoschke, K. Schrüfe, J. Holz, C. Pacha, T. Schulz, M. Ostermayr, A. Olbrich, G. Georgakos, E. Ruderer, W. Hansch and D. Schmitt-Landsiedell, In *IEDM Techn. Dig.*,195 (2004).
27. Th. Nirschl, St. Henzler, J. Fischer, A. Bargagli-Stoffi, M. Fulde, M. Sterkel, P. Teichmann, U. Schaper, J. Einfeld, C. Linnenbank, J. Sedlmeir, C. Weber, R. Heinrich, M. Ostermayr, A. Olbrich, B. Dobler, E. Ruderer, R. Kakoschke, K.

Schrüfer, G. Georgakos, W. Hansch, D. Schmitt-Landsiedel, In *Proc. ESSDERC*, 173 (2005).

28. F. Mayer, C. Le Roy, J.-F. Damlencourt, K. Romanjek, F. Andrieu, C. Tabone, B. Previtali and S. Deleonibus, In *IEDM Techn. Dig.*, 163 (2008).

29. M. Fulde, A. Heigl, M. Weis, M. Wirnshofer, K. von Arnim, Th. Nirschl, M. Sterkel, G. Knoblinger, W. Hansch, G. Wachutka and D. Schmitt-Landsiedel, In *Proc. 2nd Int. Nanoelectronics Conf. – INEC*, 579 (2008).

30. S.H. Kim, H. Kan, C. Hu and T.-S. Liu, In *VLSI Techn. Dig.*, 178 (2009).

31. W.Y. Choi, B.-G. Park, J.D. Lee and T.-J. King Liu, *IEEE Electron Dev. Lett.*, *28*, 743 (2007).

32. T. Krishnamohan, D. Kim, WS. Raghunathan and K. Saraswat, In *IEDM Techn. Dig.*, 947 (2008).

33. K.K. Bhuwalka, M. Born, M. Schindler, M. Schmidt, T. Sulima and I. Eisele, *Jpn. J. Appl. Phys.*, **45**, 3106 (2006).

34. S.O. Koswatta, M. Lundstrom and D.E. Nikonov, *IEEE Trans. Electron Dev.*, **56**, 456 (2009).

35. Y. Gao, T. Low and M. Lundstrom, In *VLSI Techn. Dig.*, 180 (2009).

36. S.O. Koswatta, S.J. Koester and W. Haensch, In *IEDM Techn. Dig.*, 909 (2009).

37. J. Knoch and J. Appenzeller, *IEEE Electron Dev. Lett.*, 31, 305 (2010).

38. S.O. Koswatta, S.J. Koester and W. Haensch, *IEEE Trans. Electron Dev.*, **57**, 3222 (2010).

CHAPTER 3

MATERIALS-1

28

SiGe and Ge on Insulator Wafers

N. Daval, C. Figuet, C. Aulnette, D. Landru, C. Drazek, K. K. Bourdelle,
E. Guiot, F. Letertre, B-Y Nguyen, C. Mazure

SOITEC Parc Technologique des Fontaines, 38190 Bernin, France

This paper discusses the options to boost performances of PMOS
in Fully Depleted CMOS via introduction of a SiGe compressively
strained layer, specifically studying thickness uniformity of such
SiGe layers.
In a second part we discuss GeOI substrates as a substrate of
choice for Ge PFET / IIIV NFET co-integration for future CMOS
nodes, and its possible manufacturing using condensation method.

Introduction

Since 90nm technology Germanium (Ge) element has become increasingly
popular in the CMOS processing for enhancing transistor performance, especially
enhancing hole mobility for P-type transistors. The main driver has been the embedded
SiGe in the source/drain region and its extraordinary boost on PFET drive current [1].
More recently Ge has enabled band engineering with respect to Silicon for Vt tuning
[2,3] in addition to channel engineering for mobility enhancement. Looking into the
future the need for SiGe alloys or pure Ge is increasing, as it is contemplated as a seed
for IIIV material growth [4], or even the replacement of Si by Ge in the channel to take
advantage of the high electron and hole mobilities [5]. Today the manufacturing reality
shows us that all Ge needs can be fulfilled by epitaxy during the processing of the devices
[6]. In this paper we will introduce the Dual Channel substrate having a strained SiGe
layer grown on top of a SOI substrate. Starting with this wafer and thru condensation
process one can produce a uniform SiGe layer suitable for Fully Depleted applications.

Strain options for planar FDSOI

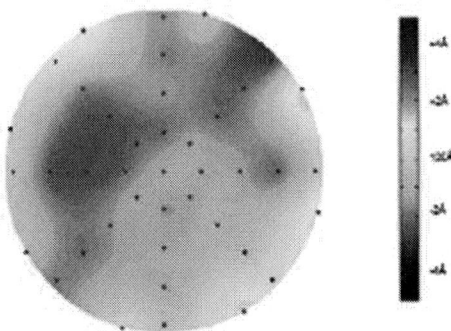

Figure 1 – Extremely uniform SOI thickness for Fully Depleted applications with 4.1A range.

Under nominal conditions for node 20nm, simulations predict 25 mV of V_T variation per 1 nm of SOI thickness variation, and this is backed up with experimental data: reader should refer to Ali Khakifirooz's paper [7] for details. Therefore, SOI thickness uniformity specification has been set at 1nm for the maximum range, allowing a maximum of 25mV variation due to thickness non-uniformity. Wafers are available with such specification with different buried oxide thickness options: the most popular are 145nm and 25nm.

Figure 1 shows a SOI layer thickness map with a range of 0.41nm. FDSOI availability is described in details by Schwarzenbach et al [8].

Published data on FDSOI devices have confirmed extremely good electrostatic control of the device [9,10] with gate length of 25nm they show DIBL inferior to 100mV/V. Undoped channel together with reduced vertical field greatly benefit the carrier mobility, and reported device performances are already very good for Low Power applications. Such good performance data, together with greatly improved electrostatics opening up the scaling path towards future nodes, give current FDSOI devices a strong edge for an introduction in manufacturing at node 20nm and will be extendable to node 15/14nm.

At node 11nm, in addition to conventional stressors, FDSOI technology can take full advantage of wafer level stressors like:

- sSOI substrates can provide a tensile strained silicon, by transfer of a strained silicon layer grown on a donor substrate with a relaxed top SiGe layer.

- Compressive strain can be provided through pseudomorphic growth of SiGe on SOI with subsequent optional oxidation step to adjust the thickness of the SiGe layer and its Ge content– known as condensation process [11].

sSOI wafers and benefits for NFETs of biaxial strain wafer level have been widely discussed in the literature [12,13], this is not the case for compressive strained SiGe option.

Dual Channel wafers

Trying to provide a compressively strained SiGe layer for PFET, one has the choice between growing a blanket layer that is continuous on the whole surface or use selective epitaxy during the device process getting SiGe only where it is needed. Given the very tight uniformity requirement for FD application we will discuss the blanket case, knowing that the selective epitaxy option can only be more complex adding loading effect as a new source of non-uniformity.

Figure 2 (left) shows the schematic structure of a Dual Channel wafers. It has a SiGe layer grown onto an SOI wafer and a silicon capping layer on top. The SOI layer has the option to be a tensile strained silicon layer. Figure 2 (right) shows a TEM cross-section of a Dual Channel wafer. This TEM shows a wafer having 12nm of 2 GPa strained silicon, with a 60% SiGe layer of 10nm and a Silicon capping of 4nm. The relaxation being negligible for that wafer the compressive strain comes from the lattice difference of SiGe

30% (template of the sSOI substrate) and the SiGe 60%, and is equal to 1.37 % lattice strain (equivalent to 2 GPa). The silicon cap has the same strain as the SOI layer.

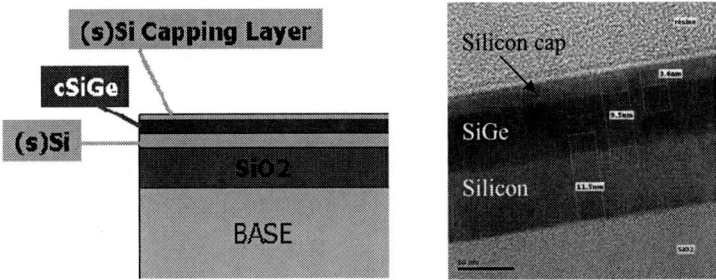

Figure 2 – (left) Drawing of the Dual Channel structure (right) TEM picture of Dual Channel wafer

With 5 different starting materials: SOI and sSOI with strain induced by relaxed SiGe 20%, 30%, 40% and 50%, it is possible to plot the strain level of the SiGe layer as a function of its Ge content. This is represented by the solid lines on the Figure 3.
Figure 4 and Figure 5 show experimental strained data points obtained on SOI and sSOI 20% starting materials and measured by Raman spectroscopy. One can see relaxation occurs when the strain drops even if the Ge content is increased.

Figure 4 shows that changing starting material from SOI to sSOI with the same growth temperature allows greater Ge contents before reaching the relaxation threshold.

Figure 5 illustrates that lower growth temperatures enable SiGe layer with higher strain levels. Decreasing the growth temperature from 650C to 550C changes the maximum strain achievable from 1.4% to 1.8% thanks to an increased critical layer thickness at lower growth temperature [14].

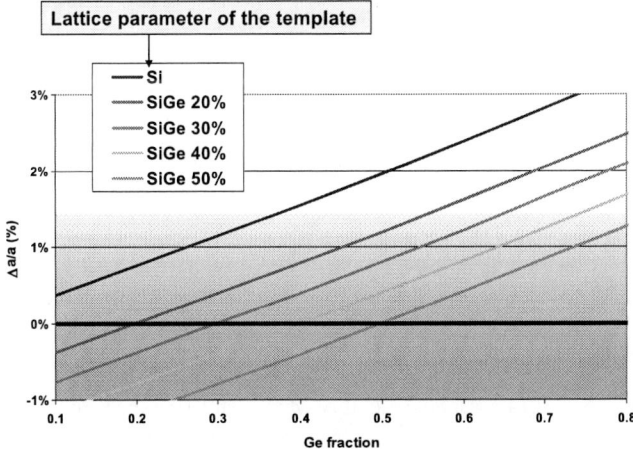

Figure 3 – SiGe strain as a function of lattice parameter of the template and its Ge content.

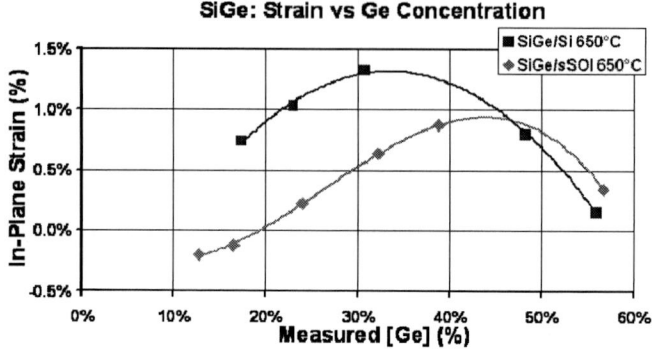

Figure 4 – Measured SiGe strain in Dual Channel structures for 2 different strain configuration in the starting material: SOI and sSOI 20%.

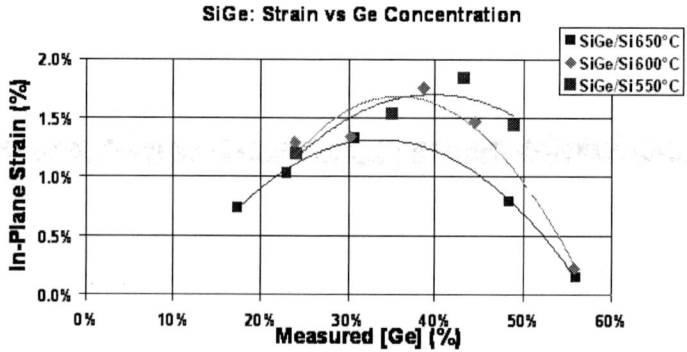

Figure 5 – Measured SiGe strain in Dual Channel structures with SOI as starting material and for different growth temperature

Such Dual Channel substrates have been used to make devices and have demonstrated high carrier mobility and showed that Ge content and strain is efficient way to tune V_T [3, 15]. Figure 6 are plots reproduced from L. Hutin [15] showing V_T in spec for 22nm HP with a single High-K/Metal stack process for short channel devices with Lg ranging from 17nm to 200nm. NFET are fabricated on 1.3GPa tensile strained silicon layer, and PFET on compressive strained SiGe layer. Those strained layers also provide long channel carrier mobility boost of 106% for NFET and up to 92% for the PFET as shown in Figure 6.

In order to evaluate the hole mobility which can be reached with such structures, we grew at low temperature (<550C) pseudomorphic $Si_{0.2}Ge_{0.8}$ on 5nm thick SOI substrates with 4nm silicon cap. SiGe layers with thicknesses ranging from 3nm up to 6nm shows increased hole mobility by a factor of 3 to 9 as compared to SOI with equivalent thickness. It is interesting to point out that the measured carrier mobility in those studies does not suffer as much as predicted from the SiGe alloy scattering and hole mobility remains high.

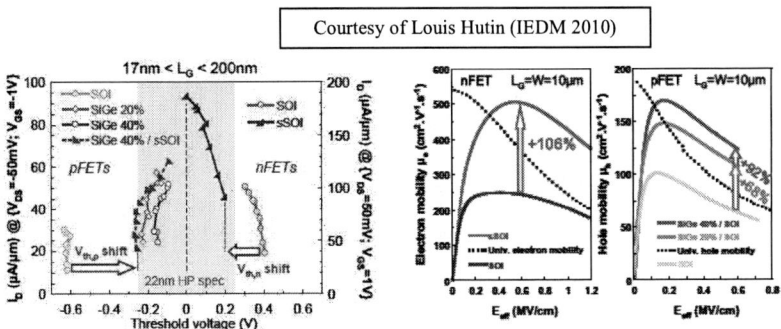

Figure 6 – Plots reproduced from L.Hutin publication (IEDM 2010) showing (left) V_T tunability with tensile strained silicon and compressive strained SiGe and (right) carrier mobility boost over SOI control in the same layers.

Condensation option

For fully depleted application it might be desirable to homogenize the Dual Channel structure in one unique SiGe layer with the thickness and Germanium content adjusted to the requirement of the application. This can be done by condensation process [11], where the thermal budget is chosen such that the Ge diffuses fast enough to have a uniform Ge concentration. The amount of formed oxide will allow adjusting the thickness of the final SiGe layer. Finally one can assume that the amount of Ge atoms is kept constant in the SiGe layer as it is "snow-plowed" by the oxidation front and stays in the SiGe layer throughout the whole process.

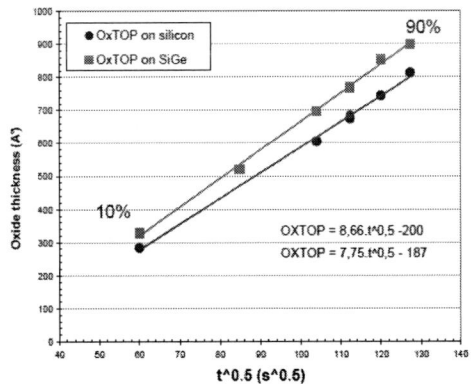

Figure 7 – Formed oxide on Dual Channel substrate and Silicon monitor – under 950C 100% O2.
10% and 90% refer to the Ge content in the condensed SiGe layer.

The main concern of using the condensation process for fully depleted technology is the uniformity of the layer thickness across the wafer. As previously mentioned under fully depleted conditions V_T depends directly on the layer thickness, setting the tight specification for SOI layer thickness for FD applications at +/- 0.5nm.
Sources of non uniformity for the SiGe layer after condensation process include the followings:

Initial SOI layer thickness range. Based on today available wafers: 0.4nm is the best wafers, 0.8nm is typical for high uniformity substrates [8].

SiGe epitaxy range in thickness and Ge composition. Epitaxy experts consider that 2 nm of range on 10nm films and 2 % points of Ge concentration across the wafers are typical. 1nm and 1 % point of Ge concentration would be achievable with great care, translating in higher maintenance costs.

Oxidation uniformity can be considered negligible in state of the art furnaces tuned for uniform oxidation.

Oxidation kinetics dependant on the Ge content. We have done experiments plotted on Figure 7, and there are also data available in the literature [16]. Under 950C dry oxidation known as conditions minimizing the oxidation rate difference for SiGe alloys with different Ge contents, we see roughly 15% faster oxidation rate for SiGe compare to Silicon. It is almost impossible to give oxidation rate as a function of Ge content as those oxidation conditions induce a Ge pile-up at the oxidation front. One can only estimate from data as reported on Figure 7, which shows oxide thickness formed on a Dual Channel wafer in comparison with a silicon monitor wafer, having the average Ge concentration in the SiGe layer going from 10% to 90% over the range of the experiment. The formed oxide is 15% thicker on the Dual Channel wafer and rather constant over the wide range (10% to 90% of Ge content) of our experiment. This tells us that under those oxidation conditions the thickness non-uniformity induced by a small Ge fraction difference is negligible. It is very likely that the Ge pile-up that forms under those conditions is determining the oxidation kinetics and is quite independent from the initial Ge fraction present in the layer.

To summarize, in the case of blanket non patterned SiGe layer solution, review of the sources of non uniformity producing a compressively strained SiGe layer suitable for FD application, we identify the thickness uniformity limiting process step being the SiGe epitaxy before oxidation. SOI starting material and condensation oxidation are second order sources of non uniformity.

In the case of selective epitaxy only in the active PMOS areas, high uniformity will be even more challenging to obtain, having to take into account the loading effect during epitaxy and the 3D effects on the edges of the patterns during the condensation oxidation.

GeOI for high mobility and IIIV growth

GeOI – Smart Cut™ transfer of a Germanium layer

Figure 8 – Picture of 200mm GeOI wafer

Looking at post-8 nm nodes, high mobility III-V material, such as InGaAs might be required to have enough performance boosters for the NFET. Such material could be either grown on Ge or transferred directly on insulator. Figure 9 shows the Smart Cut™ fabrication process of GeOI wafers, the Ge donor can be either Ge Bulk or a Ge epilayer on a Silicon wafer. The epi-layer option does open up the possibility of 300mm or even 450mm manufacturing line.

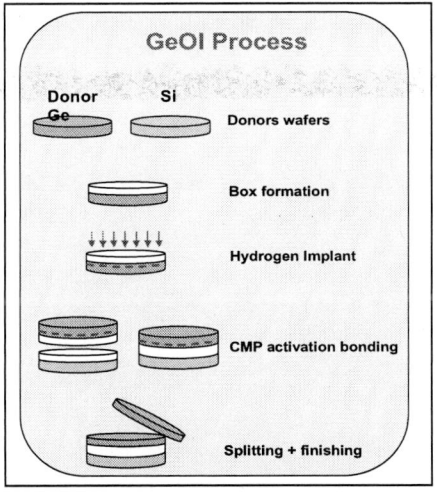

Figure 9 – GeOI fabrication process

One of the key parameter of the GeOI wafer is the Threading Dislocation Density (TDD). As shown on Figure 10, Smart Cut™ process does not change the density of dislocations between the donor and the final GeOI structure. In that case shown on Figure 10, 1E7 cm⁻² TDD corresponds to a 2µm Ge epitaxy on Silicon wafers. Lower TDD levels are accessible through thicker and graded buffers between silicon and germanium, and also the ultimate solution of starting with a Ge bulk substrate to have the lowest Ge dislocation density available.

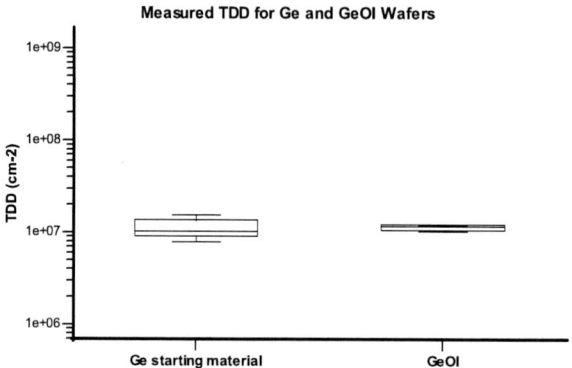

Figure 10 – TDD in Germanium layer before and after transfer

GeOI – Condensation process

We have also studied the possibility of using condensation process to manufacture GeOI substrates, enriching Ge in the SiGe film going all the way to pure Ge. With uniformity considerations as described in the previous section, we have calculated thickness and Ge concentrations introducing only a small thickness uniformity of 1.5nm. Figure 11 shows the plots.

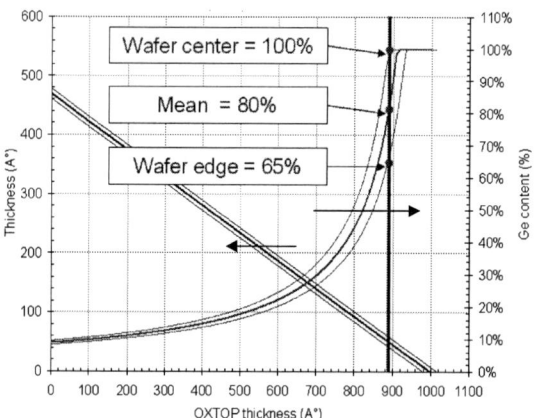

Figure 11 – Condensation process thickness and Ge concentration.

$$\%Ge = [\%Ge(0) * t_{SiGe}(0)] / t_{SiGe} \qquad [1]$$

Assuming no Ge loss during the condensation process and the Ge content given by [1], which %Ge is a Ge concentration and tSiGe is a thickness of the SiGe layer, (0) refers to the initial quantities before oxidation. The Ge concentration depends on inverse proportion of TSiGe, therefore even the small thickness non-uniformity (1.5nm) leads to a large thickness and Ge non-uniformity film after Ge condensation process.

Using condensation to manufacture GeOI wafers, one will have to accept large Ge non-uniformity. This means Ge concentration will have to be intentionally stop in the range of 80% Ge content to avoid discontinuity of the layer.

Conclusion

We have reviewed the current status of the Dual Channel and GeOI substrates for performance boost.

The SiGe channel engineering might be a very attractive mobility booster method for 11nm node and beyond, especially for FDSOI. The final SiGe thickness uniformity using Ge condensation technique will rely on the initial SiGe epitaxy step before oxidation. The thickness uniformity of the initial SiGe epitaxy film has direct impact on the final thickness uniformity after condensation. Thus it is important to obtain the excellent uniformity of the thin SiGe epitaxy selectively or blanket growth.

GeOI can be the substrate of choice for Ge and III-V future electronics. Various TDD levels (from low 1E4 to 1E7) are available depending on the application requirements.

Acknowledgement

Authors would like to thank CEA-LETI teams for their contribution in device fabrication and characterization.

References

1. K.J. Kuhn, A. Murthy, R.Kotlyar, M. Kuhn, *ECS Trans*, 33 (6) p3-17 (2010)
2. O. Reilly, *Semicond. Sci. Technol.*, 4 121-137 (1989)
3. C. Le Royer , M. Cassé, F. Andrieu, O. Weber, L. Brevard, P. Perreau, J.-F. Damlencourt, S. Baudot, C. Tabone, F. Allain, P. Scheiblin, C. Rauer, L. Hutin, C. Figuet, C. Aulnette, N. Daval, B.-Y. Nguyen, K. K. Bourdelle, *Proc. ESSDERC*, (2010)
4. T.E. Kazior , J.R. LaRochel, D. Lubyshev, J. M. Fastenau, W. K. Liu, M. Urteaga, W. Ha, J. Bergman, Choe, M. T. Bulsara, E. A. Fitzgerald, D. Smith, D. Clark, R. Thompson, C. Drazek, N. Daval, L. Benaissa, E. Augendre, *Proc. IEEE MTT-S Boston*, p1113-16 (2009)
5. S. Bedell, A. Majumdar, J. A. Ott, J. Arnold, K. Fogel, S. J. Koester, D. K. Sadana, *IEEE Electron Device Lett.,*29(7), (2008) p811-3
6. D. Sadana, S. Bedell, T. N. Adam, A Reznicek, H. He, 33 (6) (2010) p59-70
7. A. Khakifirooz, K. Cheng, P. Kulkarni, J. Cai, S. Ponoth, J. Kuss, B. S. Haran, A. Kimball, L. F. Edge, A. Reznicek, T. Adam, H. He, N. Loubet, S. Mehta, S. Kanakasabapathy, S. Schmitz, S. Holmes, B. Jagannathan, A. Majumdar, D. Yang,

A. Upham, S.-C. Seo, J. L. Herman, R. Johnson, Y. Zhu, P. Jamison, Z. Zhu, L. H. Vanamurth, J. Faltermeier, S. Fan, D. Horak, H. Bu, D. K. Sadana, P. Kozlowski, D. McHerron, J. O'Neill, B. Doris, W. Haensch, E. Leobondung, G. Shahidi, *Proc. VLSI-TSA*, (2010)

8. W. Schwarzenbach, X. Cauchy, O. Bonnin, N. Daval, C. Aulnette, C. Girard, B.-Y. Nguyen, C. Maleville, *ECS Trans*, to be published (2011)

9. Q. Liu, A. Yagishita, N. Loubet, A. Khakifirooz, P. Kulkarni, T. Yamamoto, K. Cheng, M. Fujiwara, J. Cai, D. Dorman, S. Mehta, P. Khare, K. Yako, Y. Zhu, S. Mignot, S. Kanakasabapathy, S. Monfray, F. Boeuf, C. Koburger, H. Sunamura, S. Ponoth, A. Reznicek, B. Haran, A. Upham, R. Johnson, L. F. Edge, J. Kuss, T. Levin, N. Berliner, E. Leobandung, T. Skotnicki, M. Hane, H. Bu, K. Ishimaru, W. Kleemeier, M. Takayanagi, B. Doris, R. Sampson, *Symp. VLSI Tech.*, (2010)

10. K. Cheng, A. Khakifirooz, P. Kulkarni, S. Ponoth, J. Kuss, D. Shahrjerdi, L. F. Edge, A. Kimball, S. Kanakasabapathy, K. Xiu, S. Schmitz, A. Reznicek, T. Adam, H. He, N. Loubet, S. Holmes, S. Mehta, D. Yang, A. Upham, S.-C. Seo, J. L. Herman, R. Johnson, Y. Zhu, P. Jamison, B. S. Haran, Z. Zhu, L. H. Vanamurth, S. Fan, D. Horak, H. Bu, P. J. Oldiges, D. K. Sadana, P. Kozlowski, D. McHerron, J. O'Neill, B. Doris, *IEDM Tech. Dig.*, (2009)

11. T. Tekuza, N. Sugiyama, S. Takagi, *Appl. Phys. Lett.*, Vol. 79, No. 12, (2001)

12. I. Cayrefourcq, A. Boussagol, G.K. Celler, *ECS Trans.,*3 (7) p399-410 (2006)

13. S. Flachowsky, R. Illgen, T. Herrmann, T. Baldauf, A. Wei, J. Hontschel, W. Klix, R. Stenzel, M. Horstmann. *10th ULIS* in Aachen (2009)

14. B.W. Dodson, J.Y. Tsao, *Appl. Phys. Lett.*, 51(17), 1325 (1987)

15. L. Hutin, C. Le Royer, F. Andrieu, O. Weber, M. Cassé, J.-M. Hartmann, D. Cooper, A. Béché *, L. Brévard, L. Brunet, J. Cluzel, P. Batude, M. Vinet, and O. Faynot, *IEDM Tech. Dig.*, (2010)

16. F. K. LeGoues, R. Rosenberg, B. S. Meyerson, *J. Appl. Phys.*, Vol. 65, No. 4, (1989)

Au-Catalyst Induced Low Temperature (~250 °C) Layer Exchange Crystallization for SiGe on Insulator

Jong-Hyeok Park[a], Masashi Kurosawa[a,b], Naoyuki Kawabata[a], Masanobu Miyao[a], and Taizoh Sadoh[a]

[a] Department of Electronics, Kyushu University,744 Motooka, Nishi-ku, Fukuoka, 819-0385, Japan
[b] JSPS Research Fellow,8 Ichiban-cho, Chiyoda-ku, Tokyo, 102-8472, Japan

The gold-induced crystallization technique has been investigated to achieve poly-SiGe films on insulators at low temperatures ($\leq 300^\circ$C). By annealing of the amorphous SiGe (Ge concentration: 0-100%)/Au stacked structures formed on insulating substrates, positions of the SiGe and Au layers are inverted, and the Au/SiGe stacked structures are obtained. Crystallization of the SiGe layers in the inverted samples is confirmed by the Raman scattering spectroscopy analysis. Moreover, the Raman measurements reveal that the Ge fractions in the crystallized SiGe layers are almost the same as those of the initial amorphous SiGe layers. This gold-induced layer-exchange crystallization technique of SiGe layers at a low temperature (~250°C) will be very useful to obtain poly-SiGe layers on plastic substrates, which are essential to realize flexible high-speed thin-films transistors and high-efficiency solar cells.

Introduction

Low-temperature crystallization technique of semiconductor materials such as SiGe on insulating plastic substrates (softening temperature: ~300°C) should be developed to realize the flexible high-speed thin film transistors (TFTs) and high-efficiency solar cells. In recent years, metal-induced layer-exchange crystallization techniques of Si using catalyst metals, such as Al and Ag, have been proposed to decrease the growth temperature of Si (1-3). In these techniques, poly-crystalline Si films are obtained through the layer-exchange of the Si/metal stacked structures at temperatures (~450 and ~550°C) below the eutectic temperatures (577 and 845°C) for the catalysts of Al and Ag, respectively. However, further decrease in crystallization temperatures is necessary in order to employ plastic substrates. Moreover, this technique should be developed to crystallization of SiGe films. We speculate that this layer-exchange growth occurs in the combination with metals, where the metals form eutectics with semiconductor elements. Since Au forms eutectic alloys at low temperatures with Si (363°C) and Ge (361°C), the Au-catalyst is expected to enable layer-exchange growth at very low temperatures.

In the present study, we examine the Au-induced layer-exchange growth of SiGe. Consequently, we realize the very low temperature (~250°C) growth of SiGe with the whole Ge fractions on insulators.

Experimental Procedure

Au films (thickness: 100 nm) were deposited on quartz substrates at a room temperature by the electron beam evaporation. Subsequently, a-SiGe films (Ge fraction: 0-100 %, thickness: 100 nm) were deposited on the Au layers by the molecular beam technique at a room temperature, where Si and Ge were evaporated using an electron-beam gun and a Knudsen cell, respectively. Subsequently, the samples were annealed (T_a: 150-400°C) in dry N_2 ambient to induce crystallization.

Results and Discussion

Figures 1(a)-1(d) show the depth profile of elements in the Si/Au and Ge/Au samples before and after annealing (250°C, 20 h) obtained by Auger electron spectroscopy (AES). From the profiles for the samples before annealing, it is confirmed that Si/Au and Ge/Au stacked structures are formed (Figs. 1(a), 1(c)). On the other hand, Si or Ge atoms move to the bottom layers, and Au atoms move to the top layers after annealing (Figs. 1(b), 1(d)). This indicates the layer-exchange by annealing. Similar results showing the layer exchange were observed from the samples annealed at temperatures in the range of 250-350°C.

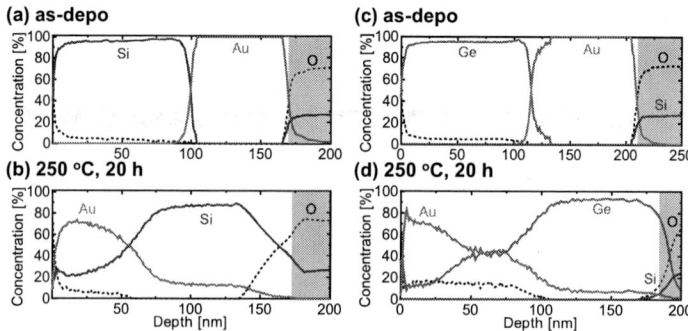

Figure 1. Concentration profile of Si, Ge, Au, and O atoms in samples for Si/Au ((a),(b)) and Ge/Au samples   ((c),(d)).

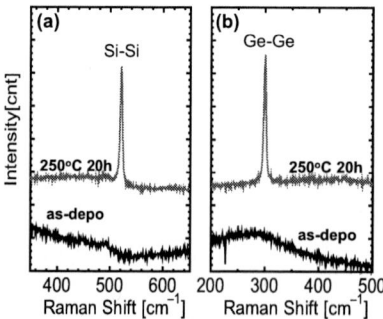

Figure 2. Raman spectra observed from the back side through the quartz substrates for samples of Si/Au (a) and Ge/Au samples (b).

In order to investigate the crystallinity of the grown layers obtained from Si/Au and Ge/Au samples, micro-probe Raman measurements were performed from the backside of the samples. The results are shown in Figs. 2(a) and 2(b), respectively. The Raman peaks due to Si-Si and Ge-Ge vibration mode are clearly visible from both the annealed samples, though no Raman peaks are observed from the as-deposited samples. These results demonstrate that the crystallization of Si and Ge through the layer exchange process is achieved at a low temperature (250°C).

To develop this technique to crystallization of SiGe films with the whole Ge fractions, we investigated growth features for the SiGe/Au samples by annealing. The energy dispersive x-ray (EDX) spectra obtained from the SiGe/Au samples before and after annealing are shown in Figs. 3(a)-3(f). In these measurements, the energy of electron beam excitation was selected as low as 3 keV to detect the elements near the surfaces (~100 nm) selectively. The results show that SiGe films with Ge fractions of about 30, 50, and 80% are formed before annealing. After annealing (250°C, 20 h), the dominant constituent of the upper layer changes to Au for all samples. This indicates that the layer-exchange occurs in those samples.

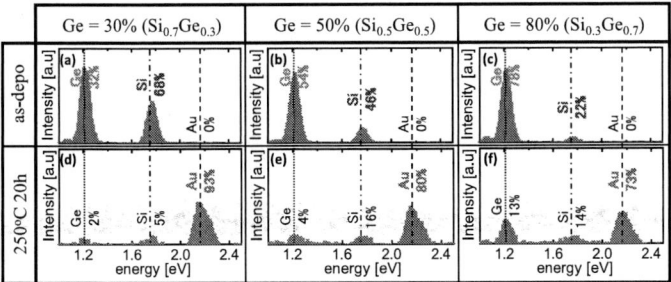

Figure 3. Elements in upper layer evaluated by EDX measurements for the SiGe/Au samples with Ge concentrations of 30, 50, and 80% before (a-c) and after annealing at 250°C for 20 h (d-f).

To investigate the crystallinity of the grown SiGe layers, we performed the Raman measurements. Figure 4(a) shows the Raman spectra of the grown layers of the samples with various initial Ge concentrations (30, 50, and 80%). The spectra of a crystal Si substrate are also shown for a reference. In the respective spectra of the SiGe samples, three peaks are observed. They are Raman peaks due to the Ge-Ge, Si-Ge, and Si-Si vibration modes in crystal SiGe. We evaluated the Ge fractions x of the SiGe layers from the positions of the Si-Si peaks using the following equation [1] reported in Ref. 4.

$$\omega_{Si\text{-}Si} = 520.7 - 66.9x \ (cm^{-1})$$
[1]

The Ge fractions (x_f) of the formed SiGe layers are summarized as a function of the Ge fractions (x_i) of the initial a-SiGe layers in Fig. 4(b). The Ge fractions of the grown layers are almost the same as those of the initial a-SiGe layers. This indicates no segregation of Si or Ge atoms occurs during the growth. Thus, this technique is useful to obtain SiGe films with controlled Ge fractions at a low temperature (~250°C).

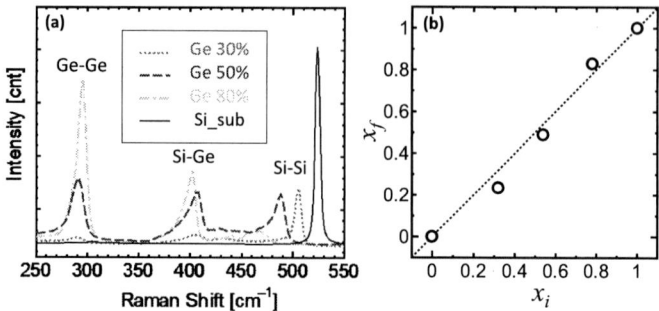

Figure 4. Raman spectra of the grown layers for the samples with various Ge concentrations (30, 50, 80%) after annealing at 250°C for 20 h (a), and Ge concentration (x_f) of the grown layers evaluated from the Raman shifts as a function of the initial Ge concentration (x_i) measured by EDX (b).

Conclusion

We have developed the gold-induced layer-exchange crystallization technique to obtain poly-SiGe (Ge fraction: 0-100%) layers on insulating substrates at a low temperatures. By annealing of the amorphous SiGe/Au stacked structures, crystallization of SiGe layers occurs through the exchange of the SiGe and Au layers at ~250°C. The Raman measurements reveal that the Ge fractions in the crystallized SiGe layers are almost the same as those of the initial amorphous SiGe layers. This gold-induced layer-exchange crystallization technique will be useful to obtain poly-SiGe layers on plastic substrates, which are essential to realize flexible high-speed TFT and high-efficiency solar cells.

Acknowledgments

A part of this work was supported by Research Foundation for Materials Science, and a Grant-in-Aid for Scientific Research from the Ministry of Education, Culture, Sports, Science and Technology of Japan.

References

1. O. Nast, T. Puzzer, L. M. Koschier, A. B. Spoul, and S. R. Wenham, *Appl. Phys. Lett.* **73**, 3214 (1998).
2. M. Kurosawa, N. Kawabata, T. Sadoh, and M. Miyao, *Appl. Phys. Lett.* **95,** 132103 (2009).
3. M. Scholz, M.Gjukic, and M. Stuzman, *Appl. Phys . Lett.* **94**, 01208 (2009).
4. F. Pezzoli, E. Bonera, E. Grilli, M. Guzzii, S. Sanguinetti, D. Chrastina, G. Isella, H. von Känel, E. Wintersberger, J. Stangl, and G. Bauer, *Materials Science in Semiconductor Processing.* **11**, 279 (2008).

ECS Transactions, 35 (5) 43-50 (2011)
10.1149/1.3570775 ©The Electrochemical Society

Strain Nano-Engineering: SSOI as a Playground

O. Moutanabbir[a,b], A. Hähnel[a], M. Reiche[a], W. Erfurth[a], A. Tarun[b], N. Hayazawa[b], and S. Kawata[b], F. Naumann[c], M. Petzold[c]

[a] Max Planck Institute of Microstructure Physics, Weinberg 2, D 06120 Halle (Saale), Germany
[b] Nanophotonics Laboratory, RIKEN Advanced Science Institute, Hirosawa, Wako, Saitama 351-0198, Japan
[c] Fraunhofer Institute for Mechanics of Materials, Walter-Hülse-Strasse 1, D 06120 Halle (Saale), Germany

> Ultrathin strained silicon-on-insulator (SSOI) has been in the limelight of device scientists and engineers as one of the materials that can possibly extend the lifetime of the current silicon technology. In this paper, we show that, beyond this technological interest, SSOI also provides a rich platform to explore and explain a number of fundamental phenomena. In particular, ultrathin SSOI substrates are exploited to elucidate basic nanomechanical properties of silicon. More precisely, the bending and local lattice rotation associated with free surface-induced relaxation upon nanoscale patterning are investigated using micro-Raman scattering, high resolution transmission electron microscopy, and nano-beam electron diffraction. The observed morphological changes in SSOI nanostructures cannot be explained by the classical Stoney's formula or related formulations developed for nanoscale thin films. Instead, a continuum mechanical approach is employed to describe these observations through three-dimensional numerical calculations of relaxation-induced lattice displacements. The use of SSOI to implement novel nanoscale devices is also discussed.

Introduction

Strain is omnipresent in semiconductor science and technology. In particular, the influence of strain on both electronic and excitonic band gap structure of group-IV semiconductors has sparked a surge of interest as a strategy to engineer enhanced or novel electronic and optoelectronic devices (1, 2). For instance, the strain influence on silicon band gap has been exploited to build higher performance transistors (1). Here the strain breaks the sixfold degeneracy in the elliposoidal valley of the conduction band as well as the degeneracy between heavy and light hole bands. This translates to a reduction of the intervalley scatterings in the conduction band and of the charge carrier effective transport mass leading to an enhancement in the mobility. In the current technology, the deposition of SiGe on specific active regions beside the channel is currently used to *locally* engineer the strain in metal-oxide-semiconductor field-effect transistors (1). The strains induced in this case are typically uniaxial. This method shows, however, a limited

43

efficiency with the technology scaling. Moreover, these local strain engineering processes can hardly be applied for the emerging architectures such as FinFETs and gate-all-around nanowire devices. In this landscape, building transistors directly from a *globally* strained Si material emerges as a potential alternative. This can be achieved using tensile strained silicon-on-insulator substrates (SSOI), which combine the benefits of strained Si and silicon-on-insulator technologies in addition to the circumvention of difficulties associated with SiGe layers, *viz*; high leakage current, Ge diffusion, and enhanced *n*-type dopant diffusion.

Besides the aforementioned technological interest in electronics, ultrathin SSOI also provides a rich playground to study the influence of strain on the basic properties of silicon nanoscale systems. In this work, we present unprecedented insights into the nanomechanical properties of silicon unveiled through the investigation of bending and local lattice rotation associated with free surface-induced relaxation upon nanoscale patterning of ultrathin SSOI substrate. We employed micro-Raman scattering, high resolution transmission electron microscopy (HRXTEM), and nano-beam electron diffraction (NBED) to address these phenomena. Our experimental data are augmented with detailed three-dimensional finite element calculations of the strain behavior in SSOI nanoscale structures.

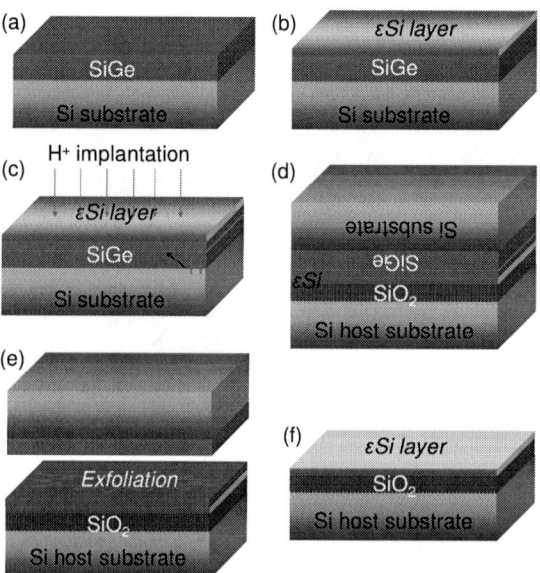

Figure 1: Schematic illustration of the process flow employed in the fabrication of ultra thin strained Si layer directly on oxide (or SSOI) by using thin layer transfer. (a) Growth of relaxed $Si_{1-x}Ge_x$ virtual substrate; (b) Growth of biaxially tensile strained Si on $Si_{1-x}Ge_x$ virtual substrate; (c) Hydrogen ion implantation into the grown heterostructure; (d) Bonding of the hydrogen-implanted onto a SiO_2/Si substrate; (e) Thermal annealing-induced layer exfoliation around the hydrogen implantation depth; (f) Thin strained Si directly on SiO_2/Si obtained after the removal of the residual $Si_{1-x}Ge_x$.

Fabrication of SSOI

The SSOI wafers used in this study were fabricated by using wafer bonding and a thin layer transfer method as illustrated in Figure 1. The tensile strain is generated by the heteroepitaxial growth of a Si thin film on a relatively thin (~500 nm) $Si_{1-x}Ge_x$ buffer layer, which was relaxed by helium ion implantation and thermal annealing. Both the strained and buffer layers were grown using reduced pressure chemical vapor deposition. After the growth, the obtained heterostructure was subject to hydrogen ion implantation under ion-cut optimal conditions. The implanted wafer was then bonded onto a handle wafer consisting of a Si substrate with a ~140 nm-thick SiO_2 layer on the top. Thermal annealing at an intermediate temperature (~500 °C) induces sub-surface microcracking, which leads to the exfoliation around the implantation depth provided that the bonded interface is stable enough. A strained Si thin layer on insulator is obtained after the removal of the residual $Si_{1-x}Ge_x$ buffer layer. The amount of the strain can be adjusted via Ge content in the buffer layer. Strained Si layers with a thickness as low as 4 nm are fabricated. Thicker wafers are usually obtained by additional homoepitaxy. Figure 2 displays an image of a 200 mm SSOI wafer with a thickness of ~20 nm.

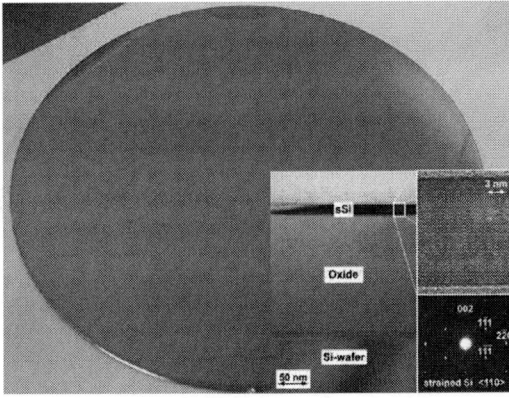

Figure 2: A 200 mm SSOI wafer fabricated using thin layer transfer process. Inset: Cross-sectional transmission electron microscopy (XTEM) image of a ~20 nm-thick strained Si on oxide. High resolution XTEM image and electron diffraction pattern of the strained layer are also shown. 300 mm SSOI wafers are also available.

SSOI nanoscale structures: Free surface-induced strain redistribution

Ordered arrays of strained Si nanowires and nano-islands were fabricated from SSOI substrates using electron beam lithography and dry reactive ion etching (RIE). The RIE process was performed at a pressure of 6 mTorr using a mixture of SF_6 (100 sccm) and O_2 (5 sccm) with a rate of ~2 nm s^{-1} at a temperature of -110 °C. Figure 3(a) displays a representative scanning electron microscope image of the array of strained silicon nanowires (30 nm × 1000 nm). Figure 3(b) shows a SEM image of rectangular nanostructures having a width of ~100-120 nm and a length of 400 nm. The corresponding cross-sectional TEM image is shown in Figure 3(c). The strain in the obtained nanostructure was investigated by NBED using a probe C_s-corrected FEI-Titan 80-300 microscope operating at 300 kV with a 20 μm 2nd condenser aperture, which defines both the semi-convergence angle of 0.3 mrad and the full width at half maximum (FWHM) of the illuminating electron beam of about 3 nm in the μProbe STEM mode.

The strain states in SSOI nanostructures were also probed using micro-Raman scattering spectroscopy in backscattering geometry in a LabRam HR800 UV spectrometer operating with a He-Cd ultraviolet (UV) laser line having a wavelength of 325 nm.

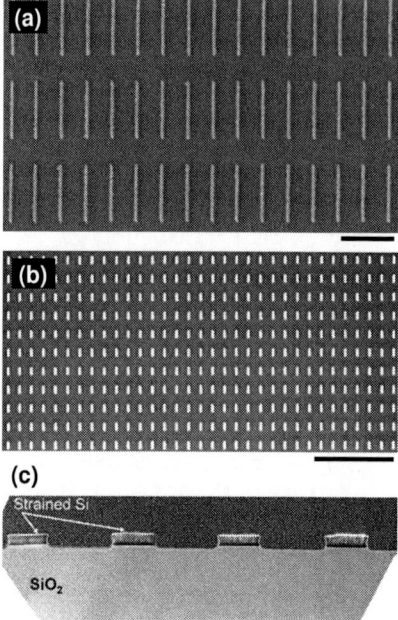

Figure 3: Scanning electron microscope image of an array of SSOI nanostructures investigated in this work: (a) 30 nm × 1000 nm nanowires. The scale bar denotes 1 μm; (b) 120 nm × 400 nm islands. The scale bar denotes 4 μm. (c) XTEM image of the array of 120 nm × 400 nm SSOI nanostructures.

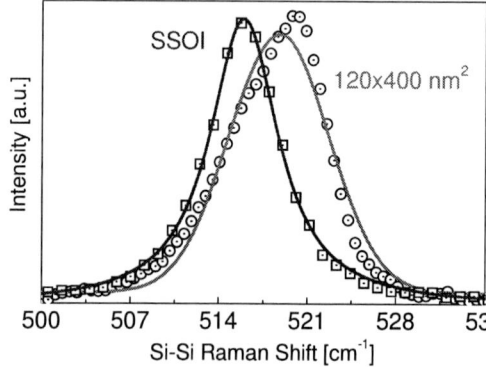

Figure 4: UV micro-Raman spectra of SSOI substrate (squares) and 120 nm × 400 nm strained Si nanostructure (circles). The solid lines correspond to *voigt* function fits.

It is expected that nanoscale patterning –a crucial step in device processing– induces partial and nonuniform relaxation of the strain in SSOI-based nanoscale devices (see Ref.

3 and references therein). To appreciate this phenomenon of strain relaxation for 120 nm × 400 nm nanostructures, Figure 4 displays the UV-Raman spectrum recorded for these nanostructures (circles). For the sake of comparison, the spectrum of SSOI substrate is also shown (squares). We note that the Si-Si peak of the unpatterned film is centered at ~515.9 cm^{-1} corresponding to an in-plane biaxial tensile strain of ~0.6 % (i.e., $\varepsilon_{xx} = \varepsilon_{yy} =$ ~0.6 %). Expectedly, the patterning induces a relaxation as it can be deduced from the ~3 cm^{-1} upshift of the Si-Si mode of the patterned nanostructures with respect to the Si-Si mode of the initial SSOI substrate. This upshift indicates that the sum of post-patterning in-plane strain components ($\varepsilon_{xx} + \varepsilon_{yy}$) has dropped to ~0.47 %. Additionally, we also note that the Si-Si mode of the SSOI nanostructures becomes relatively broader possibly due to the inhomogeneous distribution of the strain upon patterning.

More insights into the morphological properties of the investigated nanostructure were obtained through HRXTEM and NBED analyses. Figure 5(a) displays a typical HRXTEM image of a ~20 nm-thick strained nanostructure. A close inspection shows that the edges of the nanostructure have actually moved up, whereas the center has moved inwards making the nanostructure concave. This bending is clearly visible in the close-up image taken from the center of the nanostructure (Figure 4(b)). The measured angle between the vertical atomic planes ($2\bar{2}0$) at the edges along the transverse dimension is found to be 1.5°. Similar angle was found from NBED patterned recorded for a 200 nm × 200 nm strained nanostructures. Figure 6 displays NBED patterns measured at the two edges of a single nanostructure. The superposition of the diffraction patterns (inset) gives clear evidence of the local rotation of the lattice as a result of edge-induced relaxation. The observation of concave nanostructure supports the scenario of patterning-induced relaxation suggested earlier (3). It is noteworthy, as described in (3), that no significant bending is observed at the interface with the oxide (figure 4(b)). Moreover, under our experimental conditions, no bending is detected along the longitudinal dimension. Note, however, that due to their negligible dimensions as compared to the underlying substrate, the bending of SSOI nanostructures can not be described using the classical Stoney's formula or related formulations developed for nanoscale thin films, which assume a constant curvature in the bent structure. As we demonstrated recently (3), this assumption is not valid in the present case.

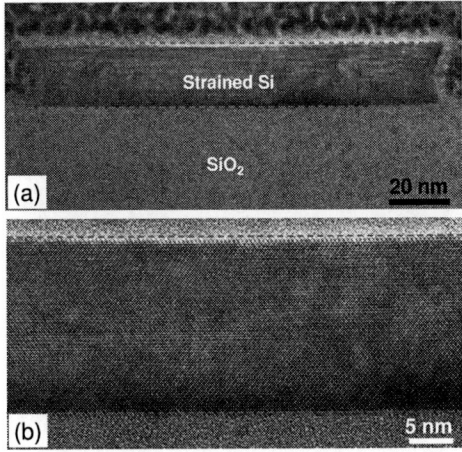

Figure 5: (a) High resolution XTEM image of a patterned 120 nm × 400 nm strained Si nanostructure directly on oxide. (b) A close-up image of the center of the 120 nm × 400 nm strained Si nanostructure. The dashed lines are guide to the eye showing that the upper side of the nanostructure is bent inward.

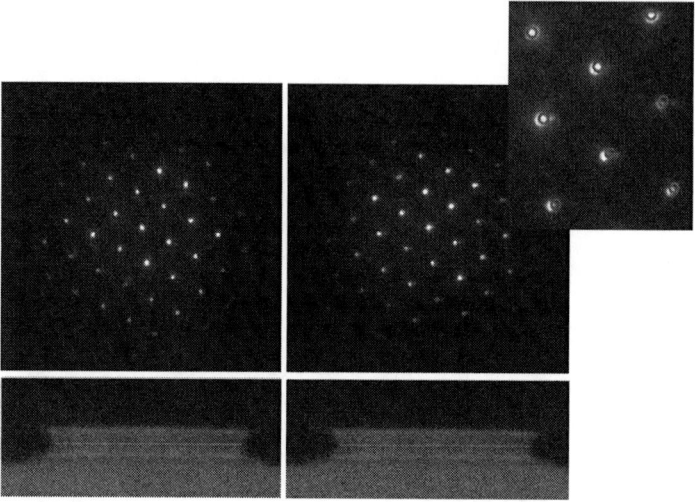

Figure 6: NBED patterns recorded for a 200 nm × 200 nm nanostructure. The patterns are measured at the two edges of the nanostructure as indicated in the corresponding XTEM images shown in the bottom. Inset: a zoom-in image of the superposed patterns where the red circles are the diffraction spots from the pattern in the left.

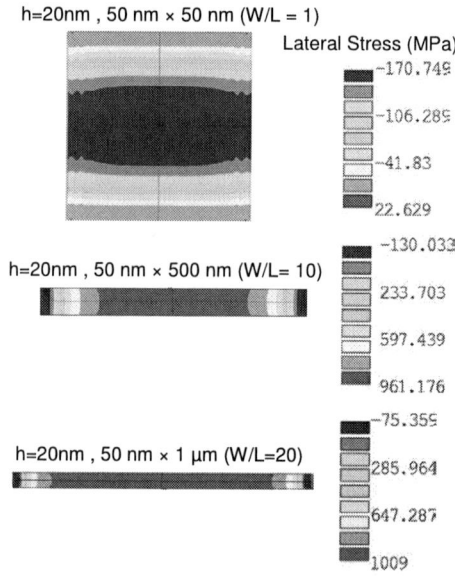

Figure 7: Results of the simulation of the first principle lateral stress as a function of the device dimensions. The starting material is a 20 nm-thick SSOI under a biaxial tensile stress of 1 GPa.

Strain nanoengineering in Si nanowires

The phenomenon of edge-induced relaxation described above is very sensitive to the geometry and size of the patterned nanostructures. This effect was investigated systematically mainly using continuum mechanics-based modeling. Figure 7 displays a representative set of our detailed 3D finite element simulations of the strain distribution in 20 nm-thick nanostructures with different width/length (W/L) ratios. We note that the residual post-lithography strain is very sensitive to the geometry of the device. Indeed, for an initial stress of 1 GPa and W = 50 nm, we found that the residual stress along the longitudinal direction increases with W/L. At W/L = 1, we notice a strong relaxation of the strain. Moreover, some regions become under a compressive strain in agreement with Raman data (4). For nanowire-like systems at W/L = 10, the initial stress is preserved in large portion along the longitudinal direction and is relaxed in the perpendicular direction. A more homogeneous strain profile and a lower strain relaxation at the edge is found for W/L = 20. Because the strain is preserved in the longitudinal direction and relaxed in the orthogonal direction, this indicates a transition from an initial biaxial to uniaxial strain.

Strained Si Nanowire

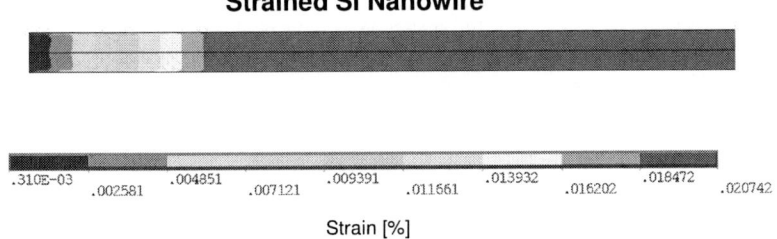

| .310E-03 | .004851 | .009391 | .013932 | .018472 |
| .002581 | .007121 | .011661 | .016202 | .020742 |

Strain [%]

Figure 8: Strain map in a quantum-sized strained Si nanowire fabricated by patterning a 5 nm-thin SSOI with an initial biaxial strain of 2%.

The interesting observation that emerges from the data in Figure 7 is that within a single nanowire the edge appears to fully relaxed or under a low compressive strain, whereas the center remains under high tensile strain. This phenomenon combined with quantum size effect can provide the basis of novel nanoscale devices. Indeed, built on the fact that the electronic structure of nanomaterials can be tuned (e.g., by quantum confinement and topological symmetries), this strain nanoengineering can be an effective strategy to potentially control the intrinsic properties of a single chemically homogenous nanowire with the aim to induce charge separation without doping or introducing an interface with other material. In fact, recent *ab initio* calculations demonstrate that local engineering of strain along the axis of a quantum-sized Si nanowire can induce LUMO and HUMO states to be localized in separate regions (5). This effect results mainly from the crystal structure of the Si nanowire and quantum confinement effects strengthen the energy level offsets. Consequently, in a partially strained Si nanowire, the frontier energy levels of the strained part are simultaneously higher and lower than the unstrained part, effectively forming a type-II junction. The experimental realization of this homojunction faces, however, important challenges in the fabrication, characterization, and control of strain-engineered quantum-sized Si nanowires. One of the key issues is to generate large strains, up to 2%, in specific areas within the nanowire. Figure 8 describes the simulated strain map in a quantum-sized Si nanowire fabricated by patterning a 5 nm-thin SSOI with an initial biaxial strain of 2%. The nanowire has a width of 5 nm and a length of 1 μm. For the clarity only a segment of the map is shown (not the whole nanowire). It is clear from the calculated map that the strain can be engineered locally within the nanowire. Here a 2% strain is preserved along the length in the region far from the edge. Close to the edge the strain is fully relaxed. Thus the control of the edge-induced relaxation in SSOI-based nanowire can be exploited to develop novel devices such as excitonic photovoltaic cells through the realization of the aforementioned strain-induced junction.

Conclusion

We have addressed nanopatterning-induced bending and local lattice rotation in ultrathin strained Si nanostructure directly on oxide. This bending results from free surface-induced relaxation of the tensile strain, which provokes an upwards displacement of the edges parallel to an out-of-plane contraction of the center leading to strained nanostructure with a concave shape. We have also demonstrated that this phenomenon is sensitive to the geometry of the patterned nanostructures. The use of SSOI as a playground to engineer the strain in Si nanowires and its exploitation to implement innovative devices was briefly discussed.

Acknowledgments

We are grateful to U. Doss, N. Zakharov, and H. Blumtritt for their technical help. This work was supported by the German Federal Ministry of Education and Research in the framework of the DECISIF project (contract no. 13 N 9881) and the nanostress project.

References

1. M. Chu et al., *Annu. Rev. Mater. Res.* **39**, 203 (2009).
2. J. Michel et al., *Nature Photonics* **4**, 527 (2010).
3. O. Moutanabbir et al., *Nanotechnology* **22**, 045701 (2011).
4. O. Moutanabbir et al., *Applied Physics Letters* **97**, 053105 (2010).
5. Z. Wu et al., *Nano Letters* **9**, 2418 (2009).

Lateral-liquid phase epitaxy of (101) Ge-on-insulator
from Si template by metal-induced crystallization

Masashi Kurosawa,[1,2] Naoyuki Kawabata,[1] Ryusuke Kato,[1]
Taizoh Sadoh,[1] and Masanobu Miyao[1]

[1] Department of Electronics, Kyushu University, 744 Motooka, Nishi-ku, Fukuoka,
819-0395, Japan
[2] JSPS Research Fellow, 8 Ichiban-cho, Chiyoda-ku, Tokyo, 102-8472, Japan

We investigate metal-induced lateral crystallization (MILC) of Si
on insulator to achieve (101) oriented Si films. Moreover, we
demonstrate the lateral liquid phase epitaxy of high quality
Ge(101) layers by using the MILC-Si films as crystal seed. This
technique will be employed to realize high-speed thin-film
transistors with Ge channel.

Introduction

Orientation-controlled single-crystalline Ge on insulating substrates is desired to achieve
high-performance thin-film transistors (TFT) with ultra-high speed operation. Recently,
we reported formation of single-crystalline Ge-on-insulator (GOI) structures by Si-Ge
mixing triggered lateral liquid phase epitaxy (L-LPE) using Si substrates as the crystal
seeds (1-3). Moreover, the Si-substrate-free method, where (001) and (111)-oriented Si
templates formed by Al-induced crystallization (4) were employed as the seed, enabled
formation of single-crystal (001) and (111)Ge on quartz substrates (5,6).

To realize ultimately high-speed transistors, GOI with the (101)-orientation should
be also developed, because hole mobility in Ge-MOS structures shows the maximum for
(101) orientation, though electron mobility is maximum for (111). In this letter, we
demonstrate the L-LPE of Ge on insulator by using the (101)-oriented Si grains formed
by the metal-induced lateral crystallization (MILC) process as the epitaxial template.

Experimental Procedure

In the experiment, p-type Si substrates with (001) orientation were used. They were
covered with SiO_2 films (thickness: 160 nm) by dry oxidation, and then a-Si layers
(thickness: 100 nm) were deposited on the SiO_2 films by molecular beam deposition
(MBD), and patterned by wet etching to form island areas (length: 300 μm, width: 20
μm). At the end of the island areas, Ni-deposited-regions (thickness: 5 nm) were formed
by a lift-off process, and annealing (550-600°C) was performed to induced MILC.
Subsequently, a-Ge layers (thickness: 100 nm) were deposited by MBD over the Si-
islands and patterned into narrow strips (length: 200 μm, width: 3 μm). Then capping
SiO_2 layers (thickness: 800 nm) were deposited by sputtering. Finally, these samples
were heat-treated by rapid thermal annealing (RTA) at 1000°C for 1 s to induce the rapid-
melting growth from the MILC Si-seeding areas.

Results and Discussion

Figure 1 shows growth characteristics of MILC in Si-islands at 550 and 600°C. Here, lateral growth length was defined as the lateral length perpendicular to the edge of Ni-supply-regions, as shown in insertion of Fig. 1. For annealing at 550°C, the lateral growth propagated linearly with the annealing time after incubation time of ~4 h. The growth speed is estimated to be ~4 μm/h. When annealing temperature is increased to 600°C, incubation time and growth speed become shorter (~6 min) and faster (~16 μm/h), respectively.

To investigate the crystal orientation of the grown Si-regions, electron back scattering diffraction (EBSD) measurements were performed. Typical results are shown in Fig. 2, which indicate the orientation of the sample surface (z-plane) and growth direction (y-plane). It is found that almost the whole region away from the Ni-supply-area (> 10 μm) is preferentially oriented to (101) orientation. Moreover, the growth direction, i.e. y-axis, is aligned to <001>.

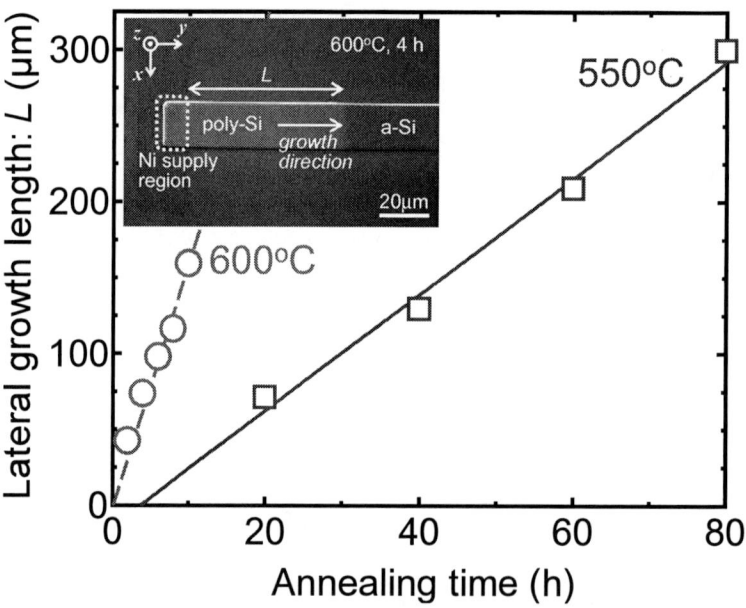

Figure 1. Lateral growth length for MILC samples at 550°C (□) and 600°C (○) as a function of annealing time. Insertion shows Nomarski optical micrograph of the sample after MILC process (600°C, 4 h).

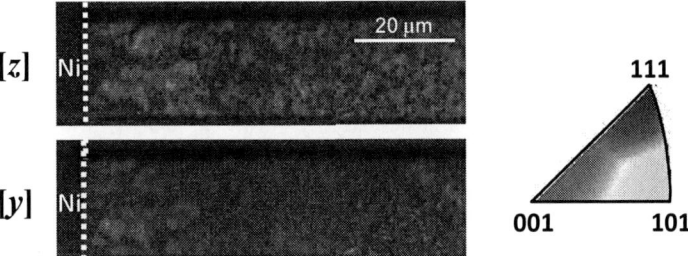

Figure 2. EBSD images showing crystal orientation of sample surface (z) and MILC grown direction (y) for the sample annealed at 600°C for 10 h.

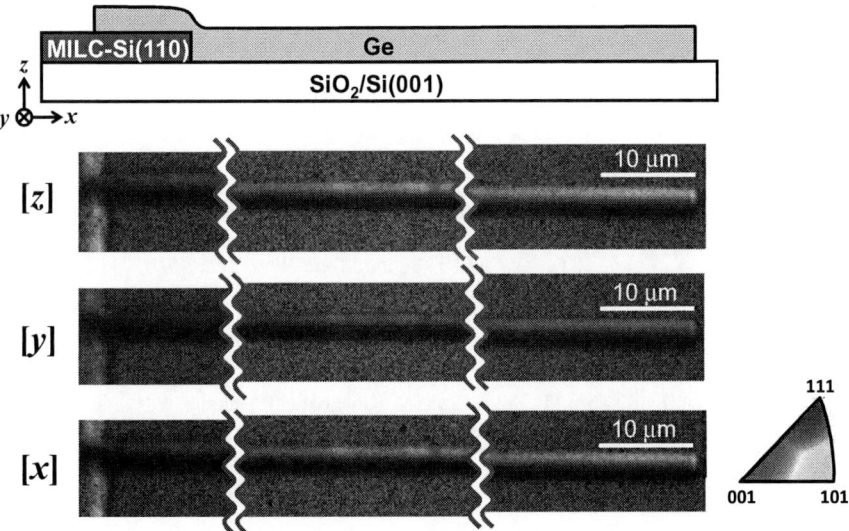

Figure 3. Schematic cross sectional view and EBSD images of the sample after RTA (1000°C, 1s).

Finally, we demonstrate the orientation controlled growth of Ge crystals by using these MILC-Si(101) islands as the template. The EBSD images of the sample after RTA are shown in Fig. 3, where the SiO$_2$ capping layer was etched off before the observation. These images show the crystal orientations along three orthogonal directions (z, y, and x directions). These results demonstrate that the epitaxial growth of GOI (101) is initiated from the MILC seed and propagates for 200 μm with keeping its orientation, where it is noticed that the lateral growth length (200 μm) is limited by the sample structures in the present study. Furthermore, cross-sectional transmission electron microscopy observations revealed no-defects in the laterally grown Ge regions. In this way, single crystalline Ge with (101) orientation has been realized on insulating layers. This technique will be useful to realize high-speed Ge-TFT.

Conclusion

Single-crystalline GOI with (101) orientation is realized by the Si-Ge mixing triggered L-LPE using MILC-Si(101) films as the crystal seeds. This Si-substrate-free method opens up the possibility of ultimately high-speed pMOS Ge transistors on insulating substrates.

Acknowledgements

A part of this work was supported by Semiconductor Technology Academic Research Center (STARC), and the Grant-in-Aid for Scientific Research from the Ministry of Education, Culture, Sports, Science, and Technology, Japan. M.K. wishes to thank to JSPS research program for young scientists.

References

1. M. Miyao, T. Tanaka, K. Toko, and M. Tanaka, *Appl. Phys. Express* **2**, 045503 (2009).
2. M. Miyao, K. Toko, T. Tanaka, and T. Sadoh, *Appl. Phys. Lett.* **95**, 022115 (2009).
3. K. Toko, T. Tanaka, Y. Ohta, T. Sadoh, and M. Miyao, *Appl. Phys. Lett.* **97**, 152101 (2010).
4. M. Kurosawa, N. Kawabata, T. Sadoh, and M. Miyao, *Appl. Phys. Lett.* **95**, 132103 (2009).
5. K. Toko, M. Kurosawa, H. Yokoyama, N. Kawabata, T. Sakane, Y. Ohta, T. Tanaka, T. Sadoh, and M. Miyao, *Appl. Phys. Express* **3**, 075603 (2010).
6. M. Kurosawa, K. Toko, N. Kawabata, T. Sadoh, and M. Miyao, *Solid-State Electronics* (accepted).

Growth-Direction Dependent Rapid-Melting-Growth of Ge-on-Insulator (GOI) and its Application to Ge Mesh-Growth

H. Yokoyama, Y. Ohta, K. Toko, T. Sadoh, and M. Miyao

Department of Electronics, Kyushu University, 744 Motooka, Nishi-ku
Fukuoka, 819-0395, Japan

Single crystal Ge-on-insulator (GOI) structures with various crystal orientations are necessary for realization of advanced high-speed and multi-functional devices. SiGe mixing triggered rapid-melting-growth of GOI is investigated as a function of seed-orientations and growth-directions. Single crystal growth of (100)-Ge strips is possible for all growth directions using (100)-oriented Si-seeds. However, rotational-growth is observed for some directions when Si-seeds with (110) and (111) orientations are employed. Such rotational-growth is completely suppressed by selecting the growth-directions deviating from the <111> direction by more than 35°. Based on this finding, growth of large mesh-patterned Ge layers with (100), (110), and (111) orientations are demonstrated.

Introduction

Single-crystalline Ge on insulator (GOI) formed on Si substrates are essential to integrate high-speed and multifunctional devices onto the Si platform. Recently, we proposed SiGe mixing triggered rapid-melting-growth, and realized (100)-oriented GOI strips (400 μm length, 3 μm width) by using Si(100) substrates as crystal seed (1). Furthermore, we optimized the sample structures and annealing conditions, which achieved the GOI(100) with high hole mobility (1040 cm^2/V s) (2-3). To expand the application fields of such GOI structures, GOI strips with various crystal orientations should be established. Since the electron and hole mobility of Ge show the highest values in (111) and (110) orientations, respectively (4), GOI structures with both (111) and (110) orientations are required to achieve ultimately high-speed complementary metal-oxide-semiconductor (CMOS) transistor circuits. In addition, recently, high quality epitaxial growth of ferromagnetic (Fe_3Si, Fe_2MnSi, Co_2FeSi et al. (5-7)) and optical (GaN et al. (8)) materials were reported on (111) Ge. Consequently, GOI crystals with various orientations are expected to be the powerful buffer layers to merge spin-transistors and opto-electronic devices with ultra-high speed MOS transistors.

In this study, rapid-melting growth for GOI with (111) and (110) orientations is examined. The detailed growth characteristics are investigated as a function of the Si-seed orientation ((111), (110), and (100)) and growth-directions. Unexpected rotational-growth of Ge strips is found depending on seed-orientations and growth-directions. Based on the consideration of bonding strength between various lattice planes, the growth conditions to prevent such rotational-growth are found. This enables single-crystal GOI strips with all crystal orientations ((100), (110), and (111)). Moreover, this orientation-control technique is applied to realize the large Ge mesh-growth with (100), (110), and (111) orientations.

Experimental Procedure

In the experiment, Si(100), (110), and (111) substrates covered with Si_3N_4 films (100 nm thickness) were employed. The Si_3N_4 films were selectively removed by wet etching to form seeding areas. Subsequently, amorphous-Ge layers (100 nm thickness) were deposited by the molecular beam deposition (MBD) technique, and they were patterned into narrow strip lines (400 μm lengths, 2-5 μm width). The sample structure is schematically shown in Fig. 1, where the angle θ of the growth direction of a strip was defined as the angle from the parallel direction to the orientation flatness ({110} plane). Then capping SiO_2 layers (800 nm thickness) were deposited by sputtering. Finally, these samples were heat-treated by rapid thermal annealing (RTA) (1000°C, 1 s). The grown layers were characterized by the electron backscattering diffraction (EBSD) analysis of the scanning electron microscopy.

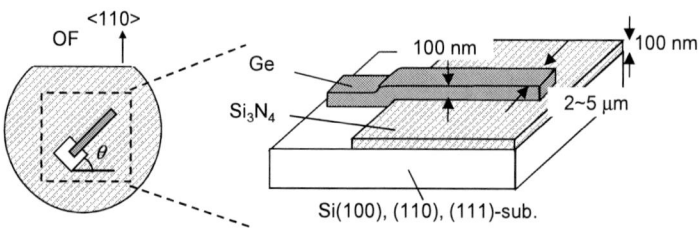

Figure1. Schematic sample structure before SiO_2 capping.

Results and Discussion

The EBSD images of the grown regions of the samples ($\theta = 0°$) are shown in Fig. 2, where the color mapping represents the crystal orientation perpendicular to the sample surface. For the samples grown on the Si(100) and (110) substrates, the crystal orientations of the Ge strips are identical to those of the Si seeding substrates (Figs. 2(a) and (b)). However, for the Si(111) substrate, the orientation gradually deviates from (111) with increasing distance from the seeding edge, and abruptly changes to (100), where a grain boundary is introduced (Fig. 2(c)). A detailed EBSD analysis showed that the crystal orientation along the growth direction was $[\bar{1}\bar{1}2]$ and did not change throughout the growth. These results indicate the rotational growth of GOI on Si(111). It is noticed that once the orientation reached to (100), it does not change any more. For other samples (Si(111), $\theta = 0°$), similar phenomena were observed. This suggests that the rotational growth is triggered by interface-energy minimization between Ge and insulator layers.

To reveal the rotational growth, the angular change in the orientations was evaluated for the samples grown on the (111) substrates with various growth directions. The maximum rotation angles in the growth (~50 μm) are summarized in Fig. 2(d), where the rotation angles show maximum values for the [$\bar{1}\bar{1}2$] and its equivalent directions, and changes with a 60° period. This period agrees with the crystal symmetry of the Ge(111) plane. This finding suggests that the rotating occurs by slipping between the (111) planes, because the bonding strength between the lattice planes is the weakest for the (111) planes in the diamond structure (9).

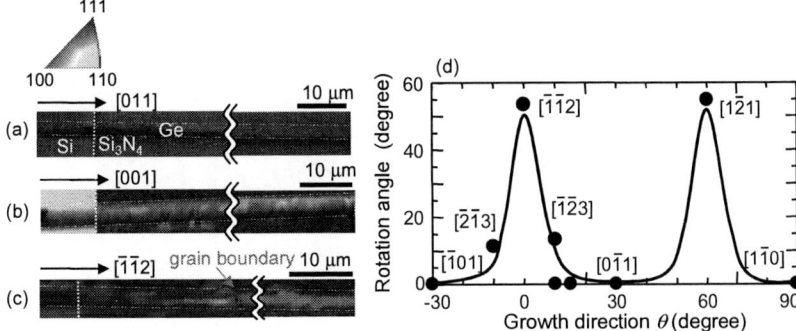

Figure 2. EBSD images of grown regions near seeding edges and at around 100 μm from seeds of samples ($\theta = 0°$) grown on (a) Si(100), (b) Si(110), and (c) Si(111), and maximum rotation angle as a function of the growth direction for samples grown on Si(111) substrates (d).

In order to examine the validity of this speculation, we explore the rotational growth for Ge strips using Si(110) seeding substrates. Figures 3(a) and 3(b) show the EBSD images of Ge strips aligned to the [$1\bar{1}2$] and [$1\bar{1}1$] directions ($\theta = 35°$ and 55°), respectively, where images obtained near the seeding edges and at 100 μm away from the seeds are compared. Rotational growth in the vicinity of <111> direction predicted in our speculation is clearly observed in these two figures. Similar experiments using Ge strips grown on Si(100) was also carried out. No rotational growth was observed for samples with various growth directions ($\theta = 0°$, 30°, 45°, 60°, and 90°), which supported the stable growth of single crystal Ge (100) strips on Si_3N_4 layers.

The maximum rotation angles of the crystal orientation of Ge surfaces are summarized as a function of the angle between the growth direction and the <111> direction. They are shown in Fig. 3(c), where data obtained from Ge strips grown on the Si (111), (110), and (100) seeding substrates are compared. It is particularly worth noting that rotating growth can be completely suppressed by selecting the growth directions deviating from the <111> direction by more than 35°. This is an important guiding principle to obtain rotation-free GOI with (100), (110), and (111) orientations.

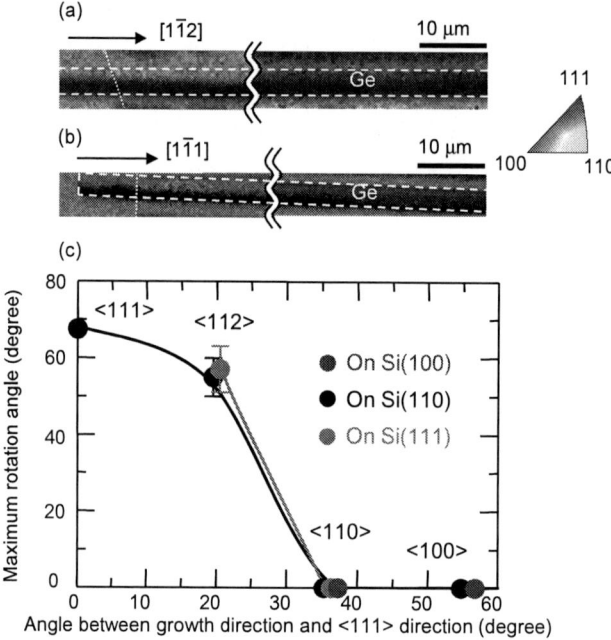

Figure 3. EBSD images for samples ($\theta = 35°$ (a), $55°$ (b)) grown on Si(110) substrates and (c) maximum rotation angle as a function of the angle deviating from <111> direction. Results obtained from Si(100), (110), and (111) substrates are summarized in (c).

In the rapid melting-growth, it is difficult to form wide GOI strips (width > 5 μm) because of Ge aggregation during the melt-back process (1). We expected that the mesh-pattern consisting of crossing narrow strips will be useful to obtain large area GOI. Thus, this orientation-control technique of Ge strips on insulator layers has been employed to obtain large mesh-patterned GOI(100), (110), and (111). In this experiments, square and hexagonal mesh-patterns are employed for Si(100) and (110), and (111) seeding substrates, respectively, to align all the growth direction of GOI strips to <110> or <100> to suppress the rotational growth. As a result, mesh patterned single crystal GOI(100), (110), and (111) structures with a large area (250μm x 500μm) are obtained, as shown in Figs. 4 (a)-4(c). On the other hand, mesh-pattern layers consisting of poly-Ge were obtained for the (110) and (111) substrates, if the growth directions of strips were near to the <111> direction. Thus, optimization of the growth directions is very important to obtain mesh patterned single crystal GOI structures. These mesh-patterned GOI structures will be useful to integrate advanced Ge-based multifunctional devices on the Si platform.

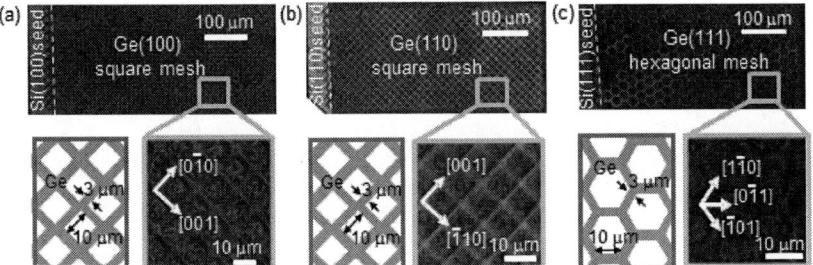

Figure 4. Schematic Ge-mesh patterns and EBSD images of grown regions on (a) Si(100), (b) Si(110), and (c) Si(111) substrates.

Conclusion

The rapid-melting-growth features of GOI initiated from Si(100), (110) and (111) substrates have been comprehensively studied as a function of lateral-growth directions. It is clarified that the growth propagates continuously keeping their seed-orientation by selecting the growth directions deviating from <111> by more than 35°. This enables single crystal GOI with various crystal orientations, i.e., (100), (110), and (111) orientations. Moreover, this orientation control technique has enabled mesh-patterned single crystal GOI(100), (110), (111) structures with a large area (250μm x 500μm). Such Ge networks with various orientation formed on insulators will be useful to realize Ge based advanced devices on the Si platform.

Acknowledgements

A part of this work was supported by Semiconductor Technology Academic Research Center (STARC), and the Grant-in-Aid for Scientific Research from the Ministry of Education, Culture, Sports, Science, and Technology, Japan.

References

1. M. Miyao, T. Tanaka, K. Toko, and M. Tanaka, Appl. Phys. Express **2**, 045503 (2009).
2. M. Miyao, K. Toko, T. Tanaka, and T. Sadoh, Appl. Phys. Lett. **92**, 022115 (2009).
3. K. Toko, M. Kurosawa, H. Yokoyama, N. Kawabata, T. Sakane, Y. Ohta, T. Tanaka, T. Sadoh, and M. Miyao, Appl. Phys. Express **3**, 075603 (2010).
4. T. Low, M. F. Li, G. Samudra, Y. C. Yeo, C. Zhu, A. Chin, and D. L. Kwong, IEEE Trans. Electron Devices **52**, 2430 (2005).
5. T. Sadoh, M. Kumano, R. Kizuka, K. Ueda, A. Kenjo, and M. Miyao, Appl. Phys. Lett. **89**, 182511 (2006).

6. K. Hamaya, H. Itoh, O. Nakatsuka, K. Ueda, K. Yamamoto, M. Itakura, T. Taniyama, T. Ono, and M. Miyao, Phys. Rev. Lett. **102**, 137204 (2009).
7. K. Kasahara, K. Yamamoto, S. Yamada, T. Murakami, K. Hamaya, K. Mibu, and M. Miyao, J. Appl. Phys. **107**, 09B105 (2010).
8. R. R. Lieten, S. Degroote, M. Leys, G. Borghs, J. Crystal Growth **311**, 1306 (2009).
9. S. M. Sze, *Physics of Semiconductor Devices*, 2nd ed. (Wiley, New York, 1981), Chap. 1, p. 11.

CHAPTER 4

ELECTRON DEVICE PHYSICS-1

62

A Simulation Comparison between Junctionless and Inversion-Mode MuGFETs

J. P. Colinge, A. Kranti, R. Yan, I. Ferain, N. Dehdashti Akhavan, P. Razavi, C.W. Lee, R. Yu, C.A. Colinge

Tyndall National Institute, University College Cork, Lee Maltings, Cork, Ireland

A new type of multigate MOSFET, called the junctionless nanowire transistor (JNT), has recently been proposed. It avoids junction formation problem and can be used to make very short-channel devices. Here, we compare the properties and performances of junctionless nanowire transistors with those of with Π-gate inversion-mode (IM) devices. The performances of silicon JNTs and IM transistors are evaluated in terms of short-channel effects, current drive, and gate capacitance. Junctionless devices are shown to have smaller short-channel effects than inversion-mode transistors with junctions. Comparison of carrier transport in the channel is made between junctionless nanowires and inversion-mode multigate field-effect transistors, and the benefits/drawbacks of bulk transport vs. surface transport are addressed.

I. Introduction

The junctionless nanowire transistor (JNT) is a heavily doped ($1\text{-}10\times10^{19}$ cm^{-3}) silicon nanowire with a Π-gate architecture. The device is basically a gated resistor that can be turned off because of the gate-to-nanowire work function difference. The doping concentration is constant throughout the device. There is, therefore, no doping concentration gradient, which greatly relaxes the processing thermal budget and facilitates the fabrication of ultrashort-channel devices. Experimental data describing the properties of long-channel n-channel and p-channel JNTs can be found in the literature (1). These devices have excellent subthreshold slope, very low off leakage current and high on/off current ratio. Here, we will describe he properties of short-channel JNTs and explore their potential for 25-10 nm node CMOS applications. The key for fabricating a JNT is to dope heavily the nanowire heavily ($1\text{-}10\times10^{19}$ cm^{-3}) such that it carries a decent current in the on state and to give it a small cross section such that it can be fully depleted in the off state. The threshold voltage depends on doping concentration, gate oxide thickness, device cross section and gate material (2). For an n-channel device it is suitable to use a material with a large workfunction such as P^{+} polysilicon or platinum, while p-channel devices require the use of a material with a small workfunction such as N^{+} polysilicon is advised. It is, however, possible to use midgap gate materials if the cross section of the device is small enough. In multigate FETs such as gate-all-around, Π-gate and Ω-gate devices it is known that, in order to avoid short-channel effects, the cross sectional dimensions (thickness and/or width) of the nanowire should, be at least twice as small as the channel length. A similar a rule of thumb applied to JNTs. Simulations, however, predict that JNTs should have less short-channel effects than MuGFETs with junctions and that gate lengths down to 3 nm should be achievable (3,4).

II. Current Drive

The devices presented here were simulated using the Atlas 3D software. Both JNTs and pi-gate inversion-mode MuGFETs have an effective gate length of 25nm and an equivalent gate oxide thickness (EOT) of 1nm. The length of the source and drain regions is 25nm. In all devices the gate material workfunction is chosen such that the off drain current at V_D=1V and V_G=0V is equal to 100 nA/um, which corresponds to the ITRS high-speed (HP) requirements. A nanowire pitch equal to $2 \times W_{Si}$ is considered for all devices. Thus if W_{Si}=10nm, the off current in an individual nanowire is equal to 100nA/um×20nm=2nA.

For an effective gate length of 25nm the current drive of the JNT is pretty similar to that of an IM MuGFETs. Figure 1 shows the on current as a function of JNT doping concentration and source and drain *underlap*. In the inversion-mode device, a source and drain *overlap* of 4 nm (on each side) is used. Thus the physical gate length of the IM device with L_{eff}=25nm is thus equal to $L=L_{eff}+2 \times overlap$=25nm+8nm=33nm. The IM device is "undoped" with N_A=5×10^{15} cm^{-3} and a S/D doping equal to 10^{20}cm^{-3}. In the JNTs, additional doping to a level of 10^{20}cm^{-3} is used in the source and drain regions to reduce their resistance. There is an underlap between these highly-doped regions and the gate, noted "*d*". The current in the JNTs is quite dependent on the underlap, *d*, with the high-doping S&D regions (Fig 1). For a JNT underlap of 10 nm (on both sides of the gate), the drive current increases with N_D. This is mostly due to the fact that the resistance of the underlap S/D regions is quite high high and is the limiting factor in the current drive. When N_D is increased, that resistance is decreased and the on current increases. For a JNT underlap of 0 nm (the heavily-doped S/D regions extend to the edges of the gate), the drive current decreases with N_D. This is due to the fact that workfunction of the gate material has to be increased with N_D in order to maintain a constant I_{off}=100nA/μm. This increases the threshold voltage, and, therefore, decreases current drive. Both effects tend to compensate each other for an underlap distance of 4 nm, as can be seen in Figure 1.

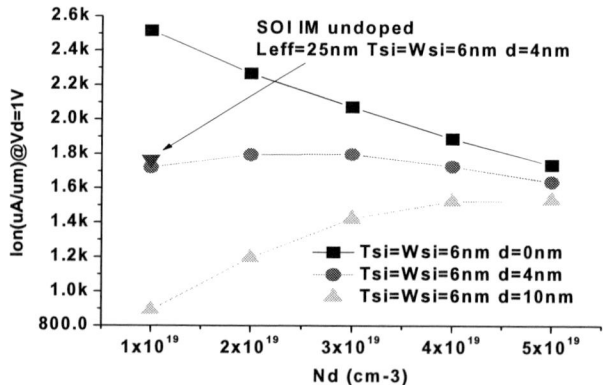

Figure 1: On current for JNTs with different N_D values and underlap values. L_g=25nm. $t_{si}=W_{si}$=6nm EOT=1nm. *d* is the S&D underlap. $V_G=V_D$=1V and I_{off}=100nA/μm @ V_D=1V, V_G=0V.

<u>Current transport</u>

Figure 2 shows the electron concentration in an IM transistor as well as in two JNTs with either a light ($N_D=10^{19}$cm^{-3}) or a heavier ($N_D=4\times10^{19}$cm^{-3}) doping concentration. In the IM device, the current is entirely confined in a surface channel where the electron concentration is higher than 10^{20}cm^{-3} (quantum effects are neglected here). In a lightly-doped JNT ($N_D=10^{19}$cm^{-3}) an accumulation layer may be formed at $V_D=1$V. In JNTs with higher doping concentrations, the current is purely a bulk current (Fig. 2).

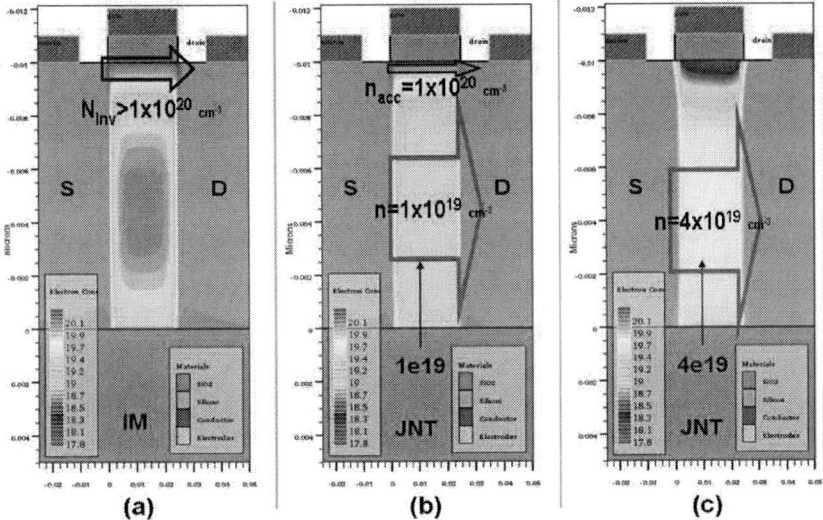

(a) **(b)** **(c)**

Figure 2: Electron concentration of IM and JNT transistors. a: IM, b: JNT with $N_D=10^{19}$cm^{-3}; c: JNT with $N_D=4\times10^{19}$cm^{-3}. S&D doping is 10^{20}cm^{-3}. $V_G=V_D=1$V and $I_{off}=100$nA/μm @ $V_D=1$V, $V_G=0$V. $W_{si}=t_{si}=10$nm; $L=25$nm, EOT=1nm. Cross section is taken in the middle of the nanowire.

It is interesting to compare the carrier density profile with the mobility profile across the devices. It is well known that bulk mobility decreases with doping concentration because of impurity scattering (5). Surface mobility in MOSFETs also decreases with the electric field in the channel, and, therefore, with any reduction of EOT (6), as shown in Figure 3.

As can be seen in Figures 2c and 2c, the current in a JNT is mostly, if not entirely, constituted by a body current that flows in the neutral (*i.e.* not depleted) center of the nanowire). Thus for a device doped to a level of 4×10^{19}cm^{-3}, the electron concentration in the channel is n= 4×10^{19}cm^{-3}. This is quite low when compared to the concentrations in excess of 10^{20} cm^{-3} found in the inversion channel of a regular MOSFET (Figure 2a). In the absence of strain enhancement, the electron mobility in the channel of an inversion-mode transistor with an equivalent oxide thickness (EOT) of 1 nm can be as low as less than 20 cm^2/Vs (Figure 4a). This is due to the high electric field (~1 MV/cm) in the channel. The current in JNT is a bulk current; the electric field perpendicular to the current flow is essentially equal to zero (7). On the other hand, impurity scattering due to the high doping concentration in the channel keeps electron mobility below ~70 cm^2/Vs.

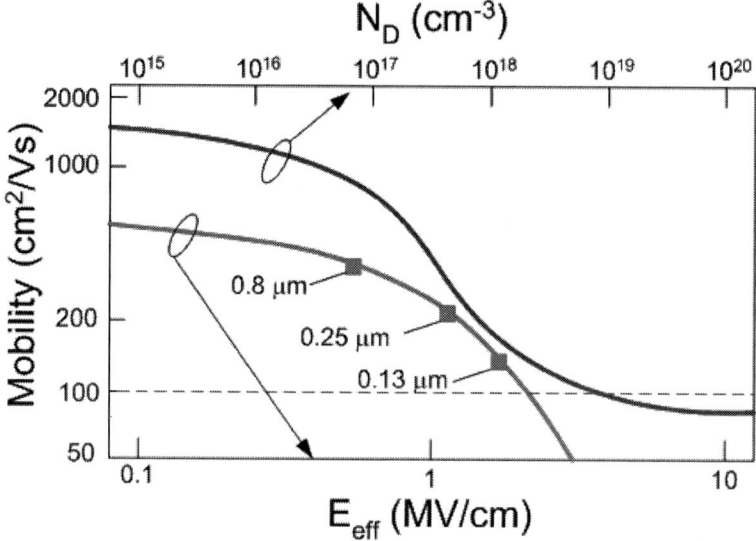

Figure 3: Electron mobility in a surface inversion channel as a function of the electric field in the channel and electron mobility in bulk silicon as a function of doping concentration (5,6). The "0.8 μm", "0.25 μm" and "0.13 μm" labels refer to the corresponding technology nodes.

Mobility

As device dimensions are further scaled down, the EOT will keep decreasing. This will further decrease surface mobility in IM MOSFETs, even in if strain-based mobility enhancement techniques are used. When JNTs are scaled down, the cross-section of the nanowire must be decreased and the doping concentration must be increased. Quite interestingly, increasing the doping concentration above 4×10^{19}cm^{-3} does not degrade mobility, and a value of $\mu_n \approx 70$ cm^2/Vs is maintained even if the doping concentration reaches 10^{20}cm^{-3} (Figure 3). This advantage of the JNT over the IM MOSFET can clearly be seen in Figure 4: the mobility *in the channel* is 10-30 cm^2/Vs in the IM device, while it is equal to 70 cm^2/Vs in the JNT. Another benefit of bulk transport is the reduction of noise. A side benefit of transport in material with high impurity scattering rate is the small mobility degradation brought about by phonon scattering when temperature is increased (8). This may be another significant advantage of the JNT over IM devices, considering that in a real environment the operating temperature of a transistor is around 100°C.

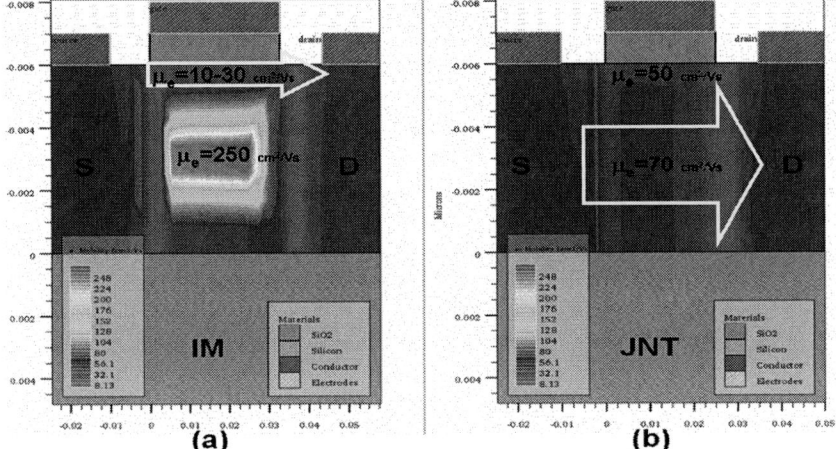

Figure 4: Electron mobility of IM and JNT transistors. a: IM, b: JNT with N_D=4x10^{19}cm^{-3}. S&D doping is 10^{20}cm^{-3}. V_G=V_D=1V and I_{off}=100nA/µm @ V_D=1V,V_G=0V. W_{si}=t_{si}=6nm; L_{eff}=25nm, EOT=1nm. Cross section is taken in the middle of the nanowire.

III. Short-channel effects

The drain-induced barrier lowering is defined as $DIBL = V_{TH}\big|_{V_D=50mV} - V_{TH}\big|_{V_D=1V}$. It has recently been has recently been acknowledged that DIBL is as a key parameter for low-voltage CMOS, as it plays an important role in determining the Ion/Ioff ratio. Achieving a low DIBL has become as important as increasing carrier mobility for improving circuit performance (9).

In a MOSFET with junctions, short-channel effects are in part due to the reduction of threshold voltage when channel length is decreased or when drain voltage is increased. This effect is due to the presence of space-charge regions in the channel associated with the source and drain PN junctions (SCE), and to the growth of the drain space-charge region with drain voltage (DIBL):

$$V_{TH} = V_{THO} - \text{SCE} - \text{DIBL} \qquad [1]$$

where V_{THO} is the long-channel threshold voltage (10,11). In a MOSFET with a physical gate length $L_{physical}$ (Figure 5c) the effective gate length is L_{eff} when the device is on, and the effective gate length is L_{SCE}, when the device is off. L_{SCE} is defined as the length of the (p-type) potential barrier in the channel that separates the source from the drain. It is smaller than $L_{physical}$ because of the extension of the charge-space regions from the source and drain PN junctions in the channel region. Since $L_{SCE}<L_{eff}$, the channel length is effectively shorter when the device is off than when it is on. When the drain voltage is increased, the DIBL effect further reduces L_{SCE} and further increases short-channel effects. In the junctionless transistor, the doping concentration is constant across the device. There are no charge-space regions where the source or drain is in contact with the channel region. Furthermore, the electrostatic "squeezing" of the channel in the off device propagates to some distance into the source and drain; as a result, $L_{eff}>L_{physical}$

when the device is off (Figure 5a). When the device is on, the "squeezing" effect is removed, such that $L_{eff} \leq L_{physical}$ (Figure 5b). As a result, L_{eff} is larger on the off state than in the on state, which greatly improves short-channel effects. This property is unique to the JNT, and is due to the fact that current blocking region is not located underneath the gate, but in the drain, at the edge of the gate (Figure 5a).

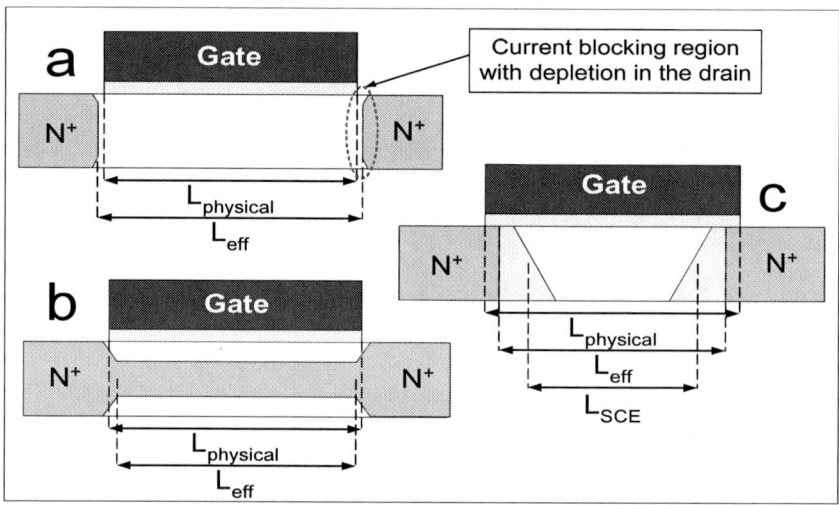

Figure 5: Different effective gate lengths in a) a junctionless transistor in the off state, b) a junctionless transistor in the on state, c) an inversion-mode transistor.

Figure 6 presents the simulated $I_D(V_G)$ characteristics of a JNT and an inversion-mode Π-gate FET. Both devices have the same dimensions. The gate length is 10 nm. The DIBL is 153 mV and 48 mV in the IM FET and the JNT, respectively. Similarly, the subthreshold slope is 84 mV/dec and 66 mV/dec in the IM FET and the JNT, respectively.

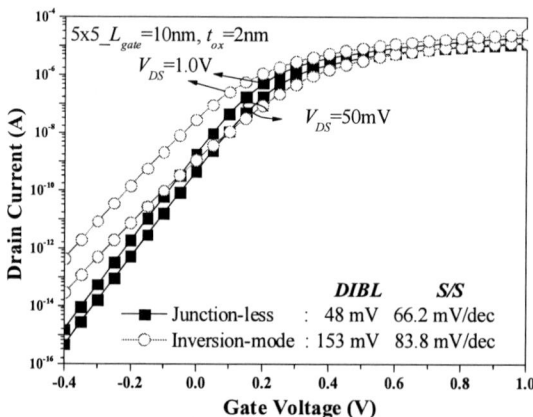

Figure 6: Simulated $I_D(V_G)$ characteristics of a JNT and an inversion-mode Π-gate FET. L=10 nm, W_{si}=5 nm, t_{si}=5 nm.

This reduction of short-channel effects have been observed in actual devices (12). Figure 7 shows the TEM picture of the cross section of a JNT, clearly showing the Π-gate architecture. The corresponding measured $I_D(V_G)$ characteristics are shown in Figure 8. The gate length is 50 nm, and the DIBL is only 7 mV. As a reference, the best DIBL values measured on FinFETs with the same gate length is on the order of 40 mV.

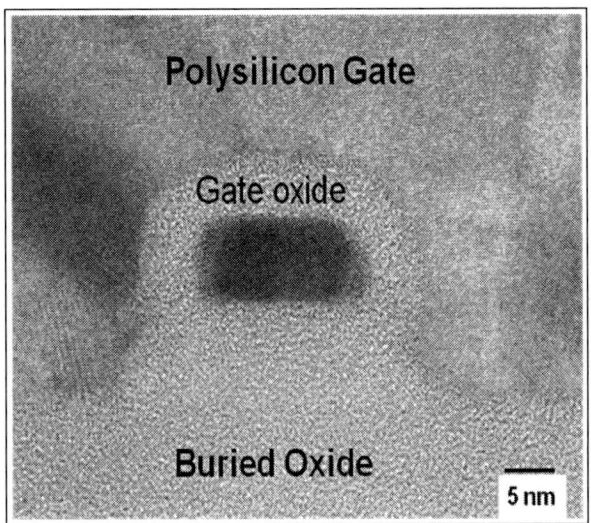

Figure 7: Cross section TEM of a JNT with L=50 nm, W_{si}=15 nm, t_{si}=8 nm and EOT=5 nm.

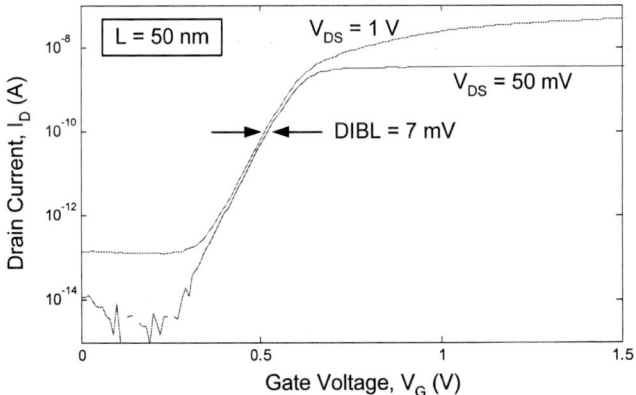

Figure 8: Measured $I_D(V_G)$ characteristics of a JNT with L=50 nm, W_{si}=15 nm, t_{si}=8 nm and EOT=5 nm.

The subthreshold slope corresponding to Figure 8 is shown in Figure 9, as a function of drain current. Slopes close to 60 mV/decade are observed at both low and high drain voltage, which is also an indication of low short-channel effects.

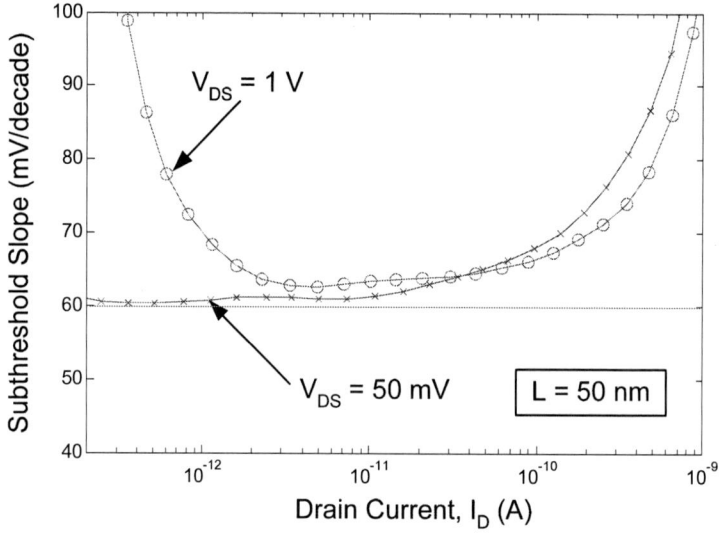

Figure 9: Subthreshold slope of the JNT presented in Figure 8; L=50 nm, W_{si}=15 nm, t_{si}=8 nm and EOT=5 nm.

IV. Gate capacitance

In an inversion-mode MOSFET, the channel is right underneath the gate oxide. For low V_{DS} values, the gate capacitance is equal to $W \times L \times C_{ox}$. In a junctionless nanowire transistor, the channel is buried in the centre of the nanowire and the gate capacitance results in the series association of C_{ox} and C_{depl}, where C_{depl} is the capacitance of the depletion region between the Si-SiO$_2$ interface and the channel. As a result the gate capacitance is lower in a JNT than in a multigate inversion-mode device of similar dimensions. It is also worth noting that the drain current in an inversion-mode transistor is directly proportional to C_{ox}. In a JNT, on the other hand, the drain current is not directly linked to C_{ox}, at least not in manner that is as straightforward as in an inversion-mode device; rather, the current is proportional to the doping concentration in the channel and to the cross section of the nanowire.

Figure 10 shows the gate capacitance $C_{gg} = C_{gs} + C_{gd}$ in inversion-mode (IM) and junctionless (JL) MuGFETs with an effective gate length, L_{eff}, of 25 nm and a cross section of 10 nm × 10 nm. $V_D = 50$ mV, $t_{ox} = 1$ nm. Two IM devices are presented: the first device has a physical gate length of 33 nm and a gate-to-source (or drain) overlap of 4 nm (noted "L_{over}=4nm"). The effective gate length is thus 33 nm - (2×4 nm) = 25 nm. The doping concentration of the source and drain is 10^{20} cm^{-3}, and the junction is abrupt. The second IM device (noted "L_{over}=0nm") has a zero overlap between the gate and the source (or drain). In this device the physical gate length is equal to the effective gate length (25 nm). The difference between the two curves is the overlap capacitance. The horizontal dashed horizontal line in Figure 10 shows the value $C_{ox} L_{eff} (W_{Si} + 2 t_{Si})$, which is the capacitance of the gate oxide running along the two sidewalls and the top of the nanowire over the effective gate length of the device (L_{eff}). The gate capacitance of the IM transistor with no source/drain overlap ($L_{over}=0$) becomes larger than $C_{ox} L_{eff} (W_{Si} + 2 t_{Si})$ at the highest gate voltage values because the simulated structure is a Π-gate device, and not a trigate transistor, which creates some additional capacitive coupling between the lateral sides of the gate and the bottom of the nanowire (*i.e.*: the effective number of gates is larger than 3).

The gate capacitance is also given for four JNTs with a physical gate length of 25 nm and doping concentrations ranging from 10^{19} cm^{-3} to 4×10^{19} cm^{-3}. The gate workfunction in each device is selected to achieve an off current of 100 nA/μm at $V_G=0$V and $V_{DS}=1$V. One can see that the JNT with the lightest doping concentration reaches accumulation at $V_G=1$V since the gate capacitance becomes equal to that of the IM device with no overlap. In the other JNTs, the surface of the nanowire is still depleted at $V_G=1$V, which results in a significant lower gate capacitance. The gate charge needs to switch a device on being

equal to $Q = \int_0^{V_{DD}} C_{gg}\, dV_G$ one can observe that Q is significantly lower in a JNT than in IM

devices, and that it decreases as the JNT doping concentration is increased.

It is interesting to note that, in some cases, C_{gd} can be zero or even negative in JNTs, due to the particularly low DIBL effect (13). This gives rise to a very low Miller capacitance, which might be of great interest for RF applications.

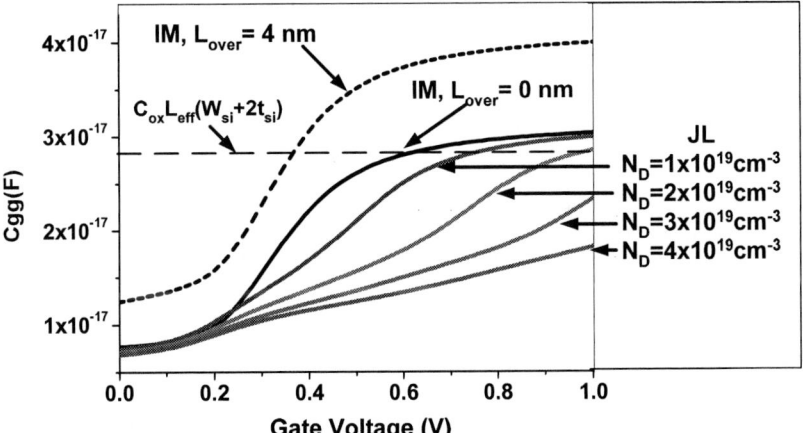

Figure 10: Gate capacitance ($C_{gg} = C_{gs} + C_{gd}$) in inversion-mode (IM) and junctionless (JL) MuGFETs with an effective gate length, L_{eff}, equal to 25 nm and a cross section of 10 nm × 10 nm. $V_D = 50$ mV, $t_{ox} = 1$ nm. The IM device with 4 nm S&D overlap ($L_{over} = 4$ nm) has a physical gate length of 25 nm + (2×4 nm) = 33 nm.

V. Conclusions

This Paper compares the properties and performances of junctionless nanowire transistors (JNTs) with those of with Π-gate inversion-mode (IM) devices. The performances of silicon JNTs and IM transistors are evaluated in terms of short-channel effects, current drive, and gate capacitance. Junctionless devices are shown to have smaller short-channel effects than inversion-mode transistors with junctions. Comparison of carrier transport in the channel is made between junctionless nanowires and inversion-mode multigate field-effect transistors, and the benefits/drawbacks of bulk transport *vs.* surface transport are addressed. JNTs have a lower gate capacitance than IM MOSFETs.

Acknowledgments

This work was supported by the Science Foundation Ireland grant 05/IN/I888: Advanced Scalable Silicon-on-Insulator Devices for Beyond-End-of-Roadmap Semiconductors. This work has also been enabled by the Programme for Research in Third-Level Institutions. This work was supported in part by the European Community (EC) Seventh Framework Program through the Networks of Excellence NANOSIL and EUROSOI+ under Contracts 216171 and 216373.

References

1. J.P. Colinge, C.W. Lee, A. Afzalian, N. Dehdashti Akhavan, R. Yan, I. Ferain, P. Razavi, B. O'Neill, A. Blake, M. White, A.M. Kelleher, B. McCarthy and Richard Murphy, *Nature Nanotechnology*, **5-3**, 225 (2010)
2. A. Kranti, R. Yan, C.-W. Lee, I. Ferain, R. Yu, N. Dehdashti Akhavan, P. Razavi and JP Colinge, *Proceedings ESSDERC*, 357 (2010)
3. C.W. Lee, A. Afzalian, N. Dehdashti Akhavan, R. Yan, I. Ferain and J.P. Colinge, *Applied Physics Letters*, **94**, 053511 (2009)
4. L. Ansari, B. Feldman, G. Fagas, J.P. Colinge and J.C. Greer, *Applied Physics Letters*, **97**, 062105 (2010)
5. C. Jacoboni, C. Canali, G. Ottaviani and A. Quaranta, *Solid State Electronics*, **20-2**, 77 (1977)
6. S.E. Thompson, M. Armstrong, C. Auth, S. Cea, R. Chau, G. Glass, T. Hoffman, J. Klaus, Z. Ma, B. Mcintyre, a. Murthy, B. Obradovic, L. Shifren, S. Sivakumar, S. Tyagi, T. Ghani, K. Mistry, M. Bohr and Y. El-Mansy, *IEEE Transactions on Electron Devices*, **51-4**, 191 (2004)
7. J.P. Colinge, C.W. Lee, I. Ferain, N. Dehdashti Akhavan, R. Yan, P. Razavi, R. Yu, A.N. Nazarov and R.T. Doria, Applied Physics Letters, **96**, 073510 (2010)
8. C.W. Lee, A. Borne, I. Ferain, A. Afzalian, R. Yan, N. Dehdashti and J.P. Colinge, *IEEE Transaction on Electron Devices*, **57-03**, 620 (2010)
9. T. Skotnicki and F. Boeuf, *Symposium on VLSI Technology Digest of Technical Papers*, 153 (2010)
10. T. Skotnicki, G. Merckel and T. Pedron, *IEEE Electron Device Letters*, **9**, 109 (1988)
11. T. Skotnicki, *Proceedings of the 30th European Solid-State Device Research Conference*, 19 (2000)
12. C.W. Lee, I. Ferain, A. Kranti, N. Dehdashti Akhavan, P. Razavi, R. Yan, R. Yu, B. O'Neill, A. Blake, M. White, A.M. Kelleher, B. McCarthy, S. Gheorghe, R. Murphy and J.P. Colinge, *Solid-State Devices and Materials Conference (SSDM)*, 1044 (2010)
13. S. Cho, I.M. Kang, R. Kim, *IEICE Electronics Express*, **7-19**, 1499 (2010)

Comparative Study of Random Telegraph Noise in Junctionless and Inversion-Mode MuGFETs

A.N. Nazarov[a], C.W. Lee[b], A. Kranti[b], I. Ferain[b], R. Yan[b], N. Dehdashti Akhavan[b], P. Razavi[b], R. Yu[b], J.P. Colinge[b]

[a] Lashkaryov Institute of Semiconductor Physics, NASU, Kiev 0328, Ukraine
[b] Tyndall National Institute, University College Cork, Lee Maltings, Cork, Ireland

> Random telegraph-signal noise (RTN) is observed in n-channel junctionless metal-oxide-silicon field effect transistors (MOSFETs) and n-channel inversion-mode MOSFETs fabricated in the same technological process on UNIBOND® SOI wafers as a function of gate and drain voltages and measurement temperature. It is shown that the RTN of the drain current in the JL transistor operating in linear mode begins to appear when the transistor starts to form an accumulation channel and has considerable lower amplitude than in the IM MOSFET. On the basis of analysis of the average charge capture and emission time of the charge from the traps responsible for the RTN the main parameters of the traps are determined.

Introduction

A new type of multigate (MuG) MOSFET, called the junctionless (JL) transistor, has recently been proposed (1). It avoids junction formation problems and can be used to make very short-channel devices (2). In such a device the channel is located inside of the Si nanowire (NW), which results in considerable smaller electric field and reduced carrier scattering inside the channel (3). This work addresses random telegraph signal noise (RTN) in JL MOSFETs and comparison with NW inversion-mode (IM) devices is made.

Experimental

Multigate silicon NW n-type MOSFETs with pi-gate architecture were fabricated on UNIBOND® silicon-on-insulator wafers (see Figures 1(a) and 1 (b)). The width and thickness of the devices is near 10 nm and the gate length is 1 μm. The gate oxide and buried oxide (BOX) thickness are 7 nm and 340 nm correspondingly. The JL MOSFET has uniform n-type doping concentration of 1×10^{19} cm^{-3} in the source, drain and channel regions (Figure 1 (c)). "Standard" IM pi-gate MOSFET have the same dimensions as the JL devices but have a p-type channel doping concentration of 2×10^{18} cm^{-3} and an n-type doping concentration of 1×10^{20} cm^{-3} in the source and drain regions (Figure 1 (d)). The JL MOSFETs have a P$^+$-polysilicon gate electrode and the IM devices have an N$^+$-polysilicon gate.

The drain current - gate voltage (I_D-V_G) characteristics and RTN in drain current were measured by Agilent B1500A Semiconductor Parameter Analyzers in temperature range from 300 to 428 K. The RTNs were measured in sampling mode of the equipment operation. The RTNs were measured at gate voltage above the threshold voltage both of the IM and JL devices, which was determined by second derivative of the I_D-V_G characteristic.

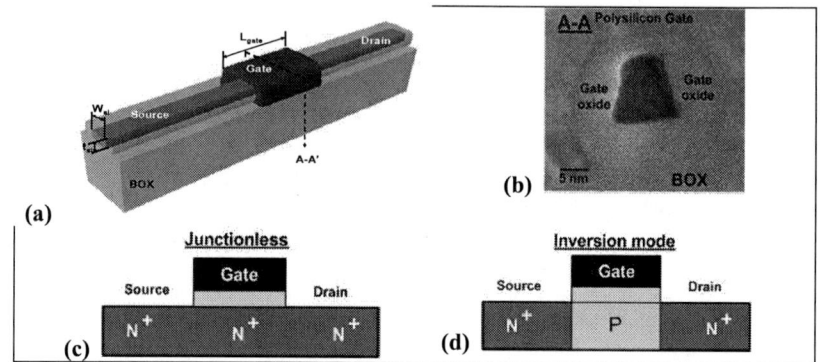

Figure 1. Schematic view of architecture of the MOSFETs (a) and HRTEM photograph of the A-A section of the device (b). Schematic view of doping in the junctionless (c) and inversion-mode (d) n-channel devices.

Results and Discussion

Dependence on Applied Gate Voltage

The Randon Telegraph Noise (RTN) of the drain current is measured above threshold voltage in linear regime and for the similar gate voltage overdrive $(V_G$-$V_{TH})$ both for IM and JL MOSFETs. The results are presented in Figure 2. The RTN can be described by two drain current levels. The time spent in the high-current (ON) state represents the capture time, τ_c, whereas the time spent in the low-current (OFF) state represents the emission time, τ_e (4). Increasing of the gate voltage results in an increase of the average emission time in both IM and JL devices, though for IM MOSFET this tendency is more pronounced (see Fig. 2 (a)). In addition the average emission time in the JL device is considerably smaller that in the IM device.

It should be noted that the magnitude of the relative RTN and flicker noise (FN) amplitude of the drain current, $\Delta I/I_{AV}$, is considerably higher in the IM devices in than in the JL devices (Figure 3(a)). Determination of the average drain current and it fluctuation for RTN and FN are presented in inset Figure 3 (a). The relative RTN amplitude of the drain current in both types of the devices depends on gate voltage overdrive following a low in $\Delta I_{RTN}/I_{AV} \approx (V_G$-$V_{TH})^{-1}$ (Figure 3(b)). A similar functional dependence can be obtained in first approximation for IM MOSFETs from the expression presented in the Reference (5):

$$\frac{\Delta I_D}{I_D} = \frac{q}{WLC_{ox}} \times \frac{g_m}{I_D} \approx \frac{q}{WLC_{ox}(V_G - V_{TH})} \times \left\{ 1 + \frac{1}{1 + \theta^{-1}(V_G - V_{TH})^{-1}} \right\}, \qquad [1]$$

where g_m is the transconductance, W and L are the electrical channel width and length, C_{ox} is the gate oxide capacitance, and θ is the mobility degradation factor. Expression [1] can also be used for AM MOSFETs operating in linear regime if the threshold voltage is replaced by the flat-band voltage. If $\theta(V_G - V_{TH})$ is small enough, the second component in the brackets can be neglected and the total amount in the brackets is equal to 1. Thus an occurrence of the inverse dependency of the relative RTN amplitude of the drain current versus applied gate voltage for the JL device is evidence of generation at the SiO$_2$-Si interface of an accumulation channel, because the dependence of RTN on the bulk channel current in the JL MOSFET is a more complicated function of gate voltage (6).

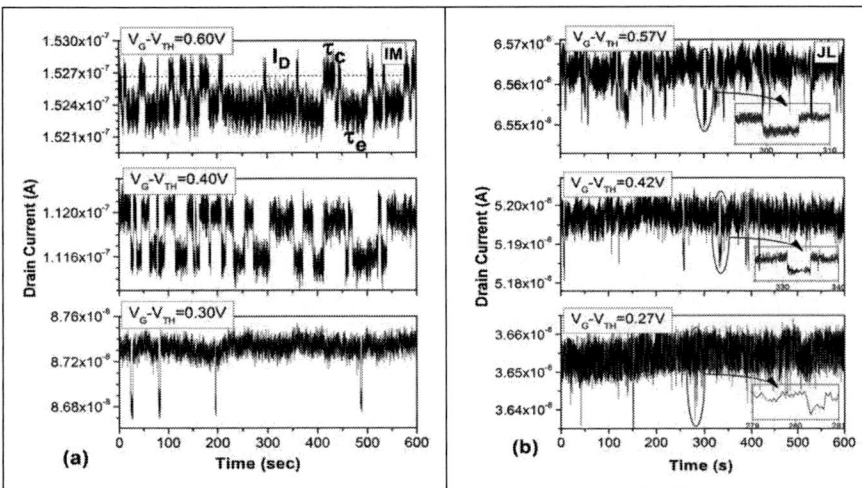

Figure 2. Random telegraph-signal noise for IM (a) and JL (b) MOSFETs for various applied gate voltages at V_D=50 mV and at room temperature.

The drain current I_D in JL MOSFET operating in accumulation mode can be expressed as (7)

$$I_D = I_{ACC} + I_{CH} = \mu_S C_{OX} \frac{W}{L}(V_G - V_{FB})V_D + \frac{q\mu_N n_C S}{L} V_D,$$ [2]

where I_{ACC} and I_{CH} denote the current components flowing through the accumulation channel and the bulk channel, respectively. μ_S and μ_N are the electron mobility in the accumulation layer and the bulk channel, respectively; n_C is the electron density in the bulk channel and S is the cross sectional area of the bulk channel. From Expression [2] it can be seen that the flat-band voltage V_{FB} can be estimated from the I_D-V_G characteristics of the JL MOSFET using the intersection point between two extrapolated straight lines corresponding to the accumulation current and the bulk current. Our estimations give us that V_{FB}=0.55V.

Measurement of the average capture time ($<\tau_c>$) and average emission time ($<\tau_e>$) in dependence on gate voltage allows us to estimate the energetic and geometrical location of the trap levels responsible for the charge trapping and creation the RTN in the gate oxide. From dependency of $\ln\left(\dfrac{<\tau_c>}{<\tau_e>}\right)$ on either $(V_G - V_{TH})$ or $(V_G - V_{FB})$ presented in

Figure 3(c) the geometrical trap depth (x_T) from the SiO_2-Si interface can be extracted by using the conventional method presented in Reference (7)

$$\frac{d\left(\ln\frac{<\tau_c>}{<\tau_e>}\right)}{dV_G} \approx \frac{q}{kT} \times \frac{x_T}{d_{ox}},$$ [3]

where d_{ox} is the thickness of the gate dielectric. The extracted x_T for the JL devices is 2.9 nm and for the IM devices ≈ 2.0 nm.

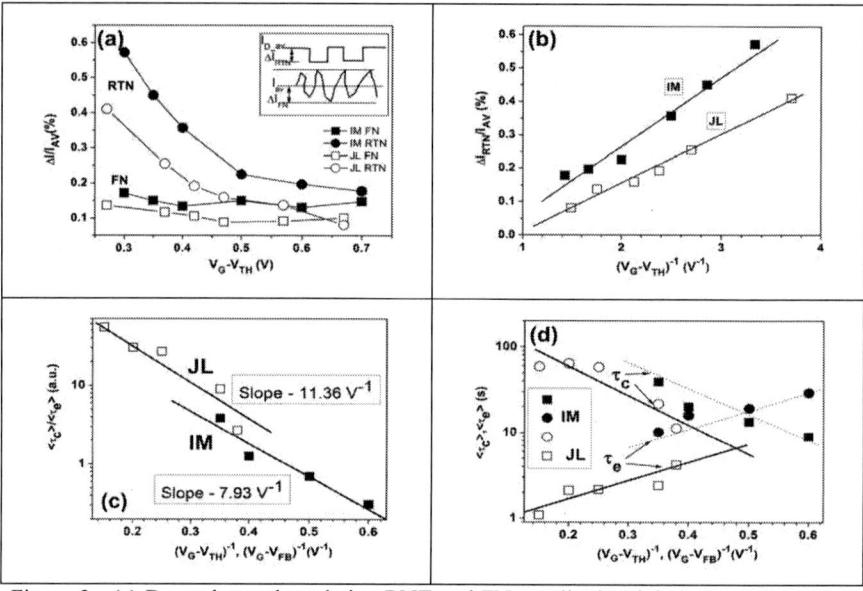

Figure 3. (a) Dependence the relative RNT and FN amplitude of drain current *vs.* gate voltage overdrive; (b) Dependence the relative RTN amplitude *vs.* the inverse of gate voltage overdrive; (c) Dependence of the ratio of average capture time to average emission time *vs.* gate voltage overdrive (for the JL MOSFET the gate voltage overdrive is V_G-V_{FB}, for the IM MOSFETit is equal to V_G-V_{TH}); (d) Average capture and emission time *vs.* gate voltage overdrive (for the JL MOSFET the gate voltage overdrive is V_G-V_{FB}, for the IM MOSFETit is equal to V_G-V_{TH}).

At the gate voltage overdrive $(V_G$-$V_{FB})^*$ for which $<\tau_c>$ and $<\tau_e>$ are equal (Figure 3(d)), using the equation of detailed balance, $\frac{<\tau_c>}{<\tau_e>} = g \exp\left(\frac{E_T - E_F}{kT}\right)$, we obtain $E_T = E_F$. Thus the following expression for the energetic location of the traps $(E_{Cox} - E_T)$ related to the conductance zone of the gate oxide for the JL device in accumulation can be deduced

$$E_{Cox} - E_T = E_{Cox} - E_F - \frac{(V_G - V_{FB})^*}{d_{ox}} x_T,$$ [3]

where $E_{Cox} - E_F$ is the energetic difference between the conductance zone of the gate oxide and the Fermi level in the silicon nanowire. Thus the $E_{Cox} - E_T$ equals to 3.05 eV for the JL device. For the IM fully depleted multigate MOSFET the estimation of the traps energetic location in the gate oxide requires the use of numerical computer simulations.

Dependence on Drain Voltage

The relative RTN and FN amplitude of the drain current in the JL devices is considerably smaller than in the IM devices at various drain voltages (Fig. 4 (a)) for the same gate voltage overdrive values $(V_G\text{-}V_{TH})$, which is sign of a more "stable" operation in the JL MOSFET than in the IM device. From the slope of the dependency of $\ln\left(\dfrac{<\tau_c>}{<\tau_e>}\right)$ on drain voltage, the geometric location the traps along of the channel related to the source, Y_T, can be estimated (5). Because the same slope of $\ln\left(\dfrac{<\tau_c>}{<\tau_e>}\right)$ vs. V_D is observed in both the JL and IM devices (Fig. 4 (b)), it can be concluded that the trap location along the channel, from source to drain, is the same for both type of the devices. It is approximately equal to 150 nm from the source.

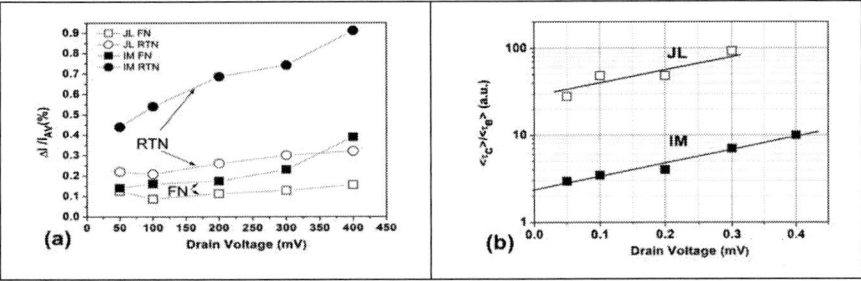

Figure 4. (a) Dependence the relative RNT and FN amplitude of drain current vs. drain voltage; (b) Dependence of the ratio of average capture time to average emission time vs. drain voltage.

Dependence on Temperature

Figure 5 (a) clearly shows that the relative RTN and FN amplitudes of drain current in the JL devices is considerable smaller than in the IM devices at a various measurement temperatures for the same gate voltage overdrive. The activation energy determined from the average emission time vs. temperature at the same gate voltage overdrive (Fig. 5 (b)) and associated with energetic location of the traps to silicon conductance zone (7) is 0.19 eV in case of the JL device and 0.77 eV in the IM device. This indicates a considerably larger bending of the energy bands in the silicon nanowire near the SiO_2-Si interface in the IM MOSFET than in the junctionless device.

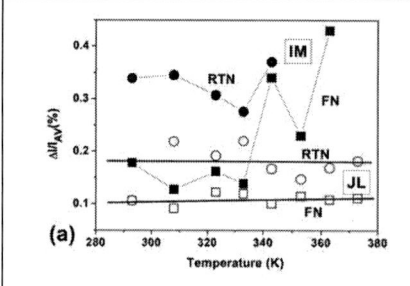 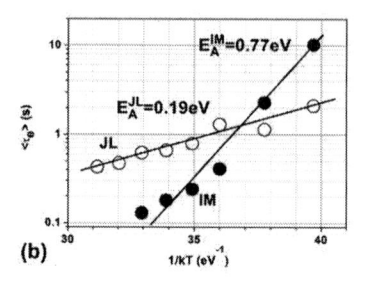

Figure 5. (a) Dependence of the relative RNT and FN amplitudes of the drain current on measurement temperature; (b) Arrhenius plot of emission time for the JL and IM MOSFET (gate voltage overdrive = 0.42V).

Conclusions

In conclusion, we have studied the random telegraph-signal noise in JL and IM NW MOSFETs and have shown that relative RTN amplitude of the drain current in JL devices is considerable smaller than that in the IM transistors as for different applied gate and drain voltages as for various measurement temperatures. Such difference can be associated with location of main current channel inside of the nanowire body in case of the JL devices.

Acknowledgments

This work was supported by the Science Foundation Ireland grant 05/IN/I888.and enabled by the Programme for Research in Third-Level Institutions. This work was supported in part by the European Community (EC) Seventh Framework Program through the Networks of Excellence NANOSIL and EUROSOI+ under Contracts 216171 and 216373, and A.N. Nazarov was supported by exchange program in frame of the Networks of Excellence NANOSIL.

References

1. C. W. Lee, A. Afzalian, N. Dehdashti Akhavan, et al. *Appl. Phys. Lett.* **94**, 053511 (2009).
2. J.-P. Colinge, C.-W. Lee, A. Afzalian, et al., *Nature Nanotechnology* **15**, 1 (2010).
3. J.-P. Colinge, C.-W. Lee, I. Ferain, et al., *Appl. Phys. Lett.* **96**, 073510 (2010).
4. K.K. Hung, P.K. Ko, C. Hu, and Y.C. Cheng, *IEEE Electron. Dev. Lett.* **11**, 90 (1990).
5. G. Ghibaudo, *Microelectronic Engineering* **39**, 31 (1997).
6. J. P. Colinge, C. W. Lee, N. Dehdashti Akhavan, et al., in *Semiconductor-On-Insulator Materials for Nanoelectronics Applications*, A.N. Nazarov, J.P. Colinge, F. Balestra, J.P. Raskin, F. Gamiz and V.S. Lysenko, Editors, p. 187, Springer-Verlag, Berlin Heidelberg (2011).
7. M. J. Kirton, and M. J. Uren, *Appl. Phys. Lett.* **48**, 1271 (1986).

ECS Transactions, 35 (5) 79-84 (2011)
10.1149/1.3570780 ©The Electrochemical Society

Hysteresis Effects in FinFETs with ONO Buried Insulator

S.-J. Chang[a], M. Bawedin[b], W. Xiong[c], J.-H. Lee[d], S. Cristoloveanu[a]

[a] IMEP-LAHC, Grenoble INP Minatec, BP 257, 38016 Grenoble, France
[b] IES, University Montpellier 2, Montpellier, France
[c] SEMATECH, Austin, Texas, USA
[d] Kyungpook National University, Daegu, Korea

FinFETs fabricated on SiO_2-Si_3N_4-SiO_2 (ONO) buried insulator are investigated for flash memory application. The Si_3N_4 layer can trap charges by tunneling at high back-gate bias. The amount of trapped charges is sensed, via gate coupling effects, by the drain current. The trapped charges in Si_3N_4 layer also induce a drain current hysteresis when the back-gate is dynamically scanned. Systematic measurements reveal that the charge trapping and drain current hysteresis are useful memory effects. The memory window depends on bias conditions and geometrical parameters.

I. Introduction

Advanced SOI devices with alternative buried insulator are studied for several applications: (i) self-heating reduction (1), (ii) strain transfer (2), (iii) fin etch definition (3) and (iv) charge storage for flash memory application (4, 5). We explore FinFETs with ONO buried insulator as innovating flash memory with remote charge trapping.

The conventional flash memory cells are based on the charge storage in a floating gate which is also used to sense the current. Until now, various structures of flash memory cells have seen suggested. For the necessary evolution of flash memory devices, cell miniaturization is one of the most import issues. Further scaling down to 22 nm technology node will be facing critical problems connected not only to the channel length reduction but also with the thinning of the tunneling oxide which may degrade the retention time. Another problem is the high value of the operating voltage: the conventional flash memory needs about 8 V to program/erase the cells. High voltage for the operation of the flash cells impedes on the co-integration with logic applications.

For these reasons, the silicon-oxide-nitride-oxide-silicon (SONOS) structure, where the nitride layer is used to store the charges, has been studied. SONOS flash memory device is attractive because the SONOS process is simpler and offers good retention characteristics due to the presence of deep trap levels in the nitride.

In this paper, we will show that the Si_3N_4 buried layer can effectively trap charges in FinFETs devices fabricated on ONO buried insulator. The hysteresis effects, induced by charge trapping/detrapping and useful as a flash memory window, will be studied as a function of back-gate bias variation. The main advantage of this device is that the roles of front and back gates are separated: charge trapping occurs in the nitride buried insulator whereas the read current is perceived by the front-gate. We report experimental results

79

which demonstrate the impacts of variable front/back gate bias and geometrical dimensions of FinFETs

II. Device fabrication

SOI wafers with oxide/nitride/oxide multi-layer buried insulator were used as starting material. The wafers were fabricated using the Smart-Cut™. The buried insulator (BOX) features 2.5 nm (SiO$_2$), 20 nm (Si$_3$N$_4$) and 70 nm (SiO$_2$), from top to bottom. The Si$_3$N$_4$ layer was sandwiched between two SiO$_2$ layers for flash memory operation. The top SiO$_2$ is very thin (2.5 nm) and enables carrier tunneling. The Si film thickness (65 nm) defined the fin height. Hydrogen annealing was used to smooth the fin sidewalls. The top gate oxide thickness grown by wet oxidation was 1.8 nm. TiSiN deposited by LPCVD was used as gate material. FinFETs with variable fin width and gate length were prepared for investigating the geometrical effects. Fig. 1 shows the structure of advanced SOI FinFETs fabricated on the ONO buried insulator. All devices operate in full depletion mode.

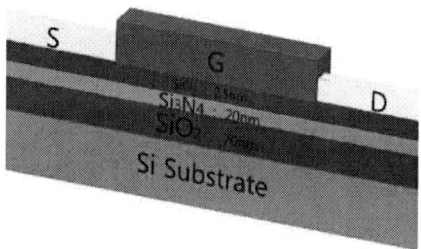

Fig. 1: Structure of advanced SOI FinFET with alternative buried insulator. The thickness of buried insulator is 2.5 nm (SiO$_2$), 20 nm (Si$_3$N$_4$) and 70 nm (SiO$_2$), from top to bottom.

III. Charge trapping by tunneling

The nitride layer can trap charges by high back-gate biasing. The charge trapping mechanism is Fowler-Nordheim (F-N) tunneling (6). Tunneling occurs from the fin body into the nitride layer through the 2.5 nm thin SiO$_2$ buried layer. Applying a strong vertical electric field, by high back-gate bias, induces a large carrier tunneling through the 2.5 nm oxide layer without damaging its dielectric properties. The trapped charges primarily change the back-channel properties, in particular the back threshold voltage. They can also modify, via coupling effects in fully depleted devices, the front-channel characteristics such as threshold voltage, mobility, subthreshold slope, etc. These characteristics are more or less affected according to the amount of trapped charges in the nitride layer. Especially, the shift of I$_D$(V$_G$) curve, resulting from charge trapping/detrapping, can be applied for flash memory devices.

Fig. 2 shows the memory effects induced by ONO trapped charge on the front-channel characteristics. I$_D$(V$_G$) curves were measured with the substrate grounded before and after ONO charging. The carrier trapping/detrapping was achieved by applying a back-gate bias V$_{BG}$ for 30 seconds before measurement. The lateral shift of I$_D$(V$_G$) curves depends on the polarity and magnitude of the programming V_{BG} bias. Front-channel threshold voltage decreases after negative back-gate bias condition, and increases after positive back-gate biasing. A negative V_{BG} bias results in, trapped holes (or detrapped electrons) in the nitride layer which increases the body potential. Therefore, front-channel threshold voltage decreases. By contrast, a positive V_{BG} bias traps electrons (or detraps holes) in the nitride layer and the body potential drops, increasing the front-channel threshold voltage. The shift of $I_D(V_G)$ curve depends on the amount of trapped charges. At $V_{BG} = \pm 50$ V (Fig. 2(b)), front-channel threshold voltage shift is larger than for $V_{BG} = \pm 30$ V (Fig. 2(a)). But, above $V_{BG} = \pm 50$ V, front-channel threshold voltage shift is saturated. Note that for $V_{BG} = -50$ V the positive charge in ONO dielectric is sufficient to activate the back-channel so that the total drain current does not switch off.

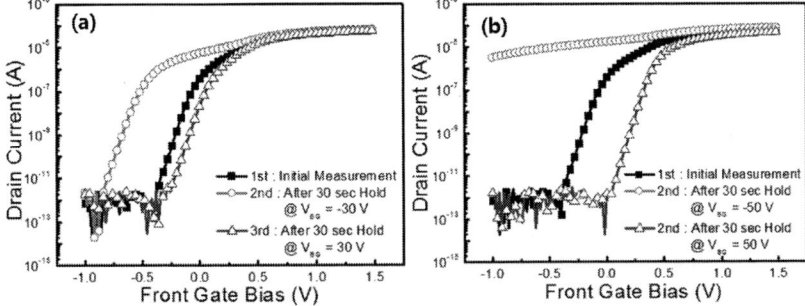

Fig. 2: Typical memory effects induced by back-gate programming. Drain current as a function of front-gate bias, measured at $V_{BG} = 0$ V after programming with (a) $V_{BG} = \pm 30$ V and (b) $V_{BG} = \pm 50$ V. $V_D = 50$ mV, $W_F = 90$ nm, $L_G = 1 \mu m$, $N_F = 20$ fingers in parallel.

As illustrated in Fig. 2, the front-channel characteristics are affected by the trapped charges in the nitride layer. A negative V_{BG}, extracting the trapped electrons or injecting holes from/into Si_3N_4 layer, leads to high drain current ('1' state). A positive V_{BG} erases the previous information and yields low current ('0' state). The difference in drain current between '0' and '1' can be used for the flash memory applications. Nevertheless, this effect is different from the usual mechanism in flash memory device because the charge is trapped in the buried insulator and is sensed by the front-channel.

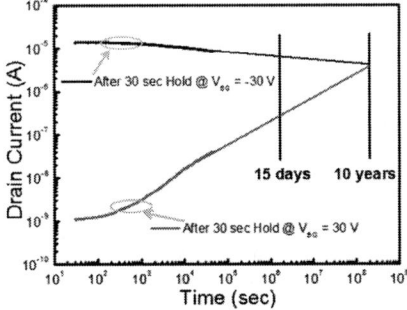

Fig. 3: Effect of trapped charges on the drain current. After charge trapping, drain current as a function of time. The trapped charges in the nitride layer are hold for a long time. $V_{FG} = 0$ V, $V_{BG} = 0$ V, $V_D = 0.4$ V, $W_F = 90$ nm, $L_G = 1$ μm, $N_F = 20$ fingers in parallel.

Evidence for the retention of the charge trapped in the nitride is obtained by monitoring the transient drain current at $V_{FG} = 0$ V and $V_{BG} = 0$ V. Fig. 3 shows that the transient effects come from the gradual release of trapped charges. However, the injected charges in the nitride layer are conserved for a long time, even through our devices were not optimized for memory applications.

IV. Drain current hysteresis

Drain current hysteresis is another consequence of ONO trapped charge. Hysteresis is observed in dynamic mode, by scanning the back-gate bias back and forth between positive and negative values. The drain current hysteresis shown in Figures 4-7, useful as a memory window, is due to the dynamic threshold voltage shift during back-gate bias scan and charge trapping. Increasing V_{BG} from a starting negative bias gradually reduces the initially trapped positive charge and increases the threshold voltage. The opposite effect (reduction of negative charge and dynamic lowering of threshold voltage) happens during the reverse scan. Fig. 4 shows the effect of front-gate bias on the drain current hysteresis. For higher front-gate bias, drain current level increases but, the back-channel threshold voltage variation, ΔV_{THB} which defines the memory window, decreases. This phenomenon can be explained by the competition between lateral and vertical coupling effects (7): the lateral gates tend to control the back-surface potential, reducing the vertical effect of the trapped charge. The controllability of the front-gate increases for higher front-gate bias, blocking the effect of trapped charges and back-gate bias. Fig. 4(b) shows the dependence of the memory window on the starting point (-50V or +50 V) of the bias scan. This demonstrates that the charge trapping efficiency is different for holes and electrons.

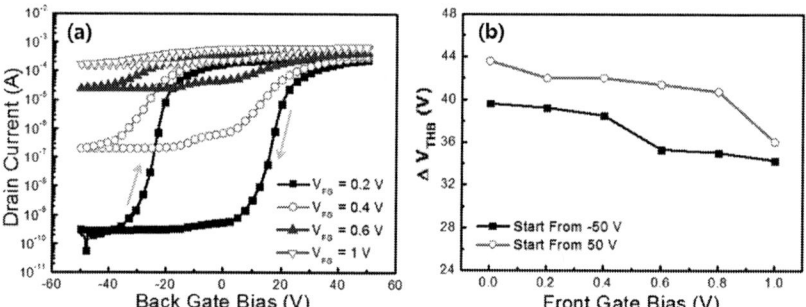

Fig. 4: Dependence of memory window on front-gate bias. (a) Drain current hysteresis as a function of back-gate and front-gate bias. (b) Back-channel threshold voltage variation (memory window) as a function of front-gate voltage and measurement starting bias value (30 sec hold time). $V_D = 0.1$ V, $W_F = 90$ nm, $L_G = 1$ μm, $N_F = 100$ fingers in parallel.

The memory window size also depends on the geometrical parameters such as gate length and fin width. For shorter device, both drain current level and ΔV_{THB} increase. As shown in Fig. 5(b), ΔV_{THB} increases rapidly under 500 nm gate length. This phenomenon is related to 3D coupling effects (7). For shorter device, the longitudinal coupling component induced by drain bias enhances the back-surface potential. This mechanism, named drain-induced virtual substrate biasing (DIVSB), is responsible for short-channel effect in SOI MOSFETs. DIVSB opposes the lateral gate control. Therefore, for shorter channel device, the longitudinal component reinforces the vertical components due to trapped charges and back-gate bias. The memory window increase for shorter channels is an outstanding asset of highly scalable FinFET/ONO flash memories.

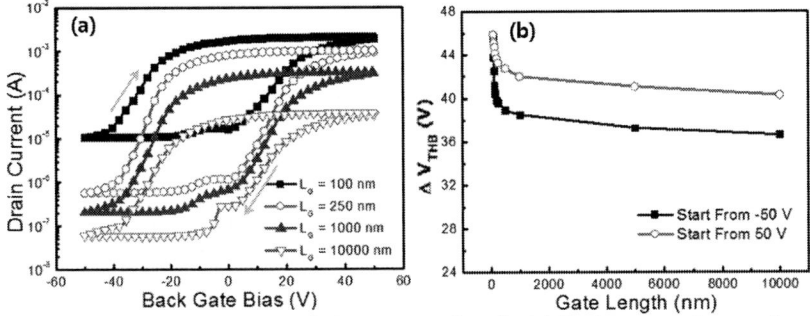

Fig. 5: Dependence of memory window on gate length. (a) Drain current hysteresis as a function of back-gate bias and gate length. (b) Back-channel threshold voltage variation as a function of gate length and measurement starting bias value. $V_{FG} = 0.4$ V, $V_D = 0.1$ V, $W_F = 90$ nm, $N_F = 100$ fingers in parallel.

The memory window size is affected by fin width as shown in Fig. 6. For wider fin width, ΔV_{THB} is larger than for narrower fins. When the fin width increases, carriers are injected into the nitride layer which has a larger area. An additional fin width effect, similar with the front-gate bias effect, is the modulation of lateral coupling. For wider fin, the controllability of the front-gate decreases. Therefore, the effect of trapped charge and ΔV_{THB} hysteresis increase.

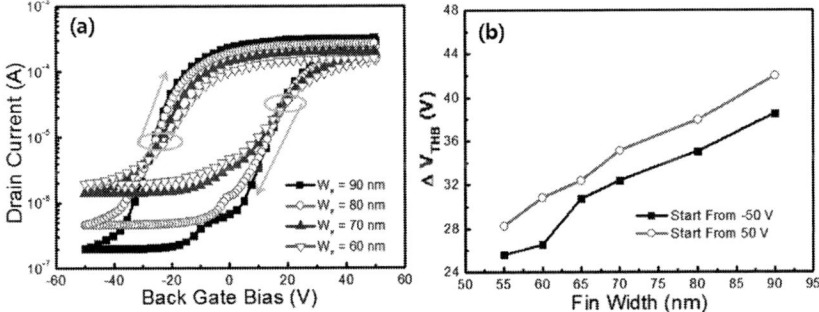

Fig. 6: Dependence of memory window on fin width. (a) Drain current hysteresis as a function of back-gate bias and fin width. (b) Back-channel threshold voltage variation as a function of fin width and measurement starting bias value. $V_{FG} = 0.4$ V, $V_D = 0.1$ V, $L_G = 1\mu m$, $N_F = 100$ fingers in parallel.

We have verified the impact of measurement conditions on the memory window. Fig. 7 shows that memory window weakly depends on delay time. A drain current hump is observed when back-gate bias is swept from inversion to accumulation. For longer delay time, the hump is decreased. This indicates that the drain current hump is a non-equilibrium mechanism.

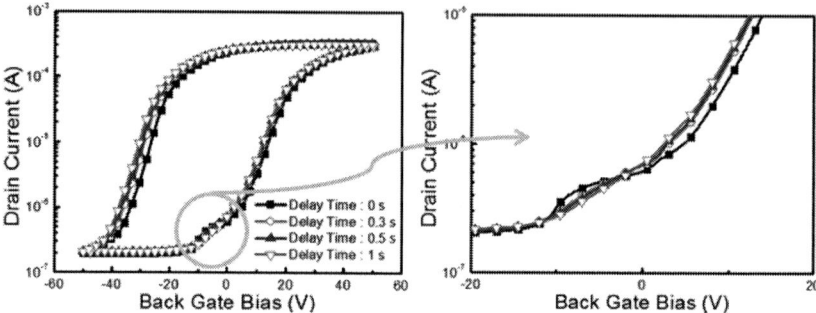

Fig. 7: Drain current hysteresis as a function of delay time. Drain current hump decreases for longer delay time. $V_{FG} = 0.4$ V, $V_D = 0.1$ V, $W_F = 90$ nm, $L_G = 1\mu$m, $N_F = 100$ fingers in parallel.

V. Conclusions

The Si_3N_4 buried layer can trap charges injected by back-gate Fowler-Nordheim tunneling. The amount of trapped charges depends on the back-gate bias. The trapped charges change the back-channel threshold voltage and, by coupling, shift the front-channel $I_D(V_G)$ characteristics. The charges trapped in the Si_3N_4 layer are maintained for a long time, which is attractive for flash memory devices. We have also shown that large drain current hysteresis (*i.e.*, memory window) is induced by dynamic charge trapping/detrapping during the variation of back-gate voltage. The memory window, useful as a flash memory, depends on the bias conditions, geometrical parameters and measurement conditions.

Acknowledgements

EUROSOI+, NANOSIL and WCU (KOSEF) organizations are thanked for support.

References

1. K. Oshima, S. Cristoloveanu, B. Guillaumot, H. Iwai and S. Deleonibus, *Solid-State Electronics*, Vol. 48, 907 (2004).
2. S. Cristoloveanu and G.K. Celler, *in Handbook of Semiconductor Manufacturing Technology*, CRC Press, London (2007).
3. P. Patruno, M. Kostrzewa, K. Landry, X. Weize, C. R. Cleavelin, H. Che-Hua, M. Ma and J. P. Colinge, *Proc. IEEE International SOI Conference*, 51 (2007).
4. R. Ranica, A. Villaret, P. Mazoyer, S. Monfray, D. Chanemougame, P. Masson, A. Regnier, C. N. Dray, R. Bez and T. Skotnicki, *IEEE Transactions on Nanotechnology*, Vol. 4, 581 (2005).
5. H. Silva and S. Tiwari, IEEE *Transactions on Nanotechnology*, Vol. 3, 264 (2004).
6. R. Bez, E. Camerlenghi, A. Modelli and A. Visconti, *Proceedings of the IEEE*, **91** 489 (2003).
7. Y. Bae, K. I. Na, S. Cristoloveanu, W. Xiong, C. R. Cleavelin and J. H. Lee, *in Annual Semiconductor Conference*, CAS 2009 51 (2009).

ECS Transactions, 35 (5) 85-90 (2011)
10.1149/1.3570781 ©The Electrochemical Society

Scaling Scheme and Performance Perspective of Cross-Current Tetrode (XCT) SOI
MOSFET for Future Ultra-Low Power Applications

Y. Omura[a,b], K. Fukuchi[a], D. Ino[b], and O. Hayashi[b]

[a] Department of Electrical, Electronics and Information Technol., Kansai University,
Suita, Osaka 564-8680, Japan
[b] Grad. School. Of Engineering Science, Kansai University, Suita, Osaka564-8680, Japan

This paper introduces a scaling scheme of the cross-current tetrode
(XCT) SOI MOSFET and preliminary results. It is demonstrated
that the XCT-SOI MOSFET is a promising solution for future
'ultra low-energy' LSIs suitable for medical applications. It is
shown that the proposed scaling scheme yields useful design
guidelines for XCT devices.

Introduction

By using the partially-depleted (PD) single-gate (SG) SOI MOSFET, one of the authors
(Omura) proposed the cross-current tetrode SOI MOSFET (XCT-SOI MOSFET) and
examined its analog performance in 1986 [1]. Scaling feasibility of similar devices has
been studied recently [2, 3]. These studies have led some people expect that XCT devices
will yield new applications such as high-voltage devices and SRAM memory cells with
high noise margin [4]. Other applications have different demands. Extremely low-power
circuits are solicited in the field of medical implants since the batteries must operate
within the human or animal body for long periods of time [5]. Although the drivability of
the XCT-SOI MOSFET is one-order lower than that of the conventional SOI MOSFET
[4], the output voltage level of the XCT-SOI MOSFET is identical to that of the
conventional SOI MOSFET. Such voltage-drive type devices are very useful in
extremely-low-power and 'low-energy' devices [5, 6] because the conventional circuit
design methodology can be basically applied without any change [7].

This paper introduces a scaling scheme of the XCT-SOI MOSFET and preliminary
results. They confirm that the XCT-SOI MOSFET is a promising solution for future
'ultra low-energy' LSIs suitable for medical applications. It is also shown that the
proposed scaling scheme yields useful design guidelines for XCT devices.

Device Structure and Features of XCT Device

Figure 1 shows a bird's eye view of the XCT-SOI MOSFET and its equivalent circuit.
The terminal configuration is the source of the XCT-SOI MOSFET characteristic [1].
The XCT device is composed of the conventional SOI MOSFET with two body contacts;
the source diffusion of the SOI MOSFET is connected through the body contact if a
metal wire is used. The source potential of the SOI MOSFET is transmitted to the body
contact (the source terminal of the parasitic JFET) as shown in Fig. 1(b). The current
flow traces the character 'α', hence the device has the name of 'cross current'. One
important aspect of the XCT device, is that it offers negative differential conductance
(NDC) in the saturation region of the drain current [1, 4]; this feature is very useful to

85

suppression of short-channel effects. The fundamental mechanism of NDC is summarized below.

(i) When the MOSFET is in the 'ON-state', the parasitic JFET works as a parasitic source resistance. This reduces the drivability of the MOSFET.

(ii) When the drain voltage of the MOSFET is raised, MOSFET operation changes from non-saturation mode to saturation mode. One of the two gate terminals of the parasitic JFET is the drain terminal of the MOSFET. In saturation mode operation, therefore, the channel width of the parasitic JFET is reduced by the rise in the MOSFET drain voltage.

(iii) As a result, the MOSFET channel current decreases as the drain voltage rises; NDC appears in the saturation regime.

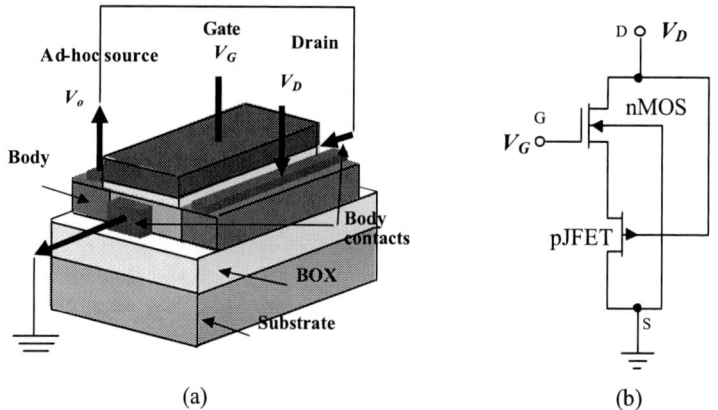

(a)   (b)

Figure 1. Schematic bird's eye view of n-channel XCT SOI MOSFET and equivalent circuit. (a) Schematic view of n-channel XCT SOI MOSFET. (b) Equivalent circuit of n-channel XCT SOI MOSFET.

Measured Characteristics of XCT Devices

This section describes the measurements conducted on fabricated XCT devices. Nominal device parameters used for fabrication are shown in Table 1. The PD SOI MOSFET structure was designed by adopting a thick SOI layer so that the parasitic JFET would work well.

Figure 2(a) shows measured I_D-V_D curves of the original SOI n-channel MOSFET with body contacts connected to the source contact. One issue with the PD SOI MOSFET structure is, generally speaking, the floating-body effect, which occurs due to the restoration of majority carriers in the Si body over time. Since the present device has body contacts, the floating-body effect and the kink effect are not seen in Fig. 2(a). The channel-length modulation effect is, however, slightly present in the saturation region of the drain current. Figure 2(b) shows measured I_D-V_D curves of the n-channel XCT device. When a high drain voltage is applied to the XCT device, NDC appears in the saturation region of the drain current [1, 4]. As the drain junction works as the one of gate electrodes of the parasitic JFET, the rise of the drain voltage reduces the channel current of the parasitic JFET on the saturation region of MOSFET. This is the primary mechanism of NDC. It is also seen that NDC becomes more significant as the gate voltage (V_G) rises. In this figure, symbols plot the simulation results yielded by the model described in [4]. The proposed model reproduces the fundamental I_D-V_D characteristics of the XCT device successfully. In other words, the advanced model proposed in [4] will be useful in conducting various device analyses. Figure 2(c) shows I_D-V_G characteristics of the original SOI MOSFET and the XCT device for comparison. It should be noted that the subthreshold swing of the XCT device is much smaller than that of the original SOI MOSFET. It should also be noted that the NDC automatically suppresses the short-channel effects.

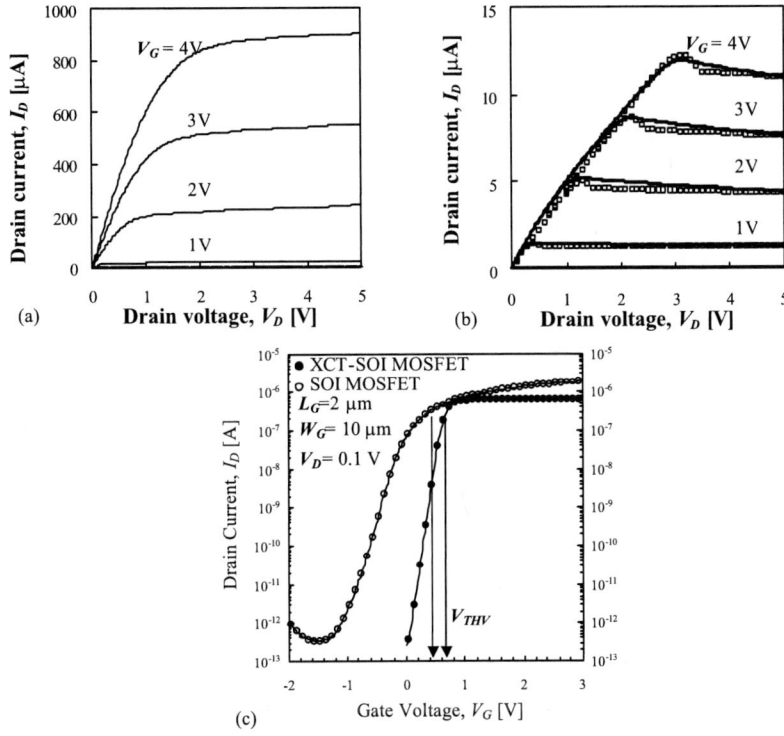

Figure 2. Experimental I-V characteristics. (a) I_D-V_D characteristics of SOI nMOSFET. (b) I_D-V_D characteristics of n-XCT SOI MOSFET. (c) I_D-V_G characteristics of SOI nMOSFET and n-XCT SOI MOSFET.

Table 1. Nominal device parameters

Device parameters	Values	[units]
Body doping, N_A	1.0×10^{16}	cm^{-3}
Gate length, L_G	2.0	μm
Channel length, L	1.5	μm
Gate width, W_G	10	μm
Gate oxide thickness, t_{ox}	30	nm
SOI layer thickness, t_{SOI}	350	nm
BOX layer thickness, t_{BOX}	300	nm

Scaling Scheme and Design Guideline

As the device dimensions considered above do not reach the modern technology level, we must investigate the performance feasibility of scaled XCT devices; presently, sub-100-nm-long channel devices should be discussed. In this study, we performed 3-D semi-classical device simulations of scaled XCT devices [8], where the hydrodynamic transport model is assumed [8].

Here we consider five XCT-SOI MOSFETs with different gate lengths; L_G= 2 μm (L= 1.5 μm), 1 μm (0.75 μm), 0.5 μm (0.38 μm), 0.25 μm (0.20 μm), and 100 nm (75 nm), where L denotes the channel length. The fundamental scaling scheme examined in this paper is shown in Table 2; this seems roughly similar to the conventional quasi-constant-field scaling scheme, but it is different from the conventional one because the scaling of the film thickness is reconsidered for the XCT device. It also shows the initial device parameters assumed. This scheme was derived from the results of many simulations. We think that this scheme is the best currently available.

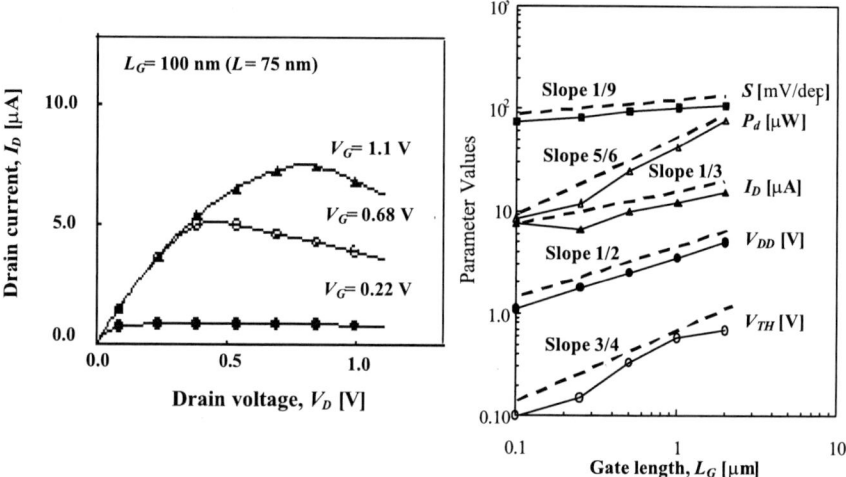

Figure 3. Calculated I_D-V_D characteristics of 100-nm-long gate n-XCT device. W_G= 0.5 μm. (3D device simulations.)

Figure 4. Scaling scheme proposed here. Fundamental data are extracted from device simulation results.

Figure 3 demonstrates 3-D device simulation results of the I_D-V_D characteristics of XCT devices with gate lengths of 100 nm. These simulation results show valid current-voltage characteristics as obtained from many trials of scaling schemes; the 3-D device simulation results strongly suggest that the XCT SOI MOSFET does support the sub-100-nm regime. We discovered that old simplified scaling schemes (e. g., constant field scaling and quasi-constant field scaling) failed to offer useful device characteristics (not shown here); this is due to the fact that the XCT-SOI MOSFET is a composite device composed of SOI MOSFET and JFET. Figure 4 reveals the simulated performance of scaled XCT-SOI MOSFETs. The significant aspects of this scaling methodology are summarized below.

(i)   Threshold voltage (V_{TH}) is scaled automatically at the rate of $(1/k)^{3/4}$. As a result, V_{TH}/V_D falls with scaling, which is not a crucial drawback in terms of the power consumption because XCT-SOI MOSFET drain current is already lower than that of conventional devices. These attributes are not achieved by the conventional bulk and SOI MOSFET.

(ii) Subthreshold swing (S) of XCT-SOI MOSFET is also slowly scaled from 107 mV/dec to 76.3 mV/dec when gate length (L_G) is scaled from 2 μm to 100 nm (details are not shown in Fig. 4); the S value scales at the rate of $(1/k)^{1/9}$. This is a great advantage for scaled devices because this behavior suppresses the standby power consumption much more than expected.

(iii) Drain current (I_D) scales at the rate of $(1/k)^{1/3}$. As a result, P_d scales at the rate of $(1/k)^{5/6}$. This scaling rate is identical to that of the conventional bulk and SOI MOSFETs that are scaled according to the conventional quasi-constant field scheme. Therefore, this scaling scheme raises the power dissipation of the circuits. However, this is not a shortcoming because the overall power dissipation of the XCT-SOI MOSFET is well suppressed by the device structure itself.

(iv) The ultra-low-power operation of scaled XCT SOI MOSFETs is very useful for future medical implant applications and some space applications because these applications do not always request high switching speeds.

Table. 2. Scaling scheme of XCT SOI MOSFET

	N_A	L_G	W_G	L	t_{ox}	t_{SOI}	t_{BOX}	V_D	V_{TH}
Scaling scheme	k	1/k	1/k	4/3k	$1/k^{1/3}$	$1/k^{1/3}$	$1/k^{1/2}$	---	
Initial values	5×10^{16} cm^{-3}	2.0 μm	2.0 μm	1.5 μm	30 nm	350 nm	300 nm	5 V	0.68 V

Perspective of Low-Energy Strategy

Finally, we demonstrate preliminary simulation results based on device characteristics of fabricated devices [9]. Figure 5 shows power dissipation (P_d) vs. delay time (t_d) characteristics. 0.1-μm gate CMOS devices in Fig. 5(b) follow the scaling scheme shown

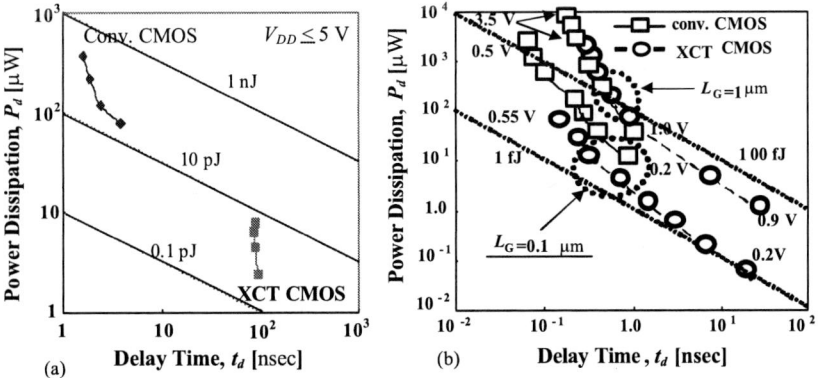

Figure 5. Simulation results of P_d vs t_d characteristics. (a) L_G= 2 μm [6], (b) L_G= 0.1 μm [10].

in Table 2. It is revealed independently of device dimension that XCT-SOI CMOS shows about 10 times delay time and about 1/100 times power dissipation in comparison to those of the conventional SOI CMOS; this results in 1/10 times operation energy of XCT SOI CMOS. It is expected that this feature of XCT-SOI CMOS is also useful to design of low-energy SRAM devices with a high static noise margin.

Summary

This paper described preliminary results on a scaling scheme for the cross-current tetrode (XCT)-SOI MOSFET. It was demonstrated that the XCT-SOI MOSFET is a very promising device for future 'ultra low-energy' LSIs that should be applied to various medical and space applications. Extensive simulation results suggest that the proposed 'modified quasi-constant-field scaling scheme' yields desirable design guidelines for 100-nm-long gate XCT devices.

Acknowledgement

Authors wish to express their thanks to Drs. Hirobumi Watanabe and Hidenori Kato (Ricoh Corp., Osaka, Japan) for the device fabrication.

References

1. Y. Omura and K. Izumi, *Ext. Abstr., 18th Int. Conf. Solid State Devices and Maerials.,* p. 715 (1986).
2. M. H. Gao, S. H. Wu, J. P. Colinge, C.Claeys and G. Declerck, *Proc. IEEE Int. SOI Conf.,* p. 138 (1991).
3. B. Dufrene, K. Akarvardar, S. Cristoloveanu, B. J. Blalock, P. Gentil, E. Kolawa, and M. M. Mojarradi, *IEEE Trans. Electron Devices,* **51**,1931 (2004).
4. Y. Azuma, Y. Yoshioka, and Y. Omura, *Ext. Abstr. Int. Conf. Solid State Devices and Materials.,* p. 460 (2007).
5. A. P. Chandrakasan, D. C. Daly, D. F. Finchelstein, J. Kwong, Y. K. Ramadass, M. E. Sinangil, V. Sze, and N. Verma, *Proc. the IEEE,* **98**, 191 (2010).
6. S. Tominaga and Y. Omura, *Abstr. 2009 IEEE Int. Meet. For Future of Electron Devices, Kansai,* p. 116 (2009).
7. Y. Omura, *Jpn. J. Appl. Phys.,* **48**, 04C071 (2009).
8. *Sentaurus,* Users Manual, 2008 (Synopsys Inc.).
9. *HSPICE,* Users Manual, 2008 (Synopsys Inc.)
10. D. Ino and Y. Omura, *Abstr. 2011 Domestic Ann. Spring Conf. (Jpn. Soc. Appl. Phys.)* (2011), No. 26p-KC-16.

CHAPTER 5

CHARACTERIZATION-1

Novel SOI Structures and Characterization Strategy

Sorin Cristoloveanu

IMEP-LAHC (UMR 5130), Grenoble INP Minatec, BP 257,
38016 Grenoble Cedex 1, France

Innovative SOI materials and devices are reviewed with special
attention to their electrical characterization. Appropriate measurement techniques and accurate interpretation of the experimental
data are necessary steps for evaluating nanosize devices.
Informative examples are selected from various structures
including ultrathin SOI and GeOI, strained MOSFETs, multiple-
gate transistors, nanowires, tunneling and memory devices.

Introduction

The family of SOI materials and devices is rapidly expanding. Not only are the Si
film and buried SiO_2 reaching the sub-10 nm range but also alternative semiconductor
films (Ge, SiGe, GaN) and buried dielectrics (ONO, diamond, glass, AlN, etc) are being
adopted.

On the device side, SOI MOS transistors combine ultrathin body, thin BOX, short
high-K/metal gates, and strain. Even more innovative devices, which take full advantage
of SOI assets, are currently explored: multiple-gate MOSFET, single-transistor floating-
body DRAM, tunnel FET, junctionless transistor, 3D nanowire FET, etc.

In this paper, we review some of these exciting developments and show that their
characterization is a rather challenging task. The evaluation of simple parameters, such as
carrier mobility, lifetime, or interface trap density, is no longer straightforward. Multi-
interface or multi-channel coupling and nanosize effects can modify both the
measurement value and the meaning of a given parameter.

Guidelines for appropriate characterization strategies and accurate interpretation of
the experimental data will be suggested. Examples are taken from various structures
including ultrathin SOI structures, strained MOSFETs, nanowires, tunneling FETs, and
memory devices.

Characterization of Advanced SOI Materials

The stringent condition for ultimate scaling is to use nanosize layers (< 10 nm). The
electrostatic integrity of short-channel devices is enhanced in Fully Depleted (FD)
structures with ultrathin body and buried oxide (BOX), which are suitable for high speed,
low-power and DRAM applications. The evaluation of the wafer properties is of
uppermost necessity for the optimization of the fabrication process.

However, the electrical characterization of semiconductor films with nanometer thickness is impossible with conventional methods because the layers are fully depleted. The only viable solution is to apply a substrate bias.

In this context, the Pseudo-MOSFET (Ψ-MOSFET), which uses the intrinsic upside-down MOS structure in SOI, is an undisputable technique as it enables *in situ* wafer inspection before CMOS device processing (1,2). The substrate plays the role of gate and two pressure probes serve as source and drain (Fig. 1a). An inversion or accumulation channel is activated at the film-BOX interface. The $I_D(V_G,V_D)$ characteristics are similar to those in fully processed MOSFETs and standard methods are used to extract parameters.

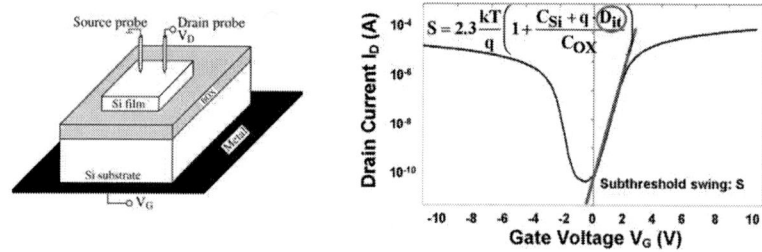

Figure 1. (a) Ψ-MOSFET configuration and (b) typical drain current variation with substrate voltage in weak inversion.

In industry, the Ψ-MOSFET serves for monitoring the quality and stability of the wafer fabrication process. At the research level, it is a very fast method for detecting defects and optimizing new materials. For example, the Ψ-MOSFET has recently been used to demonstrate the 'mobility balance' effect in GeOI: a higher Ge content improves remarkably the hole mobility but degrades proportionally the electron mobility (3). Other applications oriented to innovative structures include: measurement of Si and Ge nanowires deposited on the BOX, evaluation of alternative BOX materials (Al_2O_3, Si_3N_4, ONO, diamond …), investigation of radiation effects, etc.

Since its discovery 20 years ago, the Ψ-MOSFET method has been enriched in many respects:
- Circular source and drain contacts can be deposited or formed with Hg probes.
- Measurements at low temperature and high magnetic field provide information on the scattering mechanisms.
- Transient current or photo-current measurements indicate the carrier lifetime.
- The samples can be beveled for Ψ-MOSFET-like Spreading Resistance measurements of the resistivity profile.

The interpretation of the measurements has evolved in parallel with the layer thinning. The threshold voltage was found to increase dramatically in thinner films (4). The reason is that the numerous charges, located at the free wafer surface and associated with the native oxide, start affecting the properties of the Ψ-MOSFET channel.

If the wafer surface is passivated (by annealing, cleaning, or oxidation), the threshold voltage drops to expected values.

A similar effect leads to apparent mobility degradation with decreasing film thickness. The low-field mobility is normally determined with the Y-function, $Y = I_D/\sqrt{g_m}$, which was conceived to eliminate the influence of the gate-induced field. However, there is another component of vertical field, induced by the potential difference between the free surface charge and channel. This 'intrinsic' field is not accounted for by the Y function. It increases in thinner films leading to mobility values which are no longer 'low-field' at all. The shift along the 'universal' mobility curve explains the natural decrease in mobility with thickness. It follows that the mobility lowering is inherent and does not back up any speculation about the degradation of the film-BOX interface in thinner SOI materials.

Typical current-voltage curves are shown in Fig. 1b. The density of back interface traps D_{it} is usually deduced from the subthreshold slope. In ultrathin films, the capacitance of the film is very large, masking the capacitance of interface traps ($C_{si} \gg qD_{it}$) so that the detection limit is no longer sufficient. Low-frequency noise measurements were performed as an alternative method to evaluate the density of traps. Figure 2 shows the first noise measurements using the Ψ-MOSFET set-up with pressure contacts. The noise has a $1/f$ spectrum (Fig. 2a) and its variation with gate bias (Fig. 2b) follows the McWhorter model of carrier number fluctuations. The noise does not depend on the probe pressure but is higher than expected. This again suggests the contribution of free surface defects which may act as a source of additional noise.

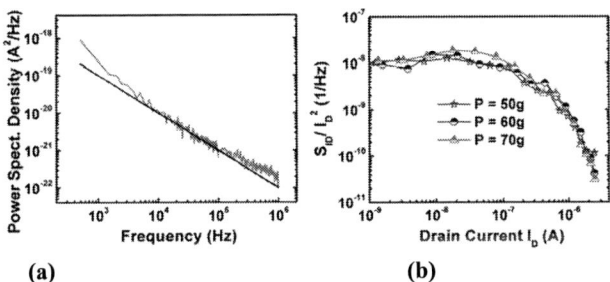

(a)   (b)

Figure 2. (a) $1/f$ noise versus frequency in Ψ-MOSFET and (b) normalized noise versus drain current showing no impact from probes pressure.

The classical Ψ-MOSFET model accounts exclusively for the film-BOX interface where the carriers flow (1,2). This simple model is accurate for film and BOX thicker than 100 nm. A 2-interface model was elaborated to depict the case of thinner Si films (5), where the properties of the top surface (passivated or non-passivated) do impact the channel. As the BOX is thinned down to 10–20 nm, the role of the substrate starts to be noticeable. The electrostatic potential drop at the substrate-BOX interface becomes comparable with the threshold voltage, leading to characterization errors. The model extension to a 3-interface formulation is more or less complicated: full analytical description or equivalent capacitance circuit.

Both approaches show that the quality of the BOX-substrate interface and its biasing state (inversion, depletion, accumulation) can also affect by coupling the properties of the channel above the BOX. Further model developments are needed in the time domain. For example, the drain current transient can be due to non-equilibrium effects occurring in the film and/or in the substrate (6).

Characterization of Advanced SOI Devices

The characterization of SOI structures is hampered by several problems: thinness of the film, presence of the BOX, three stacked interfaces, and typical defects (strain, in-depth inhomogeneity, dislocations). A number of conventional characterization methods are no longer applicable in thin films, but, in turn, novel techniques can be implemented. The MOS transistor stands as the main test vehicle. The properties of SOI structures are inferred from its static/dynamic characteristics. Parameters like mobility, threshold voltage, swing, and lifetime are determined for the front and back channels, separately or in a coupled mode. The parameter extraction is similar in SOI and bulk-Si MOSFETs and is well documented (1).

The threshold voltage and carrier mobility are extracted with the Y-function, which can be expanded to include a non-linear mobility degradation factor. Alternatively, the double-derivative method can be used to determine the threshold voltage from the position of the second derivative peak. The subthreshold slope in principle indicates the density of interface traps. However, the resolution is poor because gate dielectrics with 1 nm equivalent oxide thickness always yield nearly ideal subthreshold swing (~60 mV/decade at 300 K). An interesting solution is to measure the back-channel swing which may be more affected, via coupling mechanism, by the density of front-interface traps. Dynamic $I_D(V_G, V_D)$ measurements reveal the impact of floating-body and carrier generation-recombination mechanisms.

The following is a partial list of more refined characterization techniques.
- *Charge pumping* measures the density of interface traps. The method cannot be implemented in SOI MOSFETs with floating body. This is why body-contacted transistors or gate-controlled PIN diodes are used.
- *1/f noise* in MOS transistors originates from fluctuations in the carrier number and/or mobility. The typical variation of the noise factor shows a plateau in weak inversion which indicates the density of slow oxide traps. In small area transistors, the trapping of a single carrier is detected in the time domain as a small pulse random telegraph signal (RTS).
- *Split CV* measurements provide the effective carrier mobility as a function of electric field or inversion charge. The method can be adapted to deliver both front and back channel mobilities from a single measurement.
- *Geometric Magneto-Resistance* is the most accurate method for measuring the carrier mobility. It is implemented in short and wide transistors where the Hall effect is naturally cancelled, maximizing the magnetoresistance. The slope of the channel resistance vs. squared magnetic field yields the mobility without any assumption on the device parameters.
- *Carrier lifetime is* evaluated by measuring the forward or reverse current in gated PIN diodes.

The typical signature of all measurements described above is altered by interface coupling, when the back interface goes from inversion to depletion and accumulation. The following sub-sections contain selected examples of characterization results.

Ultrathin SOI MOSFETs

Fully depleted SOI MOSFETs benefit from the continuous search of technological boosters. State-of-the-art transistors feature ultrathin (5–10 nm) and undoped body, thin BOX (10 nm), raised and silicided source and drain, high-K dielectric, metal gate, strain and, of course, decananometer channel length.

SOI MOSFETs offer the unique option of comparing *in situ* the properties of higk-K and SiO_2 dielectrics without processing dedicated lots. Such a comparison is feasible in a single SOI MOSFET by simply probing the front and the back channels. The electron and hole mobilities are found to be systematically lower (50%) at the front interface (Si/high-K) than at the back interface (Si/SiO_2). Carrier scattering by Remote Coulomb centers, located in the high-K stack, is incriminated and documented by low-temperature measurements (7).

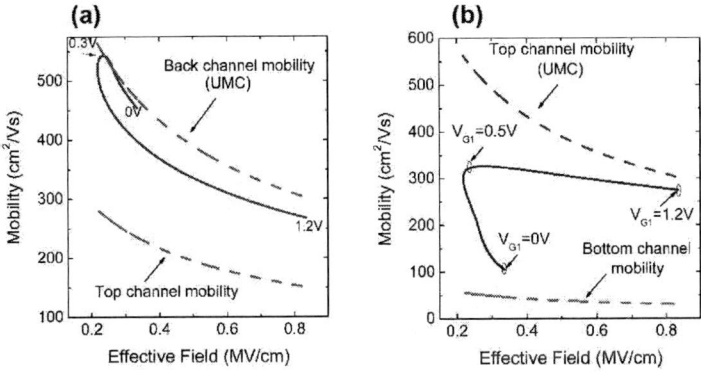

Figure 3. Unusual curves of electron mobility versus effective field achievable in SOI MOSFETs. The bottom gate is set in inversion and the front gate increases from depletion to strong inversion. Case (a): back-channel mobility is assumed to be ideal, matching the universal mobility curve (UMC), whereas top-channel mobility is 50% lower as in MOSFETs with high-K dielectric. Case (b): the front-channel mobility is ideal (UMC) whereas the back-channel mobility is 10 times lower (as in early SIMOX and SOS transistors) (after (8)).

The coupling between the front and back channels, interfaces or gates is responsible for unusual effects unknown in bulk Si CMOS. In ultrathin films, 'as-measured' front-channel properties actually include contributions from the front interface, back interface, BOX and substrate, which are not easy to isolate. The transconductance does no longer reflect the front-channel mobility as it integrates the mobility profile across the film.

Indeed, the carriers flow in the entire body, not only at the interface. This 'volume inversion' concept makes the notions of 'front-channel mobility' and 'back-channel mobility' obsolete, to be replaced by 'mobility viewed from the front or back gate'. At the experimental level, geometric magnetoresistance measurements on a device operated in double-gate and single-gate modes have demonstrated that volume inversion brings a clear gain in mobility.

Another consequence of interface coupling and volume inversion is the failure of the 'universal mobility' concept (UMC). In SOI MOSFETs, the average electric field can *decrease* as the inversion charge increases (8). For example, if the back gate is biased in inversion, increasing the front gate voltage from accumulation to inversion first reduces the vertical field. The effective mobility is even a more complicated notion because of the coexistence of multiple channels, their degree of activation, and the corresponding interface quality. Two distinct carrier distribution profiles and mobility values can correspond to the same effective field. In SOI, the average mobility can increase or decrease with the electric field, in perfect contradiction with the UMC. Figure 3 shows exotic mobility curves, totally different from UMC, which can be achieved in SOI transistors (8).

Strained SOI MOSFETs

Strain induced by the Contact Etch Stop Layer (CESL) is an efficient booster of the carrier mobility. CESL-engineered strain is tuned for maximum performance in short-channel transistors. Figure 4a shows a gain in hole mobility of 80% in 100 nm long P-channel MOSFETs (9).

The mobility dependence on channel length confirms that CESL-strain is localized at the channel extremities, as predicted by mechanical simulations (10). In a long MOSFET, there is no mobility gain because most of the channel remains unstressed. The mobility degradation for devices shorter than 100 nm is attributed to neutral defects, which are generated during the source/drain implantation or the gate stack processing and are concentrated near the source/drain junctions (11). In short transistors, the defective 'edge' regions overlap leading to an increased density of defects.

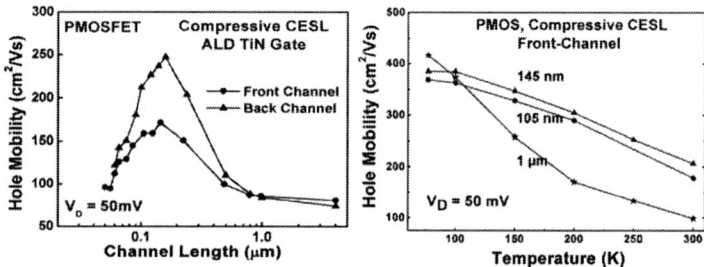

Figure 4. Hole mobility variation with (a) channel length and (b) temperature in fully depleted SOI MOSFETs with compressive CESL-strain (after (9)).

Mobility measurements at low temperature deliver additional information (Fig. 4b). The expected mobility increase at 77 K, consequence of attenuated phonon scattering, is observed only in long channels. In short MOSFETs the mobility saturates and the benefit of strain vanishes at 77 K. The mobility dependence on gate length and temperature results from competing effects of strain, neutral defects and Coulomb scattering in the source/drain depletion regions, which all are inhomogeneous along the channel. Combining these scattering mechanisms with the Matthiessen rule, it is possible to reproduce the experimental data of Fig. 4b (9).

Multiple-Gate MOSFETs

The floating body of SOI transistors can be manipulated to trigger strong hysteresis effects, useful for memory applications (12). In capacitorless single-transistor dynamic memory (1T-DRAM), the amount of majority carriers stored in the body is modulated for achieving two distinct values of drain current (states '1' and '0'). Figure 5a shows the MSD hysteresis effect used in MSDRAMs (12). The back gate V_{GB} is biased in moderate inversion and the front gate V_{GF} is scanned from depletion to strong accumulation and vice-versa. When V_{GF} is switched from 0 V to -3 V, the body experiences deep depletion and the back-channel current is suppressed. The 1-state is programmed by activating band-to-band tunneling (V_{GF} = -6 V) which supplies enough holes and completes the accumulation channel. Switching V_{GF} to -3 V results in a large electron current at the back channel because the transistor is at equilibrium.

Figure 5b shows that in Triple-Gate SOI FinFETs the MSD hysteresis is upside-down. The front-gate voltage was biased in inversion, while the back-gate voltage was swept back and forth from positive (depletion) to negative (accumulation) values. The reversal of the MSD effect indicates that the body does not experience deep depletion because the electrostatics is dominated by the front gate, which covers three sides of the fin. This explains why the front current in reverse mode is not switched off. Since the back interface is depleted (not accumulated as for the 'direct' scan), the front threshold voltage is comparatively low and the current is higher. In very narrow fins, the MSD hysteresis actually disappears due to the neutralization of the back-gate effect by the lateral gates.

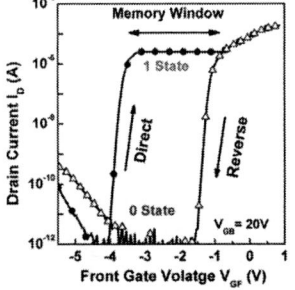 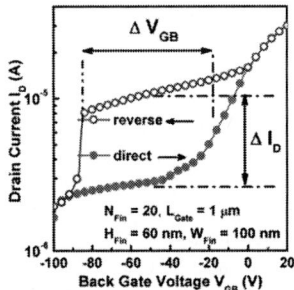

Figure 5. (a) MSD hysteresis in fully depleted SOI N-MOSFETs with double-gate action (after (12)). (b) Reversed MSD hysteresis in Triple-Gate FinFET.

Memory effects can be enhanced by replacing SiO_2 with oxide-nitride-oxide (ONO) BOX. The Si_3N_4 layer is used to store the non-volatile charge. Hysteresis effects occur when carriers are trapped/detrapped via Fowler-Nordheim tunneling or hot-carrier injection. Unlike flash memories, where the current flow and carrier trapping occur at the same interface, in ONO FinFETs the remote trapping in the BOX induces the variation of the front-gate threshold voltage. The advantage is that the carrier flow no longer disturbs the stored charge.

Nanowire MOSFETs

Gate-all-around (GAA) nanowire transistors (NWTs) are attractive candidates for future CMOS due to reduced short-channel effects. 3D-stacked Si and SiGe NWTs with high-K/metal-gate stacks were fabricated at LETI. They exhibit near-ideal subthreshold swing and very low DIBL (13). The mobility was extracted by split CV and Y-function. Combinations of substrate and gate bias enable to activate or not the two channels of the bottom nanowire situated just above the BOX. Their transport properties can thus be compared with those of the upper NWTs.

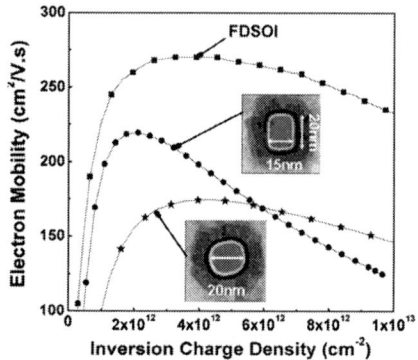

Figure 6. Effective mobility versus charge in Si nanowire FETs with circular and rectangular cross-section (after (XX)).

In all NWTs, the effective mobility drops as the width shrinks below 10 nm (13). Subband splitting, carrier/phonon confinement and technology issues can be invoked. The comparison between rectangular and circular NWTs with similar size is shown in Figure 6. A circular shape leads to marked mobility degradation at low inversion charge. In this region, the mobility is limited by Coulomb scattering involving oxide charges, high-K dipoles and interface traps. Charge pumping measurements indicate a 3 fold increase in the density of traps (14). Presumably, this is a consequence of the continuously varying surface orientation in circular NWTs. However, in very strong inversion, where surface roughness scattering prevails, the mobility is higher for the circular NWTs. A reasonable explanation is that the surface roughness has been improved during hydrogen annealing used for achieving the circular shape.

Tunneling FETs

The tunneling field-effect transistor (TFET) is similar to a MOSFET, except that the dopant types in source and drain are different. This gated PIN diode is operated in reverse mode. The current is induced by band-to-band tunneling (BTBT), which makes it theoretically possible to reach extremely low OFF current and subthreshold slope. In order to enhance the ON current, TFETs with multiple-gate structure and lower bandgap semiconductors, such as Ge and SiGe, have been proposed (15).

In symmetrical TFETs, the leakage current is large because interband tunneling occurs at either junction depending on the sign of the gate bias. This problem can be solved by inserting an intrinsic region L_{IN} separating the drain contact from the channel (Fig. 7a) (16). Since the BTBT rate is determined by the maximum electric field, it can be largely reduced at the drain side due to the increase of L_{IN}. At the source side, the peak field does not change and the tunneling rate in ON state is constant (Fig. 7b).

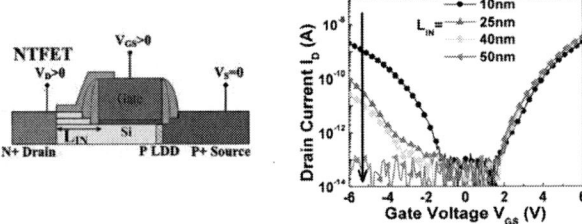

Figure 7. (a) Asymmetric N-channel TFET where tunneling occurs at the source junction. (b) Experimental current versus gate voltage characteristics for variable length of the intrinsic region (L = 400 nm, V_D = 1 V, after (16)).

A challenging aspect is to assess the precise origin of I_{ON}. In TFETs with *large* area, RTS noise (Random Telegraph Signal) is observed, revealing that only a discrete number of traps are active. In MOSFETs with comparable size, the noise is always *1/f* and RTS only appears in very small area devices (16). This result tends to confirm that the dominant current mechanism in TFETs is tunneling. Indeed, the tunneling rate is affected by the trapping process at the Si/SiO_2 interface just above the tunneling junction which is very narrow (around 10 nm).

Conclusions

Recent trends in SOI materials and devices have been reviewed. Updated techniques suitable for the characterization of novel SOI structures have been evoked and illustrated with experimental data. Their correct interpretation requires refreshed theoretical concepts and models.

Acknowledgments

Many thans are due to my students, SOI colleagues around the world and supporting organizations (Eurosoi+, WCU project and Nanosil).

References

1. S. Cristoloveanu and S.S. Li, *Electrical characterization of silicon-on-insulator materials and devices*, Kluwer (1995).
2. S. Cristoloveanu, D. Munteanu, and M.S.T. Liu, *IEEE Trans. Electron Devices*, **47**(5), 1018 (2000).
3. Q.T. Nguyen, J.F. Damlencourt, B. Vincent, L. Clavelier, Y. Morand, P. Gentil, S. Cristoloveanu, *Solid-State Electronics*, **51**(9), 1172 (2007).
4. G. Hamaide, F. Allibert, H. Hovel and S. Cristoloveanu, *J. Appl. Phys.*, **101**, 114513 (2007).
5. N. Rodriguez, S. Cristoloveanu, and F. Gamiz, *IEEE Trans. Electron Devices*, **56**(7), 1507 (2009).
6. K. Park, K. Nayak, and D.K. Schroder, *Solid-State Electronics,* **54**(3), 316 (2010).
7. L. Pham-Nguyen, C. Fenouillet-Beranger, A. Vandooren, T. Skotnicki, G. Ghibaudo, and S. Cristoloveanu, *IEEE Electron Device Letts.*, **30**, 1075 (2009).
8. S. Cristoloveanu, N. Rodriguez, and F. Gamiz, *IEEE Trans. Electron Devices*, **57**(6), 1327 (2010).
9. L. Pham-Nguyen, C. Fenouillet-Beranger, G. Ghibaudo, T. Skotnicki, and S. Cristoloveanu, *Solid-State Electronics*, **54**, 123 (2010).
10. F. Payet, F. Boeuf, C. Ortolland, and T. Skotnicki, *IEEE Trans. Electron Devices*, **55**, 1050 (2008).
11. A. Cros, K. Romanjek, D. Fleury, S. Harrison, R. Cerruti *et al*, *IEDM Tech. Dig.*, 439 (2006).
12. M. Bawedin, S. Cristoloveanu, D. Flandre, and F. Udrea, *ECS Trans.*, **19**(4), 243 (2009).
13. K. Tachi, M. Casse, D. Jang, C. Dupre, A. Hubert *et al*, *IEDM Tech. Dig.*, 313 (2009).
14. M. Casse, K. Tachi, S. Thiele, and T. Ernst, *Appl. Phys. Letts.*, **96**,123506 (2010).
15. T. Krishnamohan, D. Kim, S. Raghunathan, and K. Saraswat, *IEDM Tech. Dig.*, 947 (2008).
16. J. Wan, C. Le Royer, A. Zaslavsky, and S. Cristoloveanu, *Applied Phys. Letts.*, **97**(24), 243503 (2010)

ECS Transactions, 35 (5) 103-108 (2011)
10.1149/1.3570783 ©The Electrochemical Society

Evaluation of interface trap density in advanced SOI MOSFETs

M. Bawedin[a], S. Cristoloveanu[b], S.J. Chang[b], M. Valenza[a], F. Martinez[a], J.H. Lee[c]

[a] IES, University of Montpellier II, Montpellier, France
[b] IMEP-LAHC, Grenoble INP Minatec, Grenoble, France
[c] Kyungpook National University, Daegu, Korea

The density of traps at the top interface is difficult to assess in
advanced SOI MOSFETs with high-K dielectric and ultrathin film.
Searching for a characterization strategy, we compare various
approaches: front-channel subthreshold slope, back-channel slope,
and coupling coefficient between the front and back channels. The
back-channel slope shows the largest variation with the front trap
density. However, the resolution may not be sufficient for
MOSFETs with very thin buried oxide.

Introduction

The density of interface traps stands as a key figure of merit for SOI materials and
devices. Several advanced 'dynamic' techniques, especially elaborated for interface traps
characterization in bulk MOSFETs, have limited applicability in ultrathin SOI transistors.
For example, charge pumping, deep-level transient spectroscopy (DLTS) and high-low
frequency capacitance-conductance require special body contacts (1). All such techniques,
including low-frequency noise, suffer from interface coupling and floating-body effects
which imply the development of dedicated models.

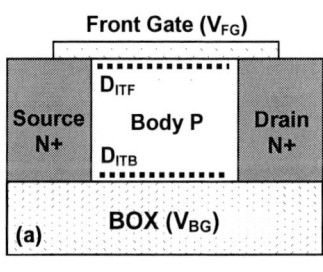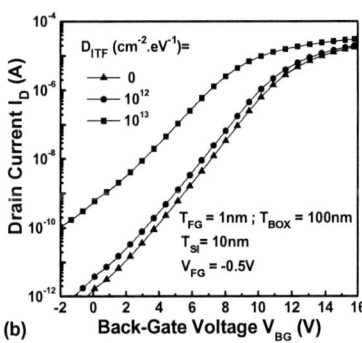

Fig.1. (a) N-MOSFET cross section and (b) simulated drain current I_D as a function of the
back-gate voltage V_{BG} for various front-interface trap densities D_{ITF}. The front-gate
insulator T_{FG}, back-gate oxide and silicon film T_{SI} are 1 nm (EOT), 100 nm and 10 nm
thick, respectively. The effective channel length and doping are 100 nm and 10^{16} cm^{-3}.
The drain V_D and front-gate V_{FG} bias are 100 mV and -0.5 V.

A pragmatic method to extract the average trap density D_{IT} at the silicon/insulator
interface uses the subthreshold swing. The swing depends on C_{IT}/C_{OX}, where $C_{IT}=qD_{IT}$
and C_{OX} is the oxide capacitance. When applied to the characterization of the front

103

interface traps D_{ITF}, this method shows poor resolution. Indeed, state-of-the-art SOI MOSFETs feature excellent swing values (~60 mV/decade, Fig.1b) and cannot be really compared in terms of interface quality. This is because the equivalent thickness (EOT) of high-K gate dielectric is around 1 nm and its capacitance (4×10^{-6} F/cm^2) masks any variation of D_{ITF} in the usual range (10^{10}–10^{12} cm^{-2}eV^{-1}). In other words, the necessary feedback from characterization to technology optimization is lost. Our goal is to explore if alternative solutions to this critical problem exist and eventually develop a reasonable strategy of characterization.

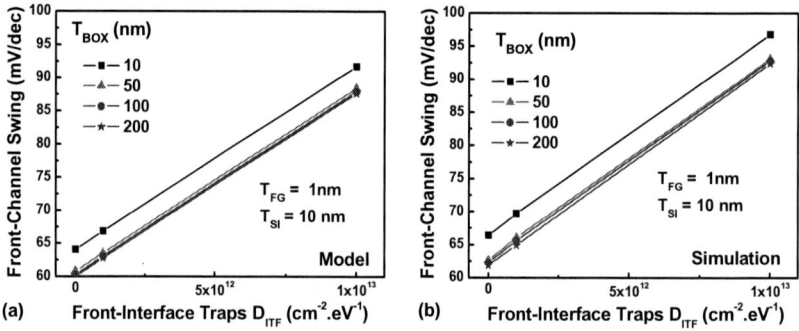

Fig.2. (a) Long-channel model and (b) simulated front-gate subthreshold slope as a function of the front-interface trap density for several BOX thicknesses T_{BOX}. The gate insulator T_{FG} and silicon film T_{SI} are 1 nm and 10 nm thick, respectively. The effective channel length and doping are 100 nm and 10^{16} cm^{-3}.

In this paper, we revisit and compare simple DC methods based on the subthreshold slope and threshold voltage extraction.

Methods and Results

The results are obtained from available long-channel transistor models (1). The adaptation of these models to the case of short (100 nm) and ultrathin SOI MOSFETs has been inspected (Fig. 2) by running 2D numerical simulations (Synopsis with quantum model) (2). The simulations systematically result in larger swings (see Fig. 2) mainly because the short-channel effects come into play and degrade the subthreshold behavior.

Method 1: Front-channel swing

The subthreshold swing is expressed as (3)

$$S_F = 2.3\frac{kT}{q}\cdot\left(1+\frac{C_{ITF}}{C_{OXF}}+\alpha_F\frac{C_{SI}}{C_{OXF}}\right) \quad \text{with} \quad \alpha_F = \frac{C_{BOX}+C_{ITB}}{C_{SI}+C_{BOX}+C_{ITB}} \qquad [1]$$

where C_{SI}, C_{OXF}, C_{BOX}, C_{ITF} and C_{ITB} are the capacitances of the fully depleted film, gate oxide, buried oxide (BOX), front-interface trap density and back-interface trap density, respectively.

In ultrathin films with high quality film-BOX interface ($D_{ITB} < 10^{11}$ cm^{-2}eV^{-1}), coefficient α_F is small and the last term can be neglected. For a reasonable front interface with $D_{ITF} < 10^{12}$ cm^{-2}eV^{-1}, the ratio C_{ITF}/C_{OXF} is also negligible and the swing is close to 60 mV/decade (Fig. 2). A change by 1 mV/decade, corresponding to about 4×10^{11} cm^{-2}eV^{-1}, is very difficult to measure. This standard method fails, as discussed above.

Method 2: Back-channel swing

The back-channel swing is expressed by interchanging the front and back interface parameters in Eq. 1. The measurement is normally performed by keeping the front channel accumulated and provides the density of traps at the back interface. In this case coefficient α_B is unity. The D_{ITB} resolution is good because the BOX is thick enough (Fig. 3). The extracted D_{ITB} value is important first for wafer quality inspection and, second, for the subsequent determination of D_{ITF}. It is worth noting that the accumulation of the front channel is no longer achievable in ultrathin Si films (< 10 nm). Indeed the coexistence of electron and hole layers, facing each other, is prohibited by the super-coupling effect (4).

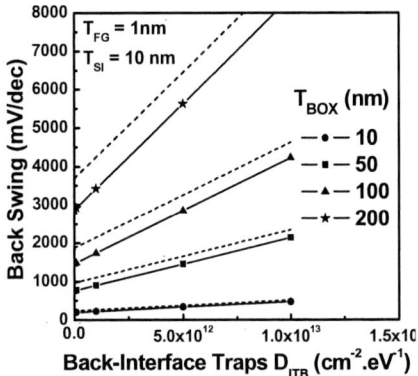

Fig. 3. Back-channel subthreshold swing S_B versus back-interface trap density for various BOX thickness. Full lines: depleted front interface; Dashed lines: accumulated front interface. $T_{SI} = 10$ nm, $D_{ITF} = 10^{12}$ cm^{-2} eV^{-1}.

When the front interface is depleted (Fig. 3), the back-channel swing is smaller and depends on D_{ITF}. The contribution of D_{ITF} is no longer summarized by C_{ITF}/C_{OXF} and is clearly revealed. Figure 4 shows the impact of D_{ITF} and BOX thickness on the back-channel swing. As the front interface changes from 'perfect' ($D_{ITF} = 0$) to 'regular' quality ($D_{ITF} = 10^{12}$ cm^{-2}eV^{-1}), the swing is degraded by 10–100 mV/decade. A thick BOX offers better resolution for D_{ITF} extraction. Since the method is *indirect*, the relative change in back swing ($\Delta S_B/S_B$) is obviously lower than $\Delta S_F/S_F$. However, it is easier to detect a change by 10–100 mV than by 1 mV. Figure 5 compares the direct method (from S_F) and indirect method (from S_B) for D_{ITF} evaluation. For very thin BOX, the accuracy of the indirect method may not be enough.

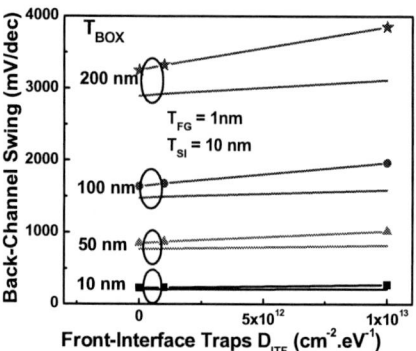

Fig. 4. Simulated (symbols) and calculated (lines) back-gate subthreshold swing S_B as a function of the front-interface trap density for several BOX thicknesses T_{BOX}. Other parameters as in Fig.1.

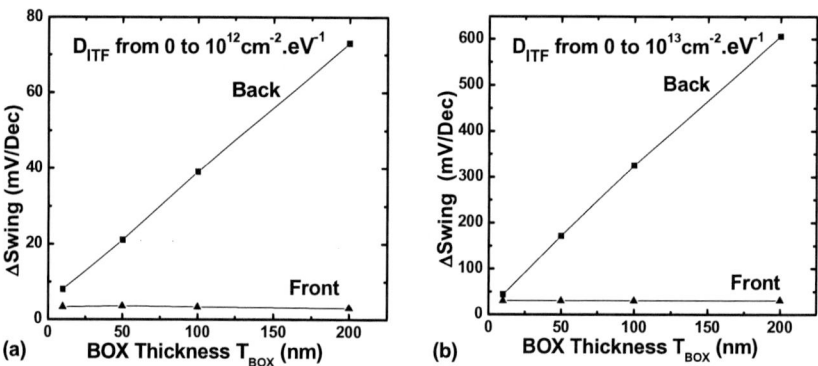

Fig. 5. Simulated back- and front-gate subthreshold swing variation (ΔS) as a function of the back-gate oxide thickness for a front-gate traps density variation ΔD_{ITF} (a) between 0 and 10^{12} cm^{-2} eV^{-1} and (b) between 0 and 10^{13} cm^{-2} eV^{-1}. Other parameters as in Fig.1.

Method 3: Coupling effect

The back-channel threshold voltage is measured as a function of the front-gate bias. When the front interface is depleted, the coupling between the two channels leads to a decrease in V_{THB} with V_{FG}. The coupling slope depends on D_{ITF} and is expressed as (5)

$$C_B = \frac{\partial V_{THB}}{\partial V_{FG}} = \frac{C_{OXF} \cdot C_{SI}}{C_{BOX} \cdot (C_{SI} + C_{OXF} + C_{ITF})} \qquad [2]$$

Figure 6 shows the coupling slope changes with D_{ITF} and T_{BOX}. This change can in turn be used to evaluate D_{ITF}. The sensitivity of this method is acceptable if the difference in thickness between the gate oxide and BOX is large enough and D_{ITF} exceeds 10^{12} cm^{-2} eV^{-1} (Fig. 7).

In practice, the coupling method is more complicated as it pre-requires very accurate threshold voltage extraction. Recommended techniques are the second derivative of the drain current or the V_{THB} definition and measurement at fixed drain current.

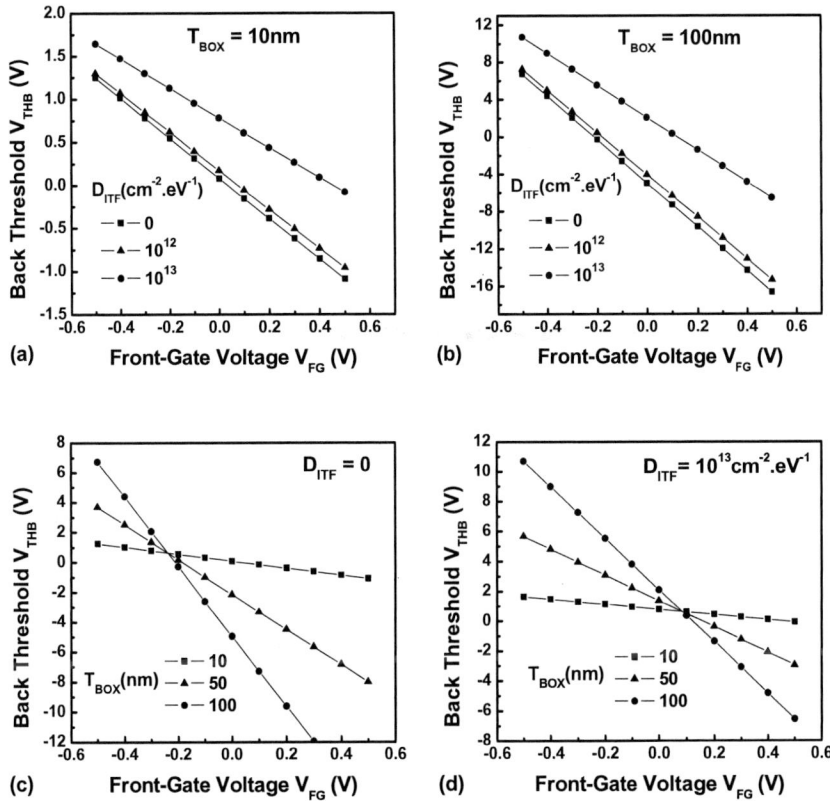

Fig. 6. Coupling curves: back-channel threshold voltage versus front-gate bias. (a) Thin BOX, (b) Thick BOX, (c) Ideal front interface, (d) Poor front interface.

From the data above, we argue that the *indirect* methods 2 and 3 can offer a more accurate evaluation of D_{ITF} than the *direct* method 1. The reason is that the back-channel swing and threshold voltage coupling coefficient also depend on the capacitances of the Si film and buried oxide, not exclusively on the C_{IT}/C_{OX} ratio.

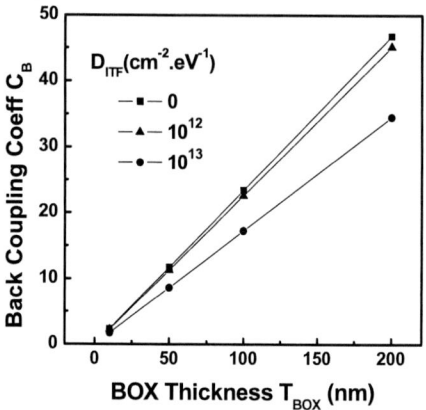

Fig. 7. Coupling coefficient dV_{THB}/dV_{FG} versus BOX thickness for various front-interface trap densities.

Conclusions

There are several possibilities to extract the density of front-interface traps from the DC characteristics of SOI MOSFETs. The measurement of the back-channel subthreshold slope appears as an efficient technique, except when the BOX is too thin. The respective merits, accuracy, and limitations of the various methods were discussed using available analytical models for long channels. The validity of the existing D_{ITF}–related analytical models in ultrathin SOI MOSFETs was examined with 2D numerical simulations. The deviations observed (for example, in Figures 2 and 4) are explained by the impact of short-channel effects. Specific SOI mechanisms such as supercoupling, volume inversion and very thin BOX are also relevant.

Acknowledgments

EUROSOI+, NANOSIL and WCU (KOSEF) organizations are thanked for support.

References

1. S. Cristoloveanu and S.S. Li, "Electrical Characterization of Silicon-On-Insulator Materials and Devices", Kluwer, (1995).
2. Synopsys TCAD, Sentaurus Workbench Advanced, Version X-2005.10.
3. B. Mazhari, S. Cristoloveanu, D.E. Ioannu and T. Caviglia, *IEEE Trans. Electron Devices*, **38**(6), 874 (1991).
4. S. Eminente, S. Cristoloveanu, R. Clerc, A. Ohata and G. Ghibaudo, *Solid-State Electronics*, **51**(2), 239 (2004).
5. H.-K Lim and J. G. Fossum, *IEEE Trans. Electron Devices*, **30**, 1244 (1983).

CHAPTER 6

POSTER SESSION

110

Humidity Effects on Substrate Bonding for Silicon-on-Glass

A. Y. Usenko

Corning, Inc., One Riverfront Plaza, Corning, New York 14831, USA

Glass-to-silicon bonding quality has been studied as a function of air relative humidity in a pre-bonding tool. Two bond-activation techniques, wet and plasma, were used. Optimum ranges of air humidity that ensure high bonding yield are determined. The bonding is finalized by annealing and types of defects appearing in the cases of low and high humidity are indicated. Results prove that humidity control is required for high yield glass-to-silicon bonding.

Introduction

Film transfer technology has been successfully applied for making silicon-on-glass substrates (1). These substrates are further used for fabricating integrated displays, systems-on-a-chip, MEMS, and other devices (2). Substrate bonding is a necessary step in the film transfer technology. While silicon wafer bonding is well developed, the bonding of glass to silicon is less investigated. As in silicon-to-silicon bonding, the glass and silicon surfaces can be prepared by either wet chemical activation, or by plasma activation. In both cases the activated surfaces are kept in air between activation and contacting of the surfaces. Air humidity affects the bonding as it determines the amount of water present on the surfaces at the moment of contacting. It is well known in bonding technology that bonding quality critically depends on the amount of water adsorbed (3). Surprisingly, the humidity-bond quality relation has not yet been quantified. Here we study glass-to-silicon bonding quality dependence from air relative humidity.

Experimental

Corning EAGLE XG® glass and standard silicon wafers were used. Both, glass and silicon were 150 mm wafers. Glass thickness was 0.5 mm and silicon thickness was 0.625 mm. The glass was cleaned and simultaneously wet activated for bonding by processing in diluted ammonia bath in an automated wet bench. The wet processed wafers were then Marangoni dried. Glass surface hydrophilicity (which is a measure of activation for bonding) was checked by wetting angle technique to be below 2 degrees (i.e., highly hydrophilic). Silicon wafers were prepared the same way, but using a diluted RCA recipe instead of ammonia. Some silicon samples were additionally activated by oxygen plasma in RIE type machine at oxygen pressure 30 mTorr, plasma power 700 W, for 2 min. Glass samples were not plasma activated, as glass gets electrically charged in the plasma and then it effectively collects particles from ambient. The glass and silicon pairs were put face-to-face into neighboring slots of a standard wafer boat and the boats were loaded into a Tenney Versa-Tenn environmental chamber. Each pair was held in the chamber for 30 minutes at the desired temperature-humidity combination to achieve equilibrium water coverage conditions on the bonding surfaces. Then the pairs were manually contacted inside of the chamber to ensure that humidity and temperature at the moment of contacting is equal to the setting of the environmental chamber. Bonding time was measured, from the moment of edge contacting to the moment when the bonding wave reaches the opposite side of the wafers. Also, initial bonding strength of the pre-bonded pairs was measured by the blade insertion technique as it is described elsewhere (4). The samples were also annealed at various temperatures and bonding strength progression with temperature was measured. After the final anneal cycle, the samples were visually analyzed for bonding defects – captured bubbles, non-bonded edges, etc.

Results

A plot of the propagation time of the bonding wave versus ambient humidity is shown in Figure 1 below. After the final anneal, the pairs pre-bonded in ambient humidity between 40 and 60% show no visible defects. The wet activated pairs pre-bonded at humidity below 40% have defects: non-bonded areas near edges. The plasma activated samples contacted in low humidity ambient had no defects. The pairs pre-bonded at humidity above 60% also have defects: non-bonded round areas (bubbles or voids) surrounded by bonded areas.

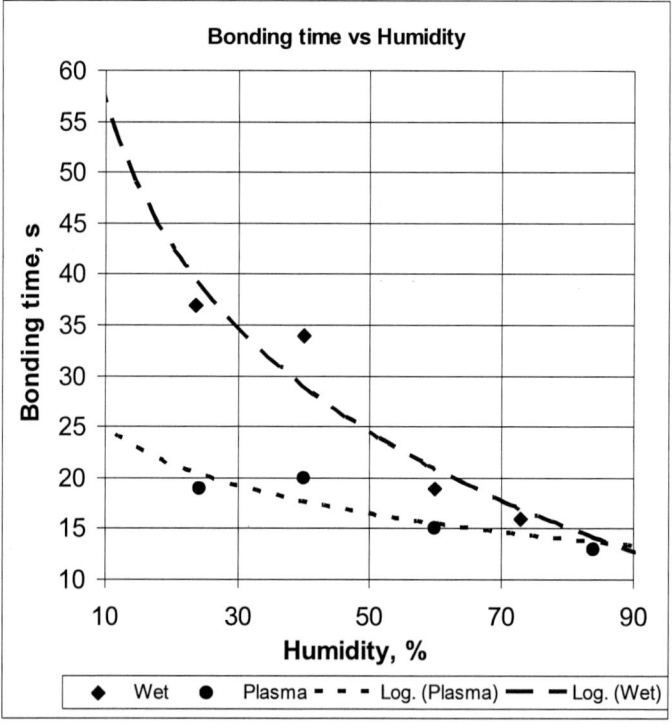

Figure 1. Bonding time as a function of humidity for wet activated pairs (diamonds, dashed line) and plasma activated pairs (circles, dotted line).

Discussion

High quality bonding in the 40-60% humidity range can be explained that surfaces are additionally covered with few monolayers of water molecules attached to OH groups via hydrogen bonds. At humidity under 40%, the water coverage is insufficient, while at humidity above 60% the water coverage is excessive. The insufficient water coverage leads to non-bonded edges. A typical example of pair bonded at low humidity is shown on Figure 2. Excessive water coverage leads to water bubble capture at the bonding interface which transforms upon annealing into void type defects typically concentrated near the end of bonding wave passage. A typical example of pair bonded at high humidity is shown on Figure 3.

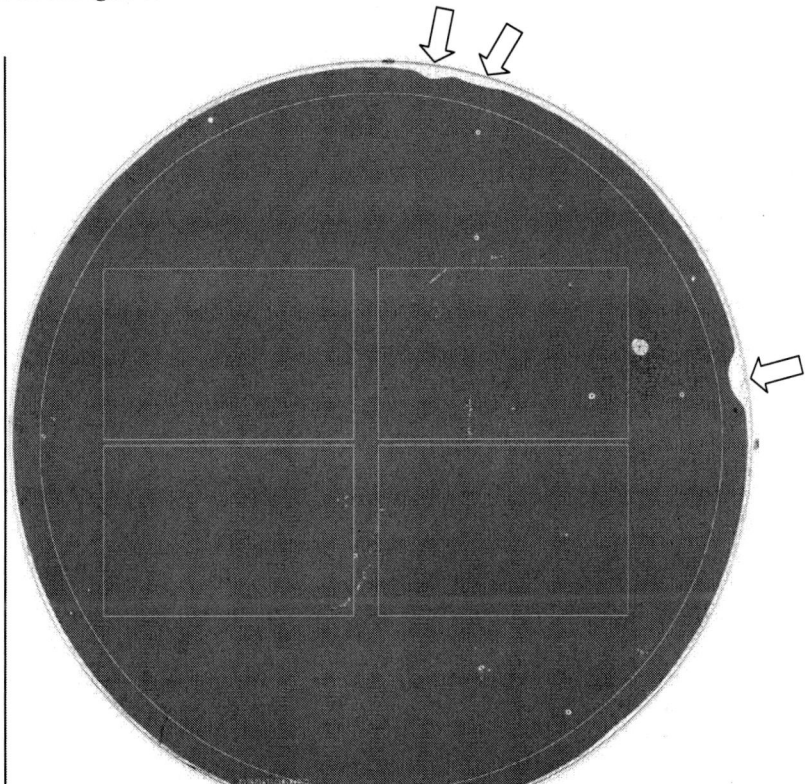

Figure 2. Non-bonded edge. Arrows show locations of the non-bonded areas. It happens while bonding at low humidity. Lack of water causes the non-bonded edges.

It can be seen from Figure 1 that there is a specific range for bonding wave propagation time. If the time is short, there is a risk of capturing water bubbles between wafers which further create defects like on Fig.3. If time is too long, it is a good indication that there is lack of water, which will likely cause defects similar to ones shown on Figure 2. Bonding wave propagation speed can be used to predict bonding quality. If the speed is 5-7 mm/s (for wet activation) or 8-10 mm/s (for plasma activation), high bonding yield is expected.

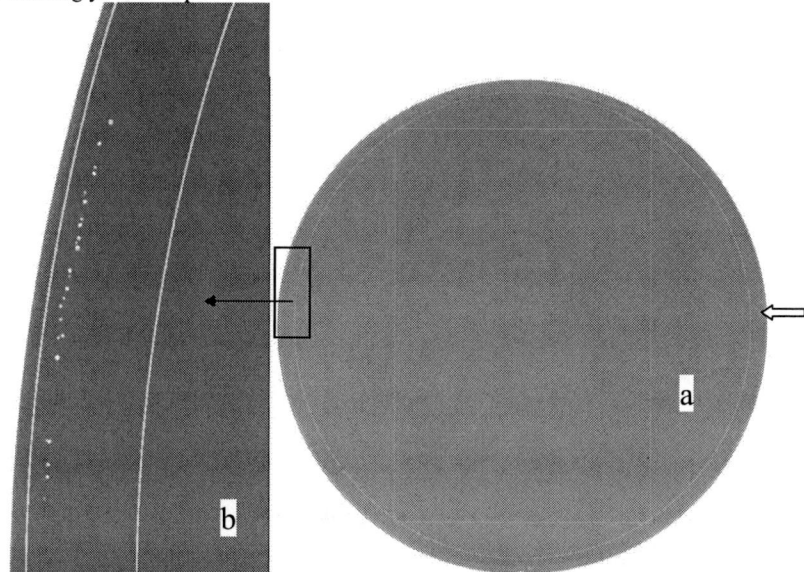

Figure 3. Line of small voids near the edge. (a) – entire bonded pair, (b) enlarged area with defects. This type of defects appears where bonding wave is about completing its way. It happens while bonding at high humidity. Excess of water causes the line of small voids. White rectangle and white circle are not defects; these are artifacts of optical measuring system. Arrow on the right side indicates a point of initial contact of the glass and silicon wafers.

Voids (also called bubbles) at bonding interface are the main type of defects created in wafer bonding. Therefore void causes, their evolution with annealing of bonded pairs, and chemistry of void formation have been extensively studied, see for example recent papers (5), (6), (7) and references therein on earlier studies. Toyoda et al. (5) describe evolution of thermally generated voids. Their pairs have no voids upon pre-bonding, but the void appear upon annealing at around 200°C, then the void size increases with anneal temperature, the voids get maximum size at around 700°C, and then the voids dissolve and eventually disappear at around 900 °C. Upon heating in 200-700 °C range, siloxane bond form, creating also water as a byproduct. The water coalesces into the bubbles thus making the voids.

In our case, there is neither any significant void size increase with temperature, nor generation of new voids upon annealing. We think, this is because we bond the silicon to glass, and glass is an efficient sink for the byproduct water. A simplest type of void is a void caused by particle trapped between the bonded surfaces. These are typically round, and the particle is visible in the middle of the void. Some of that type voids can be seen on Figure 2. Two other types of voids that appear right after contacting of surfaces being bonded are non-bonded edge Fig.2, and captured water drops Fig.3. As indicated above, the Fig.3 type voids appear only in a case of high humidity. Therefore we think that the excess of water between bonding interfaces is moved across the wafer surfaces by bonding wave. At some point, the bonding wave is not able to move further the excessive water, and water drops get trapped. This typically happens near the edge which is opposite to the point where bonding wave started. Possible reason why it typically happens right before completion of the bonding is that all the excessive water that was between the wafers is collected there. In the bonded pair Fig.3, glass and silicon wafers were contacted on right side as indicated by an arrow, bonding wave propagated across the wafers from right side to the left side, and voids were formed near the end of the bonding wave passage.

Acknowledgments

The author thanks Jeff Cites for fruitful discussions, and Ta Ko Chuang for help with experiments.

References

1. http://www.corning.com/CMS/Overview.aspx?id=27579.

2. J.S. Cites,; J.G. Couillard, and K.P. Gadkaree, in *2009 IEEE International SOI Conference*, Vol. CFP09SOI, p. 1, (2009).

3. Q.-Y.Tong, and U.Gosele, *Semiconductor Wafer Bonding*, p.82, Wiley, New York (1999).

4. T.-K. Chuang, A. Usenko, and J. Cites, in *Semiconductor Wafer Bonding 11: Science, Technology, and Applications - In Honor of Ulrich Gösele* C. Colinge, J. Bagdahn, H. Baumgart, K. Hobart, H. Moriceau, and T. Suga, Editors, p.501. The Electrochemical Society Proceedings Series, Pennington, NJ (2010).

5. E. Toyoda et al. *Japanese J. Appl. Phys.,* 48 011202 (2009).

6. X. X. Zhang and J.-P. Raskin, *J. Microelectromech. Syst.*, 14, 368 (2005).

7. M. M. R. Howlader, F. Zhang and M. G. Kibria, *J. Micromech. Microeng.*, **20** 065012 (2010).

116

Subband Structure Engineering in Silicon-on-Insulator FinFETs using Confinement

Z. Stanojevic, V. Sverdlov, and S. Selberherr

Institute for Microelectronics
TU Wien, Gußhausstraße 27-29, 1040 Wien, Austria

Splitting between equivalent valleys larger than the spin splitting energy is observed in confined electron systems, e.g. Si films grown either on SiGe substrate or Si dioxide and Si/SiGe quantum dots. Understanding the contribution of different factors in the valley degeneracy lifting is of key importance for the development of spin-based devices in Si. We demonstrate that the splitting between equivalent valleys strongly depends on the confinement direction and that it is orientation dependent. To explain the effect we use a simple but accurate two-band $\mathbf{k \cdot p}$ model for the conduction band in silicon. Our data is in good agreement with recent results obtained by first-principle calculations.

Introduction

The gradual approach of MOSFET miniaturization to saturation puts limitations on the continuation of the performance increase in logic circuits, and a search for alternative computational principles and technologies becomes necessary. The quantum computer is using quantum mechanical properties to reproduce the structure of data in its operation, which promises a substantial computational superiority over a classical computer for certain problems. The fundamental unit of quantum information is the qubit, while in a conventional computer the information is stored binary. The principle of quantum computation is based on quantum mechanical superposition and entanglement of quantum states. Qubits can be constructed from quantum states of atomic ions, quantum dots, superconducting Josephson junctions, or carrier spins.

Silicon, the material most widely used by the semiconductor industry, possesses several properties attractive for spin-based applications: weak spin-orbit interaction and predominantly zero spin nuclei. Because of these properties electron spin states in silicon should show increased stability which results in a long lifetime of spin polarized carriers. Recently (1), a possibility to inject spin polarized current into silicon was demonstrated. A coherent propagation of spin current through a silicon wafer of 350μm thickness was achieved.

The conduction band of silicon consists of six equivalent valleys. In (100) quantum wells and inversion layers the valley degeneracy is partly lifted. However, the quantum numbers of the two remaining degenerate values compete with those of the electron spin potentially threatening the stability of the qubit. In order to build a silicon based spin qubit it is necessary to lift the remaining valley degeneracies sufficiently so that the corresponding two-level system is only formed by the spin degree of freedom. In order to prevent valley degrees of freedom from interfering with spin quantum numbers, it is mandatory to develop a method for controlling the valley splitting.

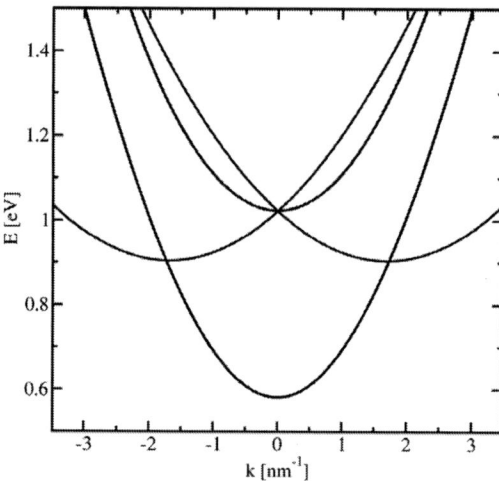

Figure 1. Subband structure in a [100] 2nm thick circular fin. The unprimed subbands are four-times degenerate. Primed subbands are centered at k_0.

The ability to build structures with atomistic precision is one of the goals of the rapidly developing nanotechnology. The control over the wave function or the spin of a single dopant is a key element in silicon quantum electronics. Recent progress in dopant engineering, coherent control over dopant states, and robust operation provides an accelerating momentum to the development of silicon quantum electronic devices. However, due to valley degeneracy, exchange coupling between two dopants in silicon is extremely sensitive to the inter-donor position (2). Controllable valley splitting opens a possibility to tune the coupling between the dopants thus relaxing the requirement for their relative displacement to be unreasonably small.

Valley splitting was discovered in the mid sixties and has been the subject of intensive investigation ever since. Recently, the theory of intervalley coupling was extended (3) to explain experimental data (4). Intervalley coupling was introduced phenomenologically at the heterostructure interfaces. The strength of the intervalley interaction parameter was then calculated by calibrating results to a tight-binding model. Estimations show that the valley splitting is in the range of 0.5meV, which is in agreement with experiments on laterally confined electron systems in point contacts. However, a much smaller valley splitting was reported in Si/SiGe heterostructures (4). In order to make the theoretical predictions qualitatively consistent with the experimental data, a quantum well slightly misaligned from (001) orientation has been considered in (3). The misalignment results in (001) atomic steps at the interface. This makes the valley splitting inhomogeneous and position-dependent. It was suggested in (3) that only the areas with high valley splitting contribute to the experimental signal. Regardless of numerous efforts, the contribution of different factors including interface disorder and quantization remains one of the greatest theoretical challenges on the path of understanding valley splitting. Recently, an experimental observation of valley splitting in Si/SiGe quantum dots was reported in (5) and an explanation of the large valley splitting at the Si/SiO$_2$ interface observed in (6) by hybridization of the valley states with

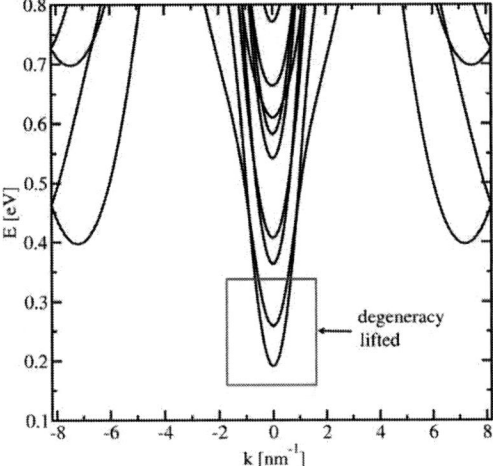

Figure 2. Subband structure in a [110] oriented 2nm thick circular fin. The two-fold degeneracy of the unprimed subbands is completely removed.

the interface states (7) was given. Control over valley splitting is one of the ingredients necessary for the successful implementation of Si spin devices.

In this work, we demonstrate the enhancement of the valley splitting in silicon fin structures and nanowires along the [110] direction. Multi-gate FinFETs are a promising alternative to bulk MOSFETs beyond the 22nm technology node.

Method

The subband structure in a confined system is computed using the $\mathbf{k \cdot p}$ model proposed in (8), which has been shown to be accurate up to 0.5eV above the conduction band edge (9). For two valleys along [001] direction the Hamiltonian is

$$H = \begin{pmatrix} \dfrac{p_z^2}{2m_l} + \dfrac{p_x^2 + p_y^2}{2m_t} + \dfrac{p_z \hbar k_0}{m_l} + U(\mathbf{r}) & \dfrac{p_x p_y}{M} \\ \dfrac{p_x p_y}{M} & \dfrac{p_z^2}{2m_l} + \dfrac{p_x^2 + p_y^2}{2m_t} - \dfrac{p_z \hbar k_0}{m_l} + U(\mathbf{r}) \end{pmatrix}, \qquad [1]$$

where $\mathbf{p} = -i\hbar\, d/d\mathbf{r}$ is the momentum operator, $U(\mathbf{r})$ is the confinement potential, m_t and m_l are the transversal and the longitudinal effective masses, $k_0 = 0.15 \times 2\pi/a$ is the position of the valley minima relative to the X point in unstrained silicon, and $M^{-1} \approx m_t^{-1} - m_0^{-1}$. The Hamiltonians for the valleys along [100] and [010] can be obtained from [1] by appropriate coordinate transformation. The resulting Schrödinger differential equation with the Hamiltonian [1] is discretized using the box integration method and solved for each value of the conserved momentum along the directions of the wave function propagation using efficient numerical algorithms available through the Vienna Schrödinger-Poisson framework (VSP).

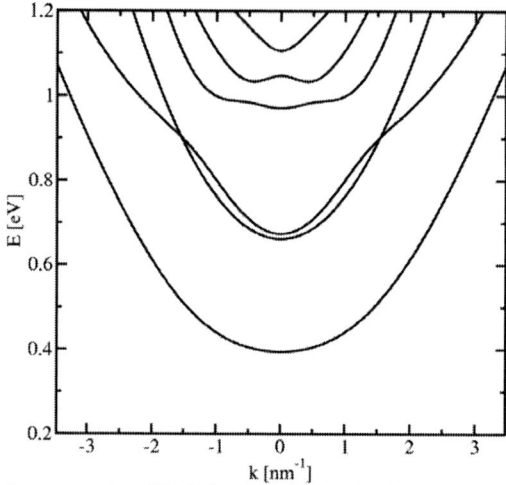

Figure 3. Subband structure in a [111] 2nm thick circular fin. The lowest subband is six-fold degenerate. Its minimum is located at the Γ-point.

Results

Fins and nanowires of [100], [110], and [111] growth orientation of circular, square, and rectangular cross sections have been investigated. The subband structure for a circular 2nm thick nanowire is shown in Fig.1-Fig.3. The circular shape of nanowires is most closely related to the shapes studied previously in (10) using a tight-binding (TB) method. The electron subband dispersion computed with the **k·p** model possesses all the important features obtained with an atomistic-based calculations, regardless its approximate character as compared to TB.

In the case of the [100] nanowire confinement causes the minima of the unprimed subbands to be projected onto the Γ-point of the one-dimensional Brillouin zone, while the minima of the primed subbands are located at $\pm k_0$ (see Fig.1). The unprimed subbands of the [100] nanowire are four-fold degenerate within the **k·p** model. In the [110] nanowire the degeneracy of the lowest subband is completely lifted as seen in Fig.2, and the lowest subband becomes non-degenerate. For the 2nm thick [111] nanowire the lowest subband minim is at the Γ-point (Fig.3), which is in agreement with TB calculations (10). The lowest subband preserves six-fold degeneracy. The degeneracy can be partly lifted in a square structure with the faces along (11-2) and (1-10) directions. Yet, the lowest subband remains four-fold degenerate.

We now turn our attention to the case of the [110] oriented nanowire shown in Fig.2. It is worth noticing that in the effective mass approximation the lowest subband must be two-fold degenerate because of the equivalency between the two valleys oriented along the [001] direction. Surprisingly, within the more accurate **k·p** model the degeneracy is removed. Similar results were obtained from the TB approach (10). The degeneracy of the lowest subband is completely lifted for a square or rectangular [110] nanowire, with the faces oriented along (001) and (-110) directions. Fig.4 demonstrates that the degeneracy is lifted in both circular and square [110] fins. Furthermore, the degeneracy is lifted for the higher unprimed subband as well. For thin fins the splitting between the

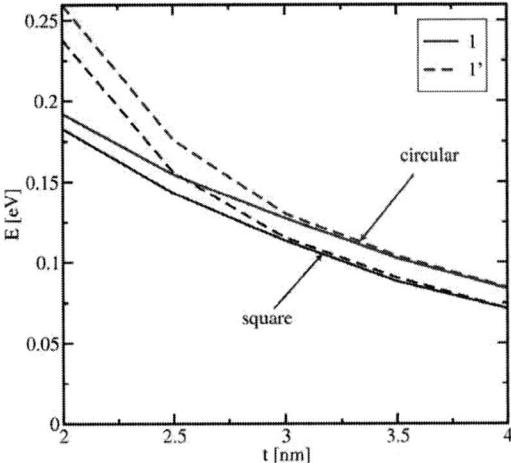

Figure 4. Dependence of the minima of the two lowest subbands on the fin thickness in circular and square [110] fins. The two-fold degeneracy is lifted in thin fins.

subbands originating from the two equivalent [001] valleys, i.e. the valley splitting, can be large. Fig.5 demonstrates good agreement between the valley splitting obtained from the **k·p** model and results of recent density functional calculations (11). For thin [110] fins valley splitting may reach a value of about one electron-volt. Thus, one can control and engineer the valley splitting between the equivalent valleys by properly designing the shape and size of silicon fins.

In order to explain such a large valley splitting let us approximate the fin confinement potential by a square well with infinite potential walls. The subband minimum is at the Γ-

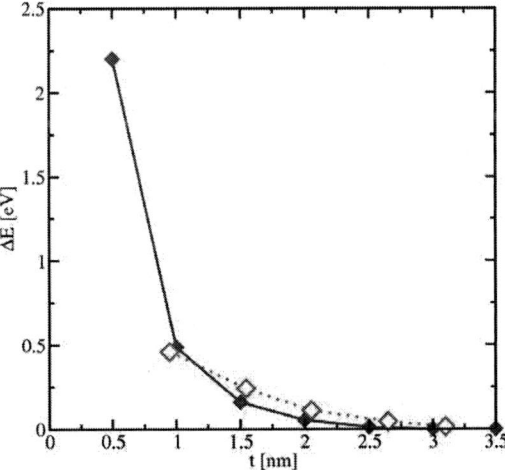

Figure 5. Comparison between the valley splitting dependences on the square fin thickness obtained with the **k·p** model [1] (filled symbols) and from density functional calculations (11).

point. At this point the off-diagonal term in the Hamiltonian [1] in a [110] fin is equal to $[\pi n\hbar/(2tM)]^2$, where n is an integer and t is the fin thickness. The non-zero value of the off-diagonal elements results in a non-parabolic dispersion along [001] direction for [001] valleys. For the quantization along the [001] direction a consideration similar to the one in ultra-thin body silicon films can be applied (9). If the value of the off-diagonal elements is not too large, the valley splitting is given by (9)

$$\Delta E_{m,n} = \left(\frac{\pi}{k_0 t}\right)^4 \frac{\hbar^2 k_0^2 m^2 n^2}{2Mk_0 t \mid 1 - (\pi n/k_0 t)^2 \mid} \sin(k_o t),$$ [2]

where m is another integer corresponding to the quantization along the [001] axis. The relative valley splitting of the first subband $E_{1,1} = \frac{\pi^2\hbar^2}{2m_l t^2}\left(1 + \frac{m_l}{m_t}\right)$ is

$$\frac{\Delta E_{1,1}}{E_{1,1}} = \left(\frac{\pi}{k_0 t}\right)^2 \frac{m_l}{Mk_0 t \mid 1 - (\pi n/k_0 t)^2 \mid (1 + m_l/m_t)} \sin(k_o t),$$ [3]

which varies strongly with the fin thickness.

Conclusion

A rigorous analysis of the subband structure in silicon fins of different shape and orientation has been performed. It is demonstrated that within the two-band $\mathbf{k}\cdot\mathbf{p}$ model the two-fold degeneracy of unprimed subbands with the same quantum number is completely lifted in [110] oriented fins. The relative splitting between the unprimed subbands is inversely proportional to the third power of the fin thickness and becomes large rapidly, when the fin thickness is decreased. The valley splitting can be reliably controlled by designing the shape and orientation of a silicon fin, which is of a great importance for developing silicon spin-based devices.

Acknowledgments

This work is supported by the European Research Council through the grant #247056 MOSILSPIN.

References

1. B.Huang, D.Monsma, I.Appelbaum, *Phys. Rev. Lett.* **99**, 177209 (2007).
2. S.Das Sarma, R.de Sousa, X.Hu, B.Koiller, *Solid-State Commun.*, **133**, 737 (2005).
3. M.Friessen, S.Chutia, C.Tahan,S.N.Coppersmith, *Phys.Rev.B* **75**, 115318 (2007).
4. S.Goswami, K.A.Slinker, M.Friesen, L.M.McGuire, J.L.Truitt, C.Tahan, L.J.Klein, J.O.Chu, P.M.Mooney, D.W.van der Weide, R.Joynt, S.N.Coppersmith, M.A.Eriksson, *Nature Physics* **3**, 41 (2007).
5. M.Borselli, R.Ross, A.Kiselev, E.Croke, K.Holabird, P.Deelman, L.Warren, I.Alvarado-Rodriguez, I.Milosavljevic, F.Ku, W.Wong, A.Schmitz, M.Sokolich, M.Gyure, A.Hunter., arXiv:1012.1363.
6. K.Takashina, Y.Ono, A.Fujiwara, Y.Takahashi, Y.Hirayama, *Phys.Rev.Lett.* **96**, 236801 (2006).
7. A.L.Saraiva, B.Koiller, M.Friesen, *Phys.Rev.B* **82**, 245314 (2010).
8. J.C.Hensel, H.Hasegawa, M.Nakayama, *Phys.Rev.* **138**, A225 (1965).
9. V.Sverdlov, O.Baumgartner, T.Windbacher, S.Selberherr, *J.Computational Electron.*, **8**, 192 (2009).
10. A.Svizhenko, P.W.Leu, K.Cho, *Phys.Rev.B* **75**, 125417 (2007).
11. H.Tsuchiya, H.Ando, S.Sawamoto, T.Maegawa, T.Hara, H.Yao, M.Ogawa, *IEEE T-ED* **57**, 406 (2010).

Single Crystal Silicon Thin Film on Polymer Substrate by Double Layer Transfer Method

J. Senawiratne[a] and A. Usenko[a]

[a] Corning Incorporated, Corning, New York 14845, USA

We report discovery of new method to transfer a single crystal silicon thin film onto a bendable polymer substrate by using layer transfer process. The method includes creation of mechanically weakened layer 100 nm - 600 nm below the Si wafer surface using boron and hydrogen ion implantations. Silicon mother wafer is then pre-bonded to glass and exfoliated. Exfoliation divides the glass-silicon assembly into weakly bound Si thin film on glass and leftover mother Si wafer. Then the silicon thin film was transferred from glass substrate into a polymer substrate.

Introduction

Silicon-on-polymer is a substrate that would enable flexible electronics, wearable computing, smart textiles, three-dimensional integrated systems, bendable displays, etc. Many electronics manufactures are looking ways to replace today's standard silicon substrates with thinner, lighter, bendable, foldable, break-free, and inexpensive silicon on polymer substrates (1). The silicon-on-polymer substrate can be made using layer transfer technique. In the layer transfer technique (2) hydrogen is implanted to separate (exfoliate) a thin Si layer onto an insulator substrate that is stable at exfoliation temperature. Successful attempts making silicon-on-polymer by layer transfer are described in literature recently (3, 4). These processes require polymer that withstand high temperature of splitting of hydrogen-implanted silicon, i.e., 400-500 °C, thus only special high temperature polymers can be used.

In this work we discovered an innovative solution to transfer single crystal silicon thin film on to a polymer substrate using what we call "double layer transfer method" which does not require high-temperature polymer substrate. Potentially, semi-fabricated (i.e. after last high-temperature processing step) integrated circuits can be transferred onto desired polymer as well using our process. The method includes two layer transfer steps: 1) transfer of Si onto a glass and 2) transfer of Si thin film from glass onto a polymer substrate.

Experimental

Standard silicon wafers 300 mm size <100> orientation, p-type, boron doped, 8 to 13 Ω-cm resistivity, 675 μm thick, prime grade were selected as mother wafers in a production run being described. First an oxide layer was grown on the wafer by thermal oxidation process at 900 °C. Then the film with oxide layer was implanted with boron. The boron doses and implantation energy were $2 \times 10^{14} - 4 \times 10^{15}$ cm^{-2} and 180 keV, respectively. Then the boron implanted Si wafers were additionally implanted with hydrogen H^+ species of energies 60 keV and doses 3×10^{16} and 4×10^{16} cm^{-2}. These implantation conditions cause exfoliation at depth of about 500 nm beneath the Si wafer surface (5, 6).

Corning EAGLE XG® glass sheet of thickness 0.5 mm was used for this experiment. The selected glass sheets were cleaned and bonded to silicon wafers with oxide layers. Standard wet activation of the silicon wafers for bonding with diluted RCA clean was used. Glass was activated for bonding with diluted ammonia bath. This removes dust and contaminants while terminating the silicon and glass surfaces with hydroxyl groups, as it is described, for example, in (7). The silicon wafers and glass sheets were rinsed in deionized water and dried after the wet processing. Hydrophilicity of the silicon and glass surfaces was tested with contact wetting angle measurements. The wetting angle was found to be below the lowest angle that is possible to measure with the setup – 2°. It indicates good bondability of the glass surface. Flatness and roughness of the glass was tested with atomic force microscopy. Glass has low roughness which is sufficient for bonding.

The pre-bonded glass-silicon assembly was placed on a hotplate facing glass side up. The glass-Si assembly was then heated from the silicon side by turning the hotplate heater ON. During the heating, the temperature of the substrate was monitored by using contact thermo couple and calibrated IR sensor. Initiation of exfoliation and splitting of wafers were monitored precisely using thermo couples. During the temperature ramping wafers were exfoliated by leaving silicon thin film on glass. Heating was turned off immediately after exfoliation. The films exfoliated at lower temperature (<380 °C) found to be weakly bound to glass substrate. The films were let cooled to room temperature and then they were glued onto polymer from the silicon side using appropriate glue (for example one can use transparent sticky tape for this purpose). Then the silicon thin film can be transferred to polymer by pealing off the polymer from the glass. The Film exfoliated at lower temperature (< 380 °C) found to be well delaminated from Si glass assembly and hence they were transferred to bendable polymer substrate without any mechanically induced defects.

Results and Discussion

Figure 1 shows the exfoliation temperature as a function of boron dose for given hydrogen dose. The exfoliation temperature reduces significantly with increasing boron + hydrogen (H+B) dose. The exfoliation temperature reduces to as lower as 280 °C for the sample which has the highest H+B dose. Compared to the hydrogen-only implantation, implantation of B and H causes platelets nucleation and micro-crack propagation to happen at lower temperature. Possibly, B atoms acts as trap centers for H atoms and hence clusters of H atoms may have trapped around them to create unstable platelets (8) and as a result of this, it can be expected lower exfoliation temperature for the film with higher B+H dose.

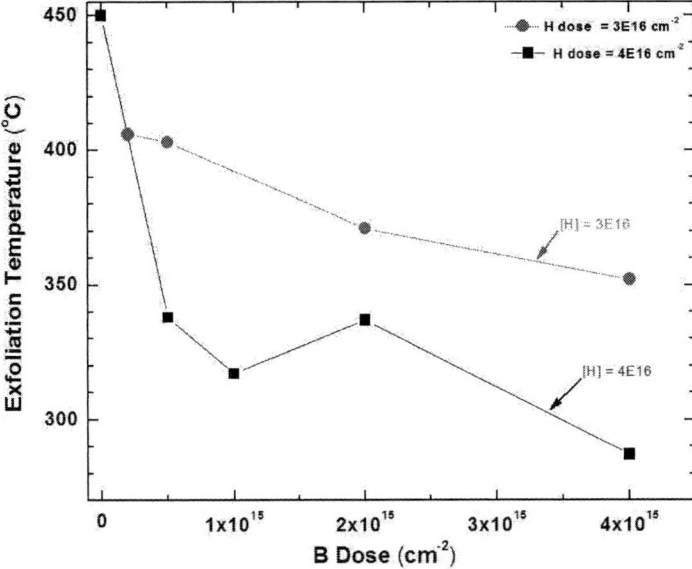

Figure 1. Exfoliation temperature as function of Boron dose for two discrete hydrogen doses: 3E16/cm² and 4E16/cm².

During the exfoliation no blistering was observed. It can be interpreted that we have strong enough bonding to prevent undesirable blistering, but weak enough for further debonding to do the second transfer, onto the polymer. Pre-bonded Si-glass assembly is initially bonded together by weak van der Waals force. During the heating these van der Waals bonds tends to converts into strong covalent bonds. Therefore, to make stronger covalent bond both exfoliation temperature and thermal budget that required for bond transformation are essential (9). Therefore, one can expects strongly bound Si on glass by increasing exfoliation temperature and heating (and thermal annealing) time. In particular, wafers exfoliated at lower temperature (< 380 °C) with faster temperature ramping results loosely bound Si thin film on glass substrate. As a result, the loosely bound Si films were able to transfer to polymer substrate very smoothly without creating any mechanically defects (see Figure 2).

We think that two causes contributed to low strength silicon-to-glass bond that further enabled our second transfer. The first cause is lowered temperature of exfoliation which is due to boron-then-hydrogen implant scheme. The second cause is silicon surface contamination during ion implantation. Implanter chambers are expected to be free of organic contamination, but in fact still there is some low contamination, possibly from roughing pump oil, polymer piping inside of the chambers, etc. As silicon wafers are cooled under the ion beam in the implanters, they might have temperature lower than the rest of chamber, thus effectively adsorbing contaminants on the wafer surface. Further, excitation from ion beam might cause polymerization of the adsorbed organic contaminants. We used a general purpose (not dedicated) ion implanter to do boron implantation, so the silicon wafer surface might also be cross contaminated with typical implant species as boron trifluoride. Polymerizing organics with fluoride can give Teflon like polymers that are very difficult to clean out, as they are hard to dissolve in RCA solution. Our preliminary TOF-SIMS data partially support this hypothesis - the TOF-SIMS shows small organic islands. At this point we cannot determine for sure, whether low bonding strength that allowed as to do double transfer is mostly due to low temperature exfoliation, or non-cleaned surface contamination also contributed to the low bonding strength. However, our process is repeatable, so there is a process window that enables the double transfer.

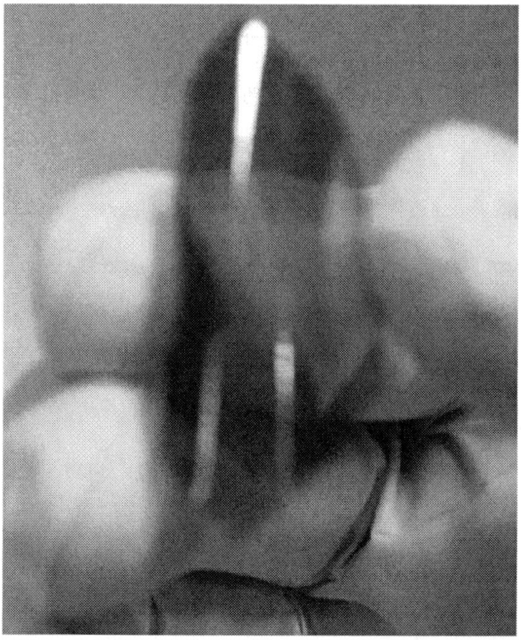

Figure 2. Picture of transferred Si thin film on a bendable polymer substrate

Crystalline quality of the transferred Si thin films on polymers was evaluated by using micro-Raman spectroscopy. As shown in Figure 3 Strong Lorentzian shaped Raman peak at 520 cm^{-1} with full with at half maximum (FWHM) of 5 cm^{-1} indicates strain free single crystal silicon on polymer substrate. The thickness of the transferred film on glass is 517 nm.

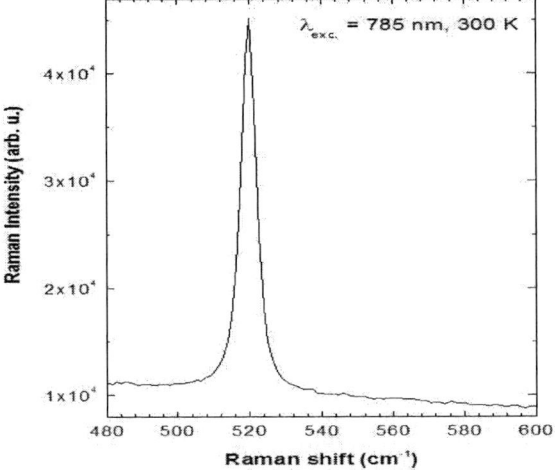

Figure 3. Strong Lorentzian shape Raman peak at 520 cm^{-1} of full width at half maximum indicated strain free single crystal Si thin film.

Conclusions

We have demonstrated successful transfer of a single crystalline silicon film onto a polymer substrate. Our version of the transfer – double transfer: first onto a temporary glass substrate, and then onto a destination polymer substrate has much less limitations to the polymer substrate compared to known methods. Particularly, it does not require polymers that withstand high temperatures. This might lead to new unique extensions of silicon based electronics.

Acknowledgments

The authors would like to thank the technical staff at Corning that provided analytical services.

References

1. www.americansemi.com/FleX.html
2. M. Bruel, "Silicon on insulator material technology" Electron. Let., **31**, 1201 (1995)
3. M. Argoud et al., "Single Crystal Silicon Film Transfer onto Polymer", ECS Trans. **33**, 217 (2010).
4. K. Byun et al., 'Single-Crystalline Silicon Layer Transfer to a Flexible Substrate Using Wafer Bonding", J. Electron. Mater, **39**, 2233 (2010).
5. http://www.srim.org
6. Q.-Y. Tong *et al*, Appl. Phys. Lett. **72**, 49 (1998)

7. Handbook of Semiconductor Wafer Cleaning Technology By Kern, Werner Noyes Publications 1993, 623 pages
8. Tong QY, Scholz R, Gosele U, Lee TH, Huang LJ, Chao YL, et al. A "smarter-cut" approach to low temperature silicon layer transfer. Appl Phys Lett. 1998;72(1):49–51
9. J. W. Lee, C. S. Kang, O.S. Song and C. K. Kim, Thin Sol. Films 394 (2001), 272 - 276

Zero Temperature Coefficient of Current Gain Cutoff Frequency and Maximum Oscillation Frequency for Various SOI and Si bulk MOSFETs

Mostafa Emam, Danielle Vanhoenacker-Janvier and Jean-Pierre Raskin

Electrical Engineering Department
Institute of Information and Communication Technologies, Electronics and Applied Mathematics (ICTEAM)
Université catholique de Louvain, 1348 Louvain-la-Neuve, Belgium

> A Zero Temperature Coefficient characteristic is presented and characterized for the first time for current gain cutoff frequency and maximum oscillation frequency of different MOSFET structures. The benefits of these points on the design of RF circuits are of first importance especially for harsh environment applications and high density RF circuits.

Introduction

The so-called Zero Temperature Coefficient (*ZTC*) point has been identified for bulk CMOS by Shoucair (1) and Prijic *et al.* (2) in both linear and saturation regions for temperatures between 27 and 200°C. According to Shoucair, the *ZTC* of drain current is usually well defined up to approximately 200°C, whereas at higher temperatures, drain-to-body leakage currents typically cause the characteristics to shift upwards on a linear scale of micro-Amperes. Later, Groeseneken *et al.* (3) and Jeon *et al.* (4) demonstrated the existence of the *ZTC* point experimentally for thin- and thick-film SOI MOSFETs, respectively. The presence of *ZTC* points in lateral asymmetric devices was shown experimentally by Emam *et al.* in (5). The analytical formulation for *ZTC* in SOI devices has been introduced by Osman *et al.* in (6). The *ZTC* point is device-geometry-independent for a given process (1) but it depends on the body bias for body-contacted SOI devices (7). Basically, there are two *ZTC* points for a transistor, one for the drain current and the other for the transconductance, and in general they have different values in linear and in saturation regions. These *ZTC* points are defined as the points at which the drain current or the transconductance remains constant as temperature varies, i.e. $\partial I_{DS}(T)/\partial T \approx 0$ or $\partial g_m(T)/\partial T \approx 0$, respectively. This can be explained as follows. The *ZTC* points are values of V_{GS} at which the reduction of the threshold voltage due to increasing temperature is counter-balanced by the reduction of the mobility, and as a result, the value of the drain current or the value of the transconductance remains constant as the temperature varies. For gate voltages lower than *ZTC*, the decrease of the threshold voltage is dominant and so the drain current increases with temperature, while for gate voltages higher than *ZTC*, the mobility degradation predominates and the drain current decreases with temperature. The *ZTC* is a very important bias point for analog designers as it corresponds to a gate voltage at which the dc performance is constant with temperature (8, 9).

In this work, a *ZTC* point is defined and experimentally presented for the current gain cutoff frequency, f_T and for the maximum oscillation frequency, f_{max}. These two *ZTC* points are shown to be valid till 250°C. Several MOSFET technologies and structures are measured to analyze these *ZTC* points. The technologies used are bulk (with a deep n-well protection) and SOI in floating-body (FB) and body-tied (BT) structures. Partially-depleted (PD) as well as fully-depleted (FD) devices are presented for the SOI technology. The bulk devices used in this work feature a deep n-well protection in which a complete isolation is provided using a deep n-well (DNW) and a n-isolation (NISO) implantation connected together to form a grounded shield for each individual transistor. This special structure imitates the behavior of a SOI device (10). In the floating-body (FB) structure; the body is not connected to any terminal whereas for the body-tied (BT) structure; the body and the source are shorted together by a silicided contact. All bulk and PD SOI devices feature 30 gate fingers of 0.13 μm-long and 2 μm-wide each connected in parallel whereas the FD SOI devices feature 48 parallel gate fingers of 0.15 μm-long and 5 μm-wide each.

ZTC point of Current Gain Cutoff Frequency

It is well known that f_T is proportional to the transconductance g_m through the relation $f_T = g_m/2\pi C_{gg}$, where C_{gg} is the total gate capacitance (gate-to-source capacitance and gate-to-drain capacitance). According to this formula, the behavior of C_{gs} and g_m with temperature impact the f_T behavior.

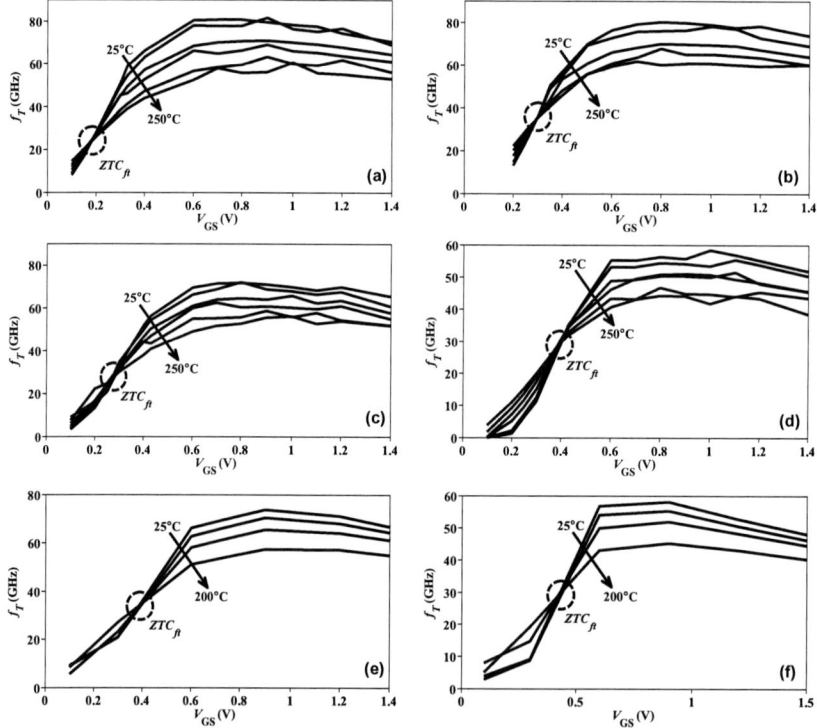

Figure 1. ZTC_{ft} for different MOSFET structures: deep n-well protected bulk in (a) FB structure and (b) BT structure, PD SOI in (c) FB and (d) BT structure, and FD SOI in (e) FB and (f) BT structure.

However, according to the extraction of the small-signal equivalent circuit using RF measurements of S-parameters (11) of several MOSFET structures, the variation of C_{gg} with temperature is very limited resulting in the appearance of a ZTC point in the behavior of f_T which coincides with the ZTC point of g_m (from dc measurements) on the V_{GS} axis.

Figure 1 shows the presence of the ZTC_{ft} for different MOSFET structures whereas Figure 2 shows the corresponding transconductance ZTC_{gm} points, both presented for the saturation region ($V_{DS} = 1.2$ V for bulk and PD SOI and 1.5 V for FD SOI). It is clear that the ZTC_{ft} and the ZTC_{gm} coincide with a negligible margin of difference. On the other hand, the behavior of C_{gg}, shown in Figure 3, does not show the ZTC behavior.

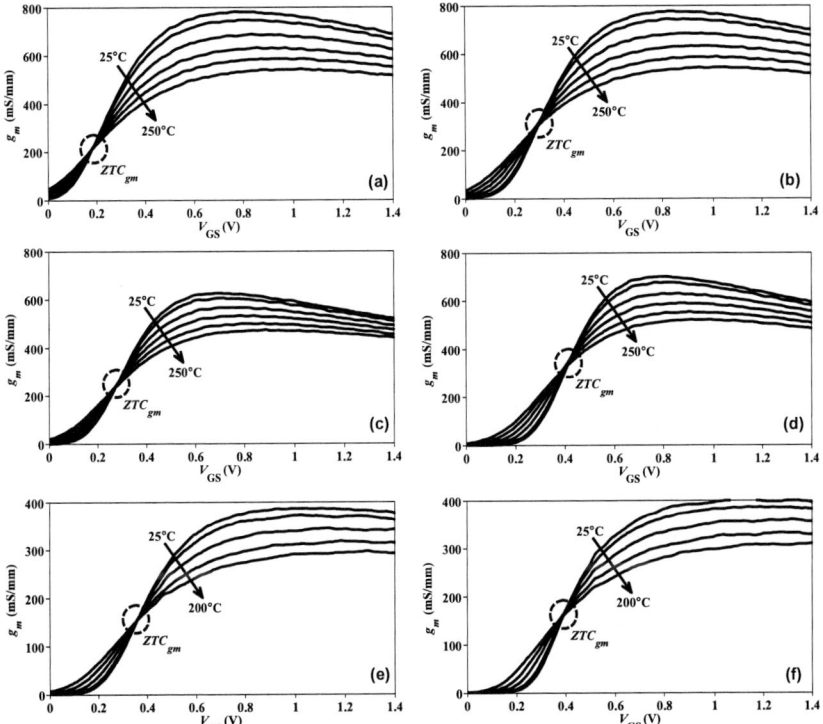

Figure 2. Transconductance g_m for different MOSFET structures: deep n-well protected bulk in (a) FB structure and (b) BT structure, PD SOI in (c) FB and (d) BT structure, and FD SOI in (e) FB and (f) BT structure.

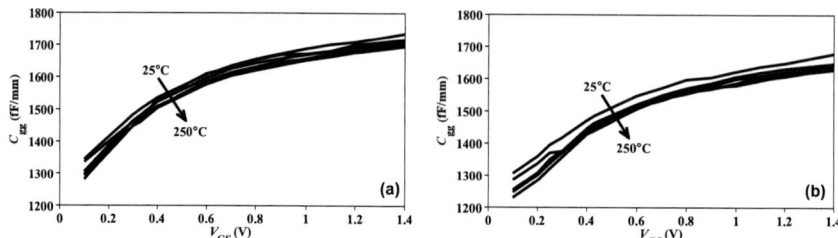

Figure 3. C_{gg} for (a) deep n-well protected bulk and (b) PD SOI both in FB structure.

ZTC point of Maximum Oscillation Frequency

The maximum oscillation frequency f_{max} also features a *ZTC* point as shown in Figure 4 for both bulk and PD SOI FB devices. The ZTC_{fmax} occurs at a value of V_{GS} lower than that of ZTC_{ft} (thus also lower than ZTC_{gm}). The expression of f_{max} is more complex [12]:

$$f_{max} = \frac{f_T}{2\sqrt{\left(R_g + R_s + R_i\right)\left(g_d + g_m \dfrac{C_{gd}}{C_{gs}}\right)}} \qquad [1]$$

The effect of the extrinsic gate and source resistances on f_{max} is quite important. On the other hand, a design at a ZTC_{fmax} point is very interesting since this occurs at a bias lower than ZTC_{ft}, hence an increasing behavior of f_T with temperature (instead of the conventional decreasing behavior) is achieved. Nevertheless, since ZTC_{fmax} occurs at very low bias values (see Figure 4), a tradeoff is unavoidable, since the absolute values of f_{max} and f_T are much lower than their maximum values at higher bias values.

It is also shown that ZTC_{fmax} points depend on the value of V_{DS}, where different values of ZTC_{fmax} points are reported for linear and saturation regions and also for different values of V_{DS} in saturation region as can be seen from the difference between Figure 4 (at $V_{DS} = 0.6$ V) and Figure 5 (at $V_{DS} = 1.2$ V). This effect is explained by the presence of another *ZTC* point, which is the ZTC_{gd} of the intrinsic output conductance g_d shown in Figure 6. As V_{DS} increases, this point moves up on the V_{GS} scale resulting in ZTC_{fmax} to move down to lower values of V_{GS}.

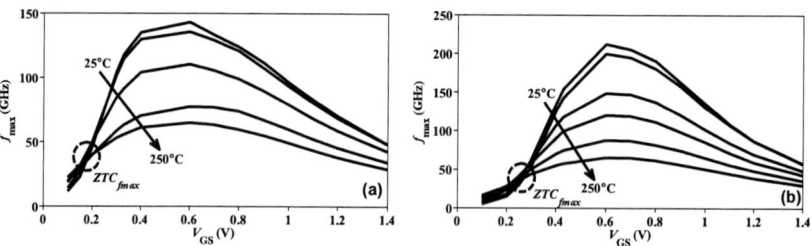

Figure 4. ZTC_{fmax} for (a) deep n-well protected bulk and (b) PD SOI both in FB structure at $V_{DS} = 0.6$ V.

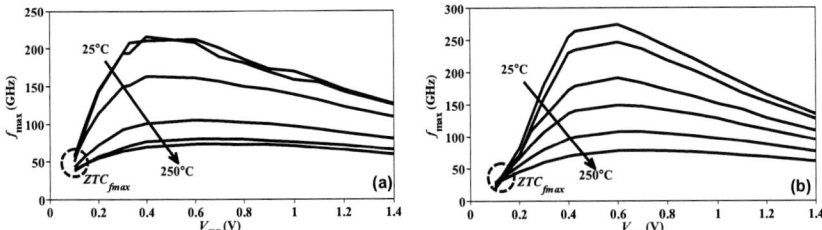

Figure 5. ZTC_{fmax} for (a) deep n-well protected bulk and (b) PD SOI both in FB structure at $V_{DS} = 1.2$ V.

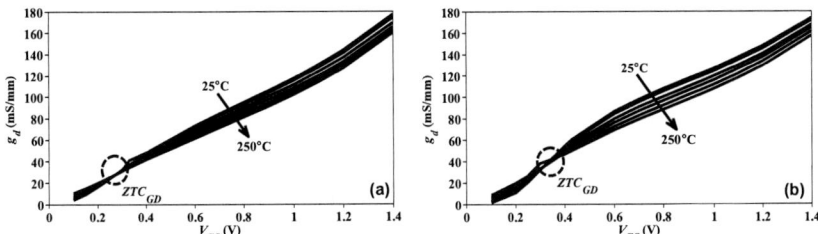

Figure 6. Output conductance g_d for (a) deep n-well protected bulk and (b) PD SOI FB both in FB structure at $V_{DS} = 1.2$ V.

Conclusion

In conclusion, the *ZTC* points for cutoff frequencies allow for an accurate and stable design of key RF circuits like the low noise amplifier (LNA) and the voltage controlled oscillator (VCO). The crucial importance of these points arises from the fact that most RF and wireless applications suffer from high temperature operating environments with the increasing number of transistors on the IC chip and the increasing frequency. Another important advantage of defining these *ZTC* points is the ability for the designer to define the region below these points where the cutoff frequencies increase, instead of decreasing, with increasing the temperature. This is mostly beneficial for low voltage low power operation.

Acknowledgments

This work has been funded by the Walloon region (Convention no. 516125).

References

1. F. S. Shoucair, *Electronics Lett.*, **25**, 1196 (1989).
2. Z. Prijic, S. S. Dimitrijev, and N. Stojadinovic, *Microelectronics Reliability*, **21**, 769 (1992).

3. G. Groeseneken, J.-P. Colinge, H. E. Maes, J. C. Alderman, and S. Holt, *IEEE Electron Device Lett.*, **11**, 329 (1990).
4. D. S. Jeon and D. E. Burk, *IEEE Trans. Electron Devices*, **38**, 2101 (1991).
5. M. Emam, A. Kumar, J. Ida, F. Danneville, D. Vanhoenacker-Janvier, and J.-P. Raskin, *Proc. IEEE Intl. SOI Conf.*, San Francisco, CA (2009).
6. A. A. Osman, M. A. Osman, N. S. Dogan, and M. A. Imam, *IEEE Trans. Electron Devices*, **42**, 1709 (1995).
7. M. E. Kaamouchi, G. Dambrine, M. S. Moussa, M. Emam, D. Vanhoenacker-Janvier, and J.-P. Raskin, *Proc. IEEE Topical Meeting on Silicon Monolithic Integrated Circuits in RF Systems (SiRF'08)*, Orlando, FL (2008).
8. B. Gentinne, J.-P. Eggermont, and J.-P. Colinge, *Electronics Lett.*, **31**, 2092 (1995).
9. M. E. Kaamouchi, M. S. Moussa, J.-P. Raskin, and D. Vanhoenacker-Janvier, *Proc. European Microwave Conference*, Munich, Germany (2007).
10. M. Emam, D. Vanhoenacker-Janvier, and J.-P. Raskin, *Proc. IEEE Topical meeting on Silicon Monolithic Integrated Circuits in RF Systems (SiRF'09)*, San Diego, CA (2009).
11. J.-P. Raskin, R. Gillon, J. Chen, D. Vanhoenacker-Janvier, and J.-P. Colinge, *IEEE Trans. Electron Devices*, **45**, 1017 (1998).
12. G. Dambrine, C. Raynaud, D. Lederer, M. Dehan, O. Rozeaux, M. Vanmackelberg, F. Danneville, S. Lepilliet, and J.-P. Raskin, *IEEE Electron Device Lett.*, **24,** 189 (2003).

ECS Transactions, 35 (5) 135-141 (2011)
10.1149/1.3570788 ©The Electrochemical Society

Research of SOI Microelectromechanical Sensors with a Monolithic
Tensoframe for High-Temperature Pressure Transducers

L. V. Sokolov

Federal State Unitary Enterprise "Institute of Aircraft Equipment"
Zhukovsky, 140185, Russia

This paper is devoted to an integrated piezoresistive pressure
sensors based on novel concept of high-temperature
microelectromechanical SOI sensors with a monolithic tensoframe.
The SOI and integrated MEMS technologies are of especial
interest for this sensors. The research enables main technology
problems due to the features of a photolithography process on a
relief surface with the high aspect ratio and resulted restrictions on
the design parameters of SOI sensor three-dimensional
micromechanical structures to be defined.

Introduction

Conventional micromechanical piezoresistive sensors [1], which contain electric p-n
junctions, have a limited operational temperature range due to the exponential
dependence of reverse current from the p-n junction temperature as known in the
semiconductor physics

$$I_o \sim n_i^2 \sim \exp[-\Delta E_F /(kT)] \qquad [1]$$

For this reason it cannot operate within a wide operational temperature range to 573K
needed Aircraft Equipment.
Piezoresistive sensors based on a polycrystalline silicon-on-isolator heterostructure,
which do not contain p-n junctions, are dramatically inferior to piezoresistive sensors
based on a monocrystalline silicon-on-isolator heterostructure in strain sensitivity.
When three-dimensional micromechanical structures are formed by chemical etching on
SOI-like wafers made with the joining method through silicon dioxide, etching defects
are often detected, such as through-holes in a thin silicon sensor membrane.
Electrical p-n junctions in the conventional silicon sensors [1] are prerequisites for the
time instability of sensors and an aircraft speed-altitude measurement system as a whole.
Anisotropic chemical etching (ACE) in KOH water solutions remains the basic
technology of solid silicon microprocessing. However, the problems of forming convex
angles with a required shape on microstructures in the silicon wafers (100) still remain.
Further development of the techniques is required to provide adequate compensation for
the convex angle undercutting process [2].
While in microprofiling of an integrated monolithic frame with the ACE technique the
inadequate compensation of the convex angle undercutting process can become the cause
of asymmetry and, therefore, unbalance for a Wheatstone measuring bridge.
The above problems can be solved, using a novel concept of high-temperature
microelectromechanical SOI sensors with a monolithic tensoframe where a dielectric
layer in an integral heterostructure is a stop layer for silicon anisotropic chemical etching
[3-5]. The research enables main technology problems due to the features of a

135

photolithography process on a relief surface with the high aspect ratio, and resulted restrictions on the design parameters of three-dimensional micromechanical structures to be defined.

Design and simulation of a SOIMT microelectromechanical sensors

The microelectromechanical piezoresistive sensor (MEMS-SOIMT) is a SOI-based membrane-type pressure sensors (chip) which measures 5.0 mm x 5.0 mm x 0.56 mm (Figure 1).

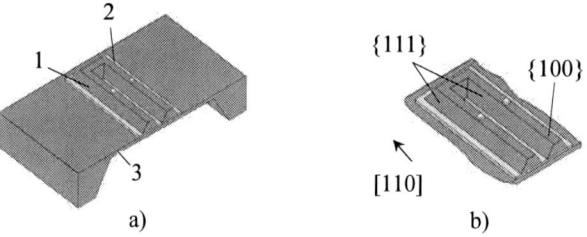

a)   b)

Figure 1.   Cross-section of SOI-MT piezoresistive Sensor (*a*): 1, 3 – Integrated Monolithic Tensoframe   and   Membrane,   2 – Thin Dielectric Layer;   (b) Crystallographic Orientation of the Tensoframe after anisotropic chemical etching of silicon (100)

The main features of the sensor include a silicon monolithic integral tensoframe is a symmetric Wheatstone bridge, an ideal isolation between the Wheatstone bridge circuit and a silicon membrane due to a thin glass layer and no p-n junctions in the structure.

Computer strain condition simulation

The research from the computer simulation of a strain condition for a SOIMT sensor structure allow the strain maxima to be defined. ANSIS had been used for simulation.

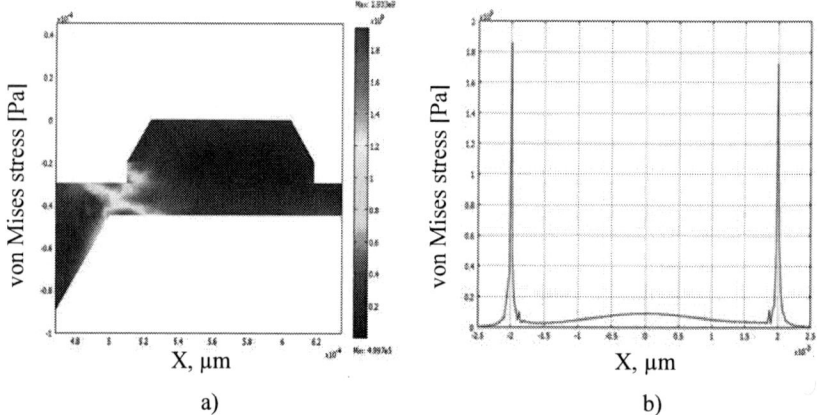

a)   b)

Figure 2. Simulation of Elastic Stress Distribution of the SOIMT Sensor for P = 50 kPa for the membrane boundares on the longitudinal section (a), the strain maxima for longitudinal section (b).

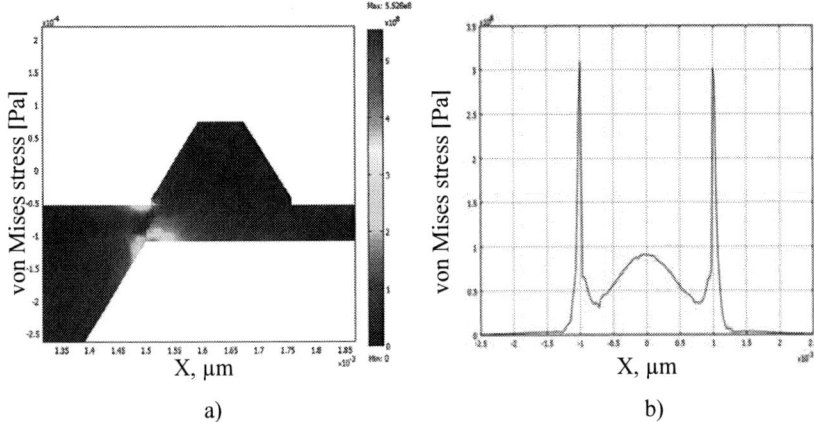

a)   b)

Figure 3. Simulation of Elastic Stress Distribution of the SOIMT Sensor for P = 2000 kPa for the membrane boundares on the longitudinal section (a), the strain maxima for longitudinal section (b).

As shown from figures 2 and figures 3 the strain maxima are located on the contour lines of a tensoframe with a membrane and within the membrane areas adjacent to the frame ends.

Investigation of a frame undercut by symmetry

SOIMT sensors have been fabricated using MEMS and SOI technologies. The investigations have been conducted using samples with the two-layer structure of the masking coating which consists of thin films of silicon dioxide and phyrolithic silicon nitride. Mask topologies have been developed to investigate the rate adjustment process during ACE of silicon (100). The topologies feature either an angular compensator configuration (Figures 4a and 4b), or the size of compensator elements (Table 1).

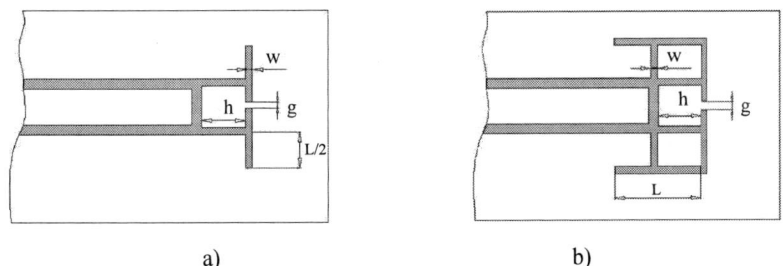

a)   b)

Figure 4. Fragments with Angle Compensators for ACE Process: with Configuration 1 (a) and Configuration 2 (b).

TABLE I. Linear Dimensions for Compensator Elements.

Configuration	Version	L, μm	L/2, μm	h, μm	W, μm
1	1-1	-	140	150	40
1	1-2	-	130	50	100
2	2-1	240	120	10	40

Manufacturing process

The simplified manufacturing process for a SOIMT microelectromechanical sensor is given in Figure 5. Etching has been performed in the 33% KOH water solution at 82 ± 2 ^{0}C with continuous solution mixing in a reactor.

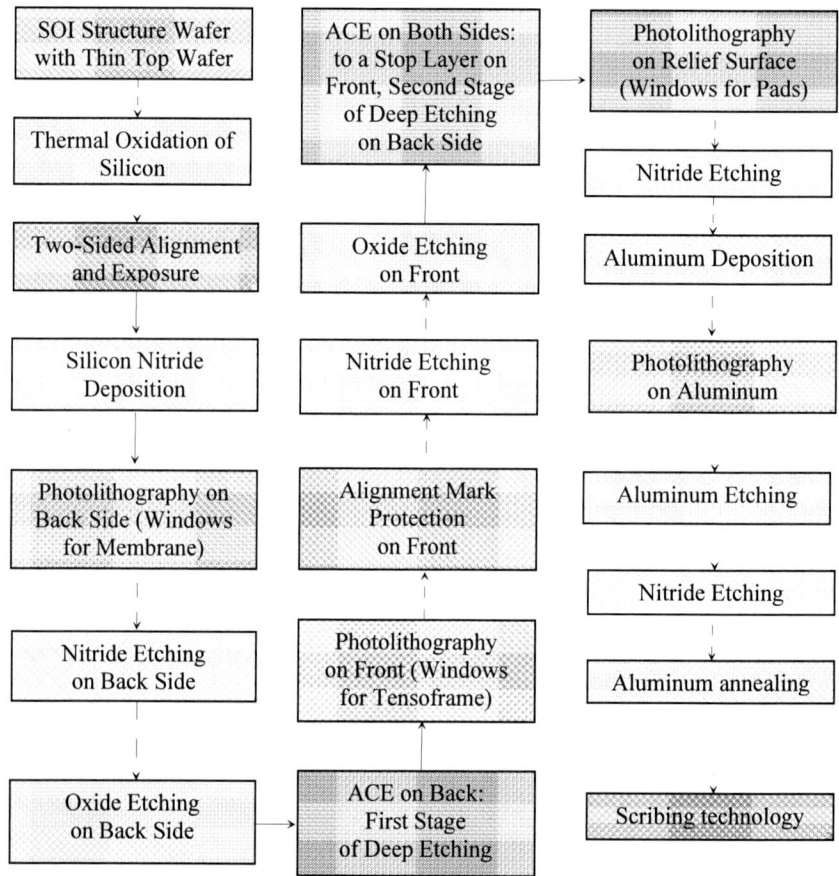

Figure 5. The simplified manufacturing route for a SOIMT microelectromechanical sensor.

Results and discussion

The quality of silicon microprocessing has been inspected a Kodak digital camera. The linear dimensions of the frame have been measured with a Karl-Zeiss optical microscope using the experimental samples to assess the frame undercut bias symmetry.

As shown on figures 6a and 6b it is obvious that the frame undercut bias depends on the height and width ratio of the angular compensators, while the other dimensions of the mask are equal (Figure 4a).

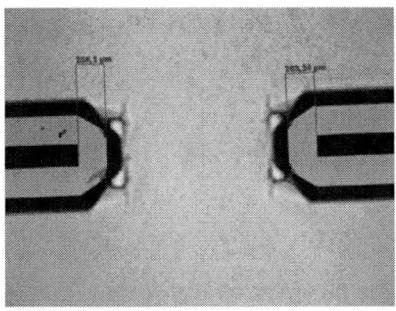 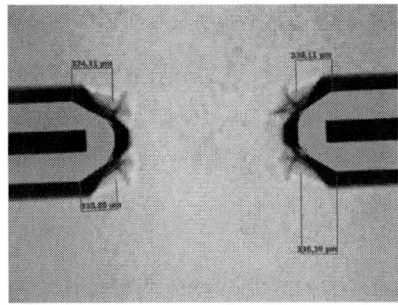

a)   b)

Figure 6. Photos of Tensoframe Sample Fragments after ACE with the Frame Undercut Bias Equal 2.73 µm from Mask Configuration (4a) and with the Frame Undercut Bias Equal 38.25 µm from Mask Configuration (4b).

If the samples are etched to a depth of 202 µm, the large frame undercut bias is obtained 38.25 µm for $h/W = 0.5$ (Figure 6a) as compared with an undercut bias of 2.73 µm for $h/W = 3.75$ (Figure 6b). Simultaneously, a slight asymmetry caused by nonuniform frame angle undercut etching is observed on both frame sides in the first case.

Well-known microelectronics technologies are used in MEMS-SOIMT sample etching, except the photolithography process on a relief surface. The feature of relief surface lithography is that a mask with holes for contact pads is superimposed on a microprofiled plate shaped as multiple 202 µm high frames on its surface.

After exposure and etching of the mask coating some contact pads are considerably smaller as compared with the specified dimensions. The possible cause is nonuniform photoresist irradiation over the entire wafer surface resulted from beam interference on a relief surface. The number of defective contact pads can be minimized by selecting exposure modes.

As a result of the ACE from the mask (Figure 4a), a 3D frame with a regular geometric shape as a symmetric Wheatstone test bridge (Figures 7a, 7b, 7c) is fabricated. Figure 7a shows a MEMS-SOIMT sample with a frame of a regular shape which is produced by using mask configuration 1 for $h/W = 3.75$ (Table 1, version 1-1). Figure 7c shows a 3D frame with the measured basic frame design parameters.

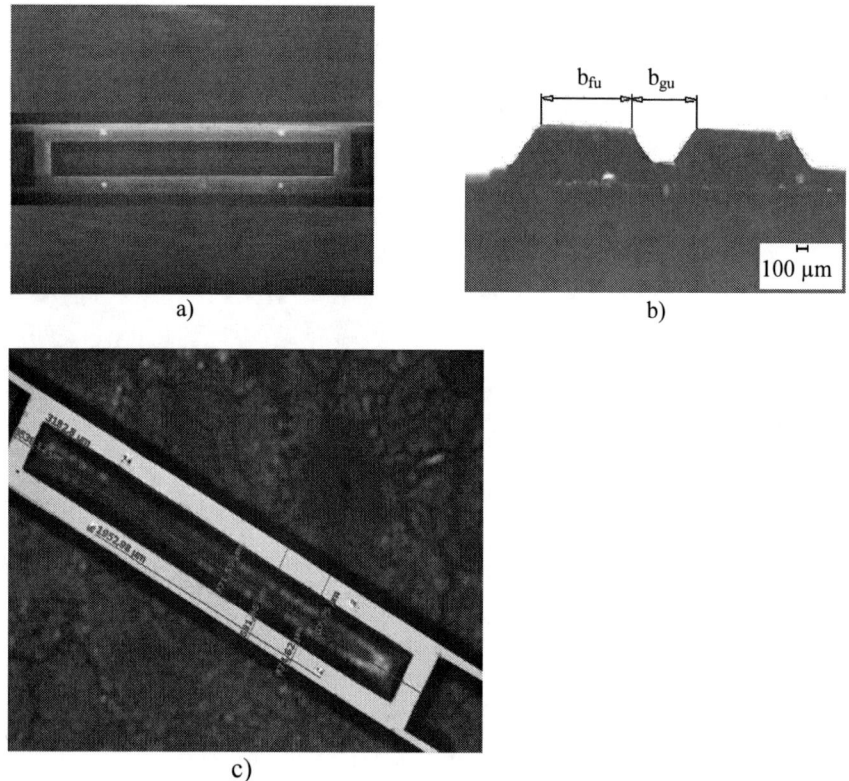

a)

b)

c)

Figure 7. Photos of MEMS-SOIMT Sample Fragments: with a frame of a regular shape (a), on a Split (b) and with Measured Tensoframe Design Parameters (c).

Comparison of the frame design parameters for the sensors (Fig. 7c) has shown that the design parameters are slightly different from the calculated ones (Table II).

TABLE II. Deviations in Frame Dimensions after Microprofiling.

Design Parameter	Dimension design, μm	Dimensions measured, μm	Δ, μm
Frame Length	3573.0	3538.12	−34.88
Frame Width, b_{fu}	200.0	150.67	−49.33
Gap, b_{gu}	327.0	371.65	+44.65

Correct geometry and required precise dimensions for a tensoframe and a membrane of a three-dimensional integral micromechanical structure have been obtained as a result of the research of the double-sided anisotropic chemical etching (ACE) process for silicon in a SOI heterostructure, using protective masks containing angle compensators, and heterostructure dielectric layer is a stop layer for silicon ACE.

Nevertheless, when the zero shift is measured, the unbalance of the Wheatstone bridge is within tens to hundreds of millivolts (for 5 VDC power supply), that supposes mechanical stresses on the sensor membrane.

Conclusions

New generation of high-temperature microelectromechanical SOI sensors based on monolithic tensoframe (MEMS-SOIMT) has been designed, fabricated and investigated.

The investigations of etch rate adjustment, convex angle protection, as well as contact lithography on the relief surface of a silicon wafer enable a new technology to be developed for the SOIMT sensors.

Correct geometry and required precise dimensions for a tensoframe and a membrane of a three-dimensional integral micromechanical structure have been obtained.

The research enables main technology problems due to the features of a photolithography process on a relief surface with the high aspect ratio, and resulted restrictions on the design parameters of three-dimensional micromechanical structures to be defined.

The research results from the computer simulation of a strain condition in a SOIMT sensor structure allow the strain maxima to be defined which are located on the contour lines of a tensoframe with a membrane and within the membrane areas adjacent to the frame ends.

The SOI technology combined with MEMS create an opportunity to push the silicon-based sensors even further.

Acknowledgments

The author of the present paper is very grateful to Professor S.P. Timoshenkov, the Moscow State Technical University for Electronic Technology, for the exploratory investigations of wafers with a silicon-on-isolator heterostructure wafes which meet tensometry requirements.

References

1. M. Elwenspoek, R. Wiegerink, *Mechanical Microsensors*, p.86, Springer-Verlag Berlin Heidelberg New York (2001).
2. O. Powell and H. Barry Harrison. *J. Micromechanics and Microengineering*, **11**, 3, (2001).
3. L.V. Sokolov, *Proceed. 9th International Symposium on Measurement Technology and Intelligent Instruments, ISMTII-2009*, **3**, p.248, S-Petersburg (2009).
4. L. V. Sokolov, *Measurement Techniques*, **52**, 9, p.947, Springer Science + Buisness Media, Inc. (2009).
5. L.V. Sokolov, *Abstracts and program of 10th ISMQC. 10th International Symposium on Measurement and Quality Control*, Osaka, Japan, p.187 (2010).

142

CHAPTER 7

ELECTRON DEVICE PHYSICS-2

144

ECS Transactions, 35 (5) 145-150 (2011)
10.1149/1.3570789 ©The Electrochemical Society

Global and/or Local Strain Influence on p- and nMuGFET Analog Performance

P. G. D. Agopian[a,b], J. A. Martino[a], E. Simoen[c] and C. Claeys[c,d]

[a]LSI/ PSI/USP - University of Sao Paulo, Sao Paulo 05508-010, Brazil.
[b]Centro Universitario da FEI, S.B.Campo 09850-901 Brazil
[c]imec, B-3001 Leuven, Belgium
[d]E.E. Dept., KU Leuven, B-3001 Leuven, Belgium

In this work, the analog performance is evaluated for tri-gate p-
and nMuGFETs processed with and without the implementation of
different global or local strain engineering techniques. For n-
channel devices, the intrinsic voltage gain showed to be worse for
strained devices when the fin is narrow. Only for wider fins the
voltage gain increases with the strain efficiency due to mobility
enhancement. Besides the voltage gain, the transconductance,
output conductance and Early Voltage are also evaluated. In spite
of the smaller impact of strain engineering, pMuGFETs show
better analog behavior for all studied parameters.

Introduction

Multiple-gate field-effect transistors (MuGFETs) have been pointed as a
promising alternative to continue the scaling at the 22 nm CMOS node and beyond. One
of the main advantages of this type of architecture is its better electrostatic control that
results in smaller short-channel effects (1,2).

In addition, the mechanical stress technique has been used to boost the carrier
mobility and consequently enhancing the devices performance. There are two types of
strain engineering techniques, i.e., the global stress (sSOI) that is obtained through the
silicon epitaxial growth over a relaxed silicon/germanium layer and local stress, that is
introduced by the use of dual contact etch stop layers (CESL), i.e., a SiN_x cap is
deposited over the gate stack, inducing either compressive (for pMOS) or tensile strain
(for nMOS) in the channel direction (3).

The global stress results in a tensile strained silicon, that in turn, causes an
increase of the electron mobility but decreases the hole mobility. Knowing that the
uniaxial compressive stress modifies the valence band structure of silicon and,
consequently, results in a hole mobility enhancement (4), sometimes the local stress
becomes more attractive because this technique can implement both tensile stress for
nMOS devices and compressive stress for pMOS in the same wafer.

In this work, strain engineering is applied to p- and nMuGFETs in order to
evaluate the analog device performance. Four different splits are analysed: standard (no
stress – reference), local strain (CESL - Contact Etch Stop Layer), global strain (sSOI)
and the last split combines the biaxial and the uniaxial stress on the same device
(sSOI+CESL). Tensile CESL is used for nMOS and compressive CESL for pMOS.

145

Devices Characteristics

All the studied devices are multiple gate FinFETs (MuGFETs) that have been processed on SOI substrates with a 150 nm thick buried oxide and having the following characteristics: fin height of 65 nm and fin width (W_{FIN}) ranging between 20 and 870 nm and channel length of 150 nm. The gate dielectric of all devices consists of 2 nm HfSiON on 1 nm SiO_2, resulting in an Equivalent Oxide Thickness (EOT) of 1.5 nm. The gate electrode is 10 nm TiN covered by 100 nm polysilicon.

Measured devices have recessed Si/Ge Selective Epitaxial Growth (SEG) source/drain (S/D) contacts in order to reduce the series resistance. Further process details can be found in (5).

Analysis and Discussion

The strained silicon influence on the transconductance in the saturation region (gm_{sat}) is shown in figure 1. One can see that the smallest gm_{sat} value for nMOS devices was obtained for the standard split and the highest gm_{sat} for devices with combined strain techniques (sSOI+CESL) due to the highest strain-induced mobility enhancement. The wider is the fin, the higher is the biaxial strain efficiency, so that for wider devices, the mobility improvement is almost exclusively due to the biaxial strain (sSOI). However, for pMOS the best gm_{sat} behavior was obtained for CESL and the worst for devices with sSOI as tensile strain in the <110> channel direction degrades the hole mobility. For the case when both techniques are applied (CESL+sSOI), the local compression from the CESL is not enough to compensate the sSOI global tensile effect.

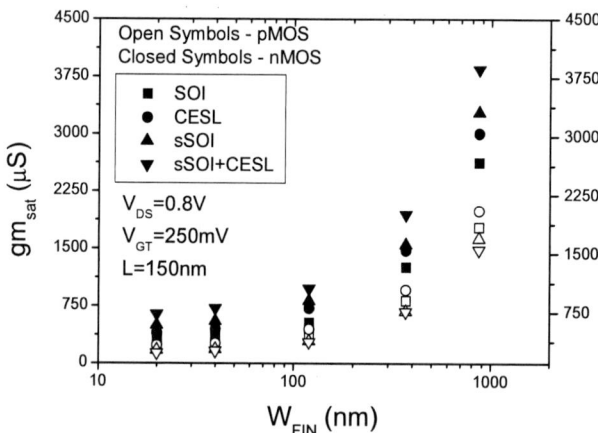

Figure 1: Transconductance as a function of W_{FIN} for different strained nMOS and pMOS devices.

The output conductance (g_D) was extracted for standard and strained n- and p MuGFETs with a gate voltage overdrive $V_{GT}=V_G-V_T=250$ mV (V_G is the applied gate voltage and V_T the threshold voltage) and a drain voltage $V_{DS} = 0.8$ V. Figure 2 shows that for the larger W_{FIN}, g_D becomes higher due to the increased susceptibility to the drain electric field. When considering the type of devices (p or nMOS) and the applied stress, it is noticed that the higher the strain efficiency the higher the g_D degradation owing to the higher mobility.

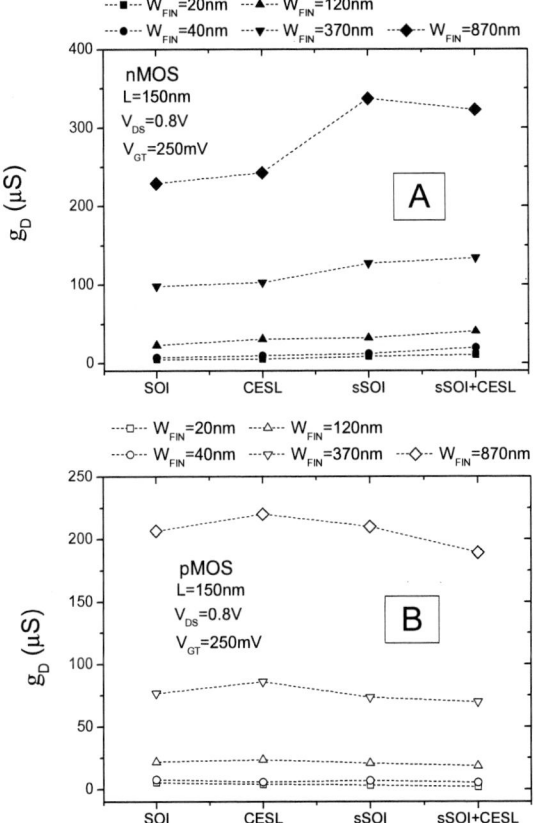

Figure 2: Output conductance behavior for nMOS (A) and pMOS (B) devices with different W_{FIN} and for different splits.

For nMOS devices (figure 2A) it is possible to observe that there is a slight increase of g_D for the CESL split. Since for narrow devices, the effective strain of sSOI in narrow fins acts only in the channel direction, the increase obtained for the sSOI split is almost the same as for the CESL one. A wider W_{FIN} corresponds with a higher sSOI efficiency, which gives rise toa pronounced increase of g_D. The worst g_D was obtained for the sSOI+CESL split for all W_{FIN}.

However, when the focus is on the pMOS devices (figure 2B), g_D showed to be less affected by the strain process. The sSOI stress results in a lower hole mobility and consequently for these devices a smaller (better) g_D was obtained.

Figure 3 presents the intrinsic voltage gain (A_V) for n- and pMOS devices with different W_{FIN} for the different splits. The intrinsic voltage gain was extracted using the experimentally obtained gm_{sat} and g_D values ($A_V=gm_{sat}/g_D$).

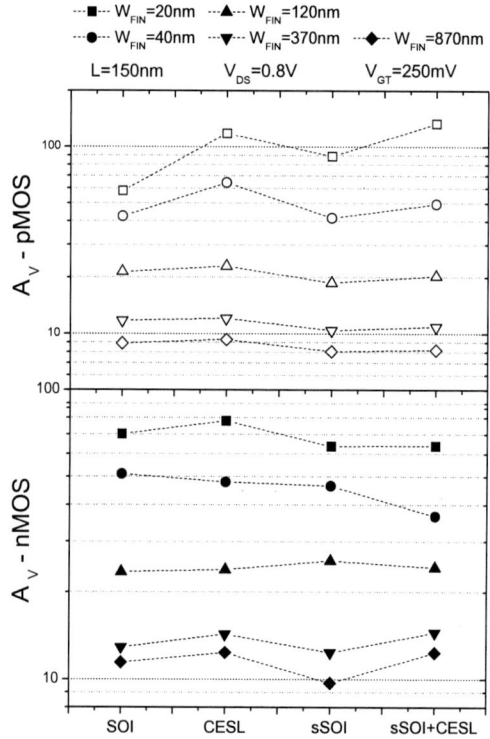

Figure 3: A_V for different splits for n- and pMOS devices with different W_{FIN}

Analyzing the intrinsic voltage gain (figure 3) for n-channel devices, it is possible to observe for narrow fins that sSOI strain causes a slight improvement in gm and a higher degradation of g_D. Consequently, the biaxially strained devices present the worse results for A_V. For larger fin devices, the biaxial strain starts to have a higher influence on the electron mobility, thus increasing gm and yielding a higher A_V for devices with both strain techniques (sSOI+CESL).

Figure 4 presents the Early voltage (V_{EA}) for both n- and pMuGFETs for several W_{FIN} for a gate voltage overdrive of 250 mV. It is known that V_{EA} increases with fin width reduction due to the better lateral gate coupling. Focusing on the strain influence on V_{EA} for nMOS devices, one can derive that the Early voltage follows the same tendency as the voltage gain (figure 4).

Making the same analysis for p-channel devices (V_{EA} and A_V), it is possible to observe that pMuGFETs with CESL present better A_V results for all W_{FIN}. In spite of the fact that CESL gives rise to an increase of gm_{sat}, the p-channel devices are less affected by the strain technology which causes a small degradation of g_D. As a result, both g_D and gm contribute to enhance the voltage gain (figure 3).

Although the Early voltage also follows the same voltage gain tendency, in this case, it is important to notice that for narrow fins the obtained V_{EA} values for p-channel devices are almost twice those obtained for n-channels (figure 4).

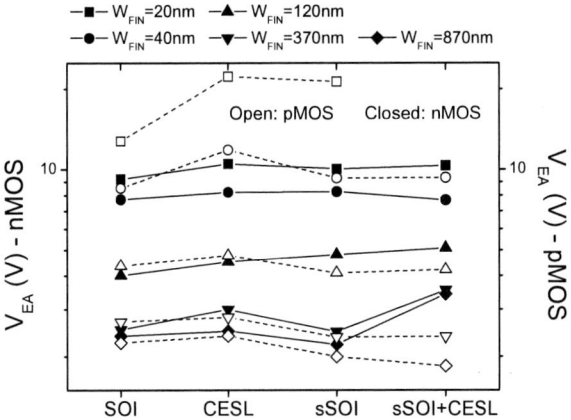

Figure 4: V_{EA} for different splits of n- and pMOS devices with different W_{FIN}

Conclusion

The analysis of p and nMuGFETs was performed focusing on the influence of strain-engineering techniques on the analog parameters. This analysis was based on the comparison of four different splits where the standard unstrained SOI was the reference.

N-channel devices show that a higher strain efficiency results in a higher g_D degradation and, consequently, a smaller A_V. Therefore, the poorest analog performance occurs for narrow fins for the split with combined strain techniques (sSOI+CESL). Only for wider fin devices, the biaxial strain starts to influence gm_{sat} and the A_V behavior improves.

The analog parameters for p-channel devices were less affected by the strain techniques. Although the uniaxial compressive strain (CESL) degrades g_D, it enhances all the other analog parameters, attaining the best analog performance.

Acknowledgments

Paula Ghedini Der Agopian and Joao Antonio Martino thank CNPq and FAPESP for the financial support for execution of this work. Part of the work has been performed within the frame of the CNPq-FWO Brazil-Flanders cooperation agreement.

References

1. J.P. Colinge, FinFET and Other Multigate Transistors, Springer, p.22 (2007).
2. J. P. Colinge, L. Floyd, A.J. Quinn, G. Redmond, J.C. Alderman, W. Xiong, C.R. Cleavelin, T. Schulz, K. Schruefer, G. Knoblinger, P. Patruno, *IEEE Electron Device Letters*, **27**, 172 (2006).
3. C. Claeys, E. Simoen, S. Put, G. Giusi, F. Crupi, *Solid-State Electronics,* **52,** 1115 (2008).
4. Y. Sun, S.E. Thompson and T. Nishida, *J. Appl. Phys.*, **101**, 104503 (2007).
5. N. Collaert, R. Rooyackers, F. Clemente, P. Zimmerman, I. Cayrefourcq, B. Ghyselen, K. T. San, B. Eyckens, M. Jurczak, and S. Biesemans, *VLSI Symp., Dig. Technical Papers,* p.52 (2006).

ECS Transactions, 35 (5) 151-156 (2011)
10.1149/1.3570790 ©The Electrochemical Society

Fin Pitch Impact on Biaxial/Uniaxial Strain Engineering of Triple-Gate Devices

M. Rodrigues[a], V. Sonnenberg[a,b], J. A. Martino[a], N. Collaert[c], E. Simoen[c], C. Claeys[c,d]

[a]LSI/PSI/USP, University of Sao Paulo, Sao Paulo 05508-010, Brazil
[b]FATEC/SP, Faculdade de Tecnologia de Sao Paulo, Sao Paulo 01124-060, Brazil
[c]imec, B-3001 Leuven, Belgium
[d]E.E. Dept., KU Leuven, B-3001 Leuven, Belgium

> This work characterizes the analog performance of SOI n-MuGFETs with uniaxial and biaxial strain configuration with respect to the influence of the fin pitch distance. Improved intrinsic gain can be achieved on strained devices independent of the pitch, mainly due to the increased transconductance. Larger fin pitch in the uniaxial strain case showed a higher gain. However, for the biaxial devices the smaller pitch resulted in a larger gain thanks to the increased Early voltage.

Introduction

Alternative MOS device architectures have been explored recently. The non-planar multiple gate (MuGFET) SOI transistor, like, e.g., triple gate, appears to be one of the most promising devices structures. The reason is their high immunity to short channel effects, excellent compatibility with the planar CMOS process and the higher gate control of the channel charges (1). Multiple-fin structures are also used in order to improve the drive current by increasing the number of fins present. In spite of that, a reduction of the carrier mobility is observed due to the different top and lateral channel crystal orientation present in a FinFET (2).

Additionally, the implementation of controlled mechanical strain has been widely used to enhance the carrier mobility and drive current (3-5). Among the strain techniques, one can distinguish the strained contact etch stop layers (sCESL) that can lead to a local tensile strain in nFET and in the case of pFET compressive strain. The strain is in this case mainly uniaxial. Another stressor technique is the substrate-induced strained material (sSOI) that is present over the whole wafer by epitaxial growth of a relaxed SiGe layer (biaxial) before deposition the Si top layer. Together with strain engineering the use of selective epitaxial growth (SEG) on source and drain is widely used in order to reduce the series resistance (6).

Finally, MuGFET structures also present an attractive behavior for analog applications with a reduced drain output conductance and an extremely large Early voltage value (7-11).

The main goal of this work is to analyze the influence of the variation of the fin pitch on the SOI triple-gate n-channel devices analog performance.

151

Devices characteristics

The FinFET devices studied were fabricated starting from a standard SOI substrate with 65nm Si film on 145nm buried oxide. The gate dielectric consists of a 1nm SiO_2 chemical oxide onto which 2.3 nm HfSiON was deposited, followed by a TiN metal gate electrode. No channel doping was applied so that the doping level of the channel is around 10^{15} cm^{-3}. The devices had 5 fins and different fin pitches. Si SEG was used on the source and drain areas to reduce the series resistance. This standard process flow (reference) was also used for the fabrication of the devices with biaxial tensile strain (1.5GPa) and the devices with uniaxial stress that was formed by an intrinsically strained 100 nm SiN_x layer deposited prior to BEOL. Different fin pitches, 0.2 and 1.0μm were considered in this analysis. The detailed device fabrication can be found in (12). Figure 1 is a schematic of the n-type SOI triple-gate showing its important geometrical parameters, the channel length (L) and the fin width (W_{fin}). Different channel lengths were considered and an effective fin width of W_{fin}=20nm.

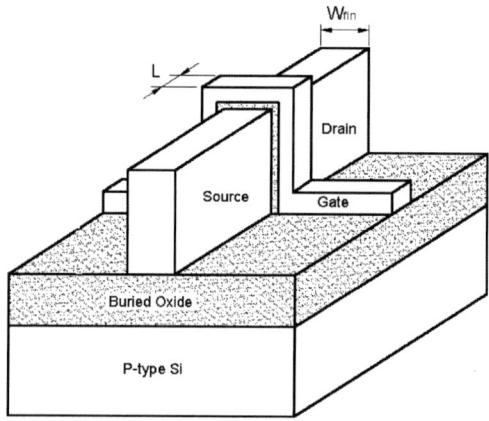

Figure 1 - Schematic of a SOI triple-gate n-channel device.

Experimental results and discussion

Figure 2 shows the transconductance as a function of the gate voltage for the different fin pitches and strain engineering methods. Reference devices show no variation in g_m for different fin pitch. However, a lower transconductance is observed for the reduced fin pitch in the case of devices with biaxial and uniaxial strain.

For the biaxial case and disregarding a series resistance variation, a possible explanation for this behavior could be that during the fin process formation, the plasma technique used may have a different impact depending on the substrate (sSOI and SOI), causing a variation of the fin width. As a result, the reduced fin pitch may present a lower fin width, resulting in this reduced transconductance. Another issue can be a possible relaxation of the biaxial strain after the fin patterning with reduced fin pitch, leading to a lower mobility and transconductance.

In the case of uniaxial strain, it seems that for the small fin pitch the CESL is degrading the transconductance. This behavior can be related with the deposition of the SiN layer that could not be filling up all the space between the narrow fins, reducing in that way the uniaxial strain influence on the transconductance.

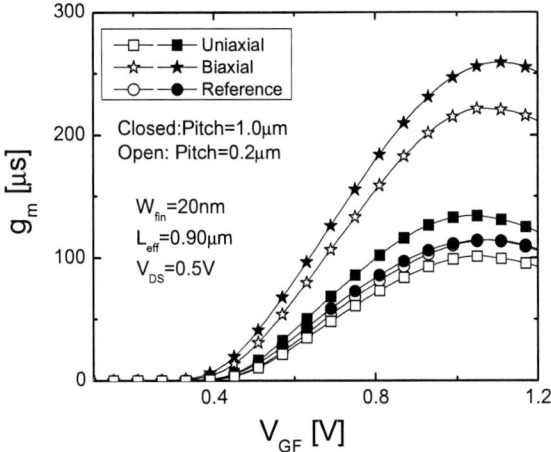

Figure 2. Transconductance as a function of the gate voltage for different fin pitches and strain engineering techniques.

The drain-induced barrier lowering (DIBL) was also extracted at V_{DS}=25mV and V_{DS}=0.5V for the different devices under study and the results are shown in Table I. According to Table I the biaxial and uniaxial strain result in smaller DIBL (better) for reduced fin pitch. This behavior is in agreement with the reduced output conductance (g_D) values presented on Table I. This phenomenon is also in agreement with the expected variation of the fin width, where lower DIBL is expected for smaller channel width (less drain electric field penetration). Considering the possible biaxial strain relaxation, the lower DIBL, gm and g_D for smaller fin pitch is due to the lower mobility. For the reference devices no variation in DIBL was observed.

TABLE I. V_T, DIBL and g_D values extracted for the different fin pitches and strain engineering methods on L_{eff}=0.90µm and W_{fin}=20nm.

Split	Pitch=0.2µm			Pitch=1.0µm		
	V_T [V]	DIBL [mV/V]	g_D [S]	V_T [V]	DIBL [mV/V]	g_D [S]
Uniaxial	0.47	47	2.25×10^{-7}	0.46	65	3.13×10^{-7}
Biaxial	0.42	38	5.89×10^{-7}	0.41	49	10.7×10^{-7}
Reference	0.47	55	7.09×10^{-7}	0.47	55	8.45×10^{-7}

The threshold voltage (V_T) was also extracted, where the biaxial strained devices present the smallest values (Table I). This behavior is related to the influence of strain that affects the material band gap (E_g). Comparing the different pitches no difference in V_T was observed.

The Early voltage ($V_{EA} \cong I_D/g_D$) was extracted through the output conductance (g_D) at a gate voltage overdrive ($V_{GT} = V_{GF} - V_T$) of 200mV, as presented in Figure 3. One can observe that V_{EA} reduces when the effective channel length shrinks as reported earlier for FinFET devices (8-9). An increase in V_{EA} is observed for strained devices and this is more pronounced for smaller fin pitch. The possible reduction in the channel width for the smaller fin pitch could be responsible for this behavior. Considering that a narrow fin width (expected for reduced fin pitch) presents a better gate control, a larger transverse electric field influence on the drain current is expected, increasing in that way V_{EA}. The reference devices didn't present any V_{EA} variation that can be related to a similar W_{fin}.

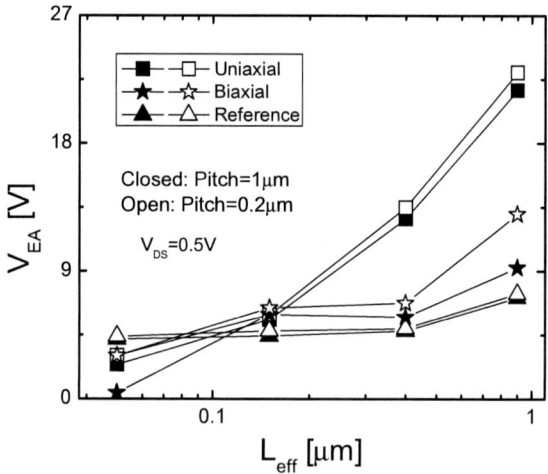

Figure 3. Early voltage as a function of the channel length for different fin pitches and strain engineering techniques.

The device intrinsic voltage gain ($A_V = V_{EA} * g_m/I_{DS}$) has been extracted and is presented in Figure 4 as a function of the channel length at $V_{DS} = 0.5V$ and $V_{GT} = 200mV$ in strong inversion ($I_{DS}/(W/L) = 2.5 \times 10^{-5} A$).

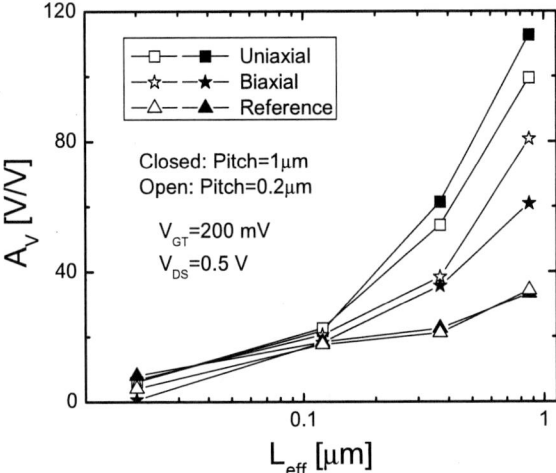

Figure 4. Calculated intrinsic gain as a function of the channel length for different fin pitches and strain engineering techniques.

Higher gain is observed for the strained devices for both pitches compared to the reference devices, due to the higher mobility. Comparing the two pitches for the uniaxially strained case, an increased gain is observed for the larger one due to the higher transconductance. However, for the biaxial devices the larger Early voltage observed for the smaller pitch is resulting in a higher gain, notwithstanding the worse transconductance. The intrinsic voltage gain for the reference triple-gate devices is not sensitive to the pitch.

Conclusions

In this paper, an analysis of the impact of the fin pitch on SOI triple-gate n-channel devices was presented. Reduced fin pitch results in a lower transconductance for the devices with biaxial and uniaxial strain compared to the reference devices. In the biaxial case two issues could be considered: a possible fin width reduction and a relaxation of the biaxial strain after the fin patterning for a reduced fin pitch. For the uniaxially strained devices, the lower transconductance could be related with a non uniform SiN_x layer deposition in the space between the narrow fins. Strained devices and a smaller fin pitch indicate a larger Early voltage. The possible fin width reduction could be leading to a better gate control yielding a higher V_{EA}. As a result, improved gain was found for uniaxially strained devices with a large fin pitch. In spite of that, biaxial devices showed higher gain for reduced pitch due to the increased V_{EA}.

Acknowledgments

Michele Rodrigues, Victor Sonnenberg and Joao Antonio Martino would like to thank to the Brazilian research-funding agencies CNPq, CAPES and FAPESP for the support for developing this work. Part of the work has been performed within the frame of the CNPq-FWO Brazil-Flanders cooperation agreement.

References

1. J. P. Colinge, *Solid State Electronics*, **48**, 897 (2004).
2. T. Rudenko, N. Collaert, S. De Gendt, V. Kilchytska, M. Jurczak, D. Flandre,. *Microeletronic Engineering*, **80**, 386 (2005).
3. T. Irisawa, T. Numata, T. Tezuka, K. Usuda, N. Sugiyama, S.-I. Takagi. *IEEE Trans. Electron Devices*, **55**, 649, (2008).
4. S.E. Thompson, M. Armstrong, C. Auth, M. Alavi, M. Buehler, R. Chau, S. Cea, T. Ghani, G. Glass, T. Hoffman, C.-H.. Jan, C. Kenyon, J. Klaus, K. Kuhn, Zhiyong Ma, B. Mcintyre, K. Mistry, A. Murthy, B. Obradovic, R. Nagisetty, R. Phi Nguyen, S. Sivakumar, R. Shaheed, L. Shifren, B. Tufts, S. Tyagi, M. Bohr, Y. El-Mansy, Y, *IEEE Trans. Electron Devices*, **51**, 1790 (2004).
5. T. Ghani, M. Armstrong, C. Auth, M. Bost, P. Charvat, G. Glass, T. Hoffmann, K. Johnson, C. Kenyon, J. Klaus, B. McIntyre, K. Mistry, A. Murthy, J. Sandford, M. Silberstein, S. Sivakumar, P. Smith, K. Zawadzki, S. Thompson and M. Bohr, In *IEEE International Electron Devices Meeting/IEDM 2003,* p.978 (2003).
6. X.J. Ning, D. Gao, P. Bonfanti, H. Wu, J. Guo, J. Chen, C.C. Shen, I.C. Chen, G. Cherng, *Materials Science and Engineering: B*, **134**, 165 (2006).
7. V. Kilchytska, N. Collaert, R. Rooyackers, D. Lederer, J.-P. Raskin, D. Flandre, *IEEE International Electron Devices Meeting/IEDM 2004*, p. 65 (2004).
8. D. Lederer, V. Kilchytska, T. Rudenko, N. Collaert, D. Flandre, A. Dixit, K. De Meyer, J.-P. Raskin, *Solid-State Electronics,* **49**, 1488 (2005)
9. J. P. Raskin, Tsung Ming Chung, V. Kilchytska, D. Lederer, D. Flandre, *IEEE Trans Electron Devices*, **53**, 1088 (2006).
10. V. Subramanian, B. Parvais, J. Borremans, A. Mercha, D. Linten, P. Wambacq, J. Loo, M. Dehan, N. Collaert, S. Kubicek, R.J.P. Lander, J.C. Hooker, F.N. Cubaynes, S. Donnay, M. Jurczak, G. Groeseneken, W. Sansen, S. Decoutere, *IEEE International Electron Devices Meeting/IEDM 2005*, p. 898 (2005).
11. M.A. Pavanello, J.A. Martino, E. Simoen, R. Rooyackers, N. Collaert, C. Claeys. *Solid-State Electronics*, **51**, 285 (2007).
12. N. Collaert, R. Rooyackers, A. De Keersgieter, F.E. Leys, F.E. Cayrefourcq, B. Ghyselen, R. Loo, M. Jurczak, *IEEE Electron Device Lett.*, **28**, 646 (2007).

Transport Properties of 3D Vertically Stacked SiGe and SiGeC Nanowires

A. Diab[a], E. Saracco[b], I. Ionica[a], C. Bonafos[c], J. F. Damlencourt[b], and S. Cristoloveanu[a]

[a] IMEP-LAHC, Minatec Grenoble-INP, BP 257, 38016 Grenoble Cedex 1, France
[b] CEA-LETI, Minatec, 17 avenue des Martyrs, 38054 Grenoble Cedex 9, France
[c] CEMES, CNRS, Univ. Toulouse, BP 94347, 31055 Toulouse Cedex 4, France

> One of the trends in microelectronics is to explore nanowire gate-all-around structures and alternative channel materials with superior properties. We investigate the electrical transport in three-dimensional (3D) vertically stacked germanium-enriched nanowires. Two starting materials have been used for the nanowire fabrication: SiGe and SiGeC. Measurements using the Pseudo-MOSFET concept show that the transport in as-grown gateless nanowires is controlled by the voltage applied on the substrate, as in a MOS transistor. Interestingly, the substrate bias effect depends on the designed geometry of the nanowire. For high width/length ratio nanowires, we show the possibility of turning on both electrons and holes channels. For nanowires with small width/length ratio, only the hole channel is visible. Transport modification with the concentration of germanium in the nanostructures is also discussed.

Introduction

Some of the nowadays challenges in microelectronics can be approached by alternatives to silicon planar integration (1). For example, a better electrostatic control and a higher immunity against short-channel effects can be achieved by using nanowire-based gate-all-around transistors (2). Nevertheless, their small diameters limit the current density in nanowire devices. To overcome this issue, nanowires can be integrated in 3D. In this context, the study of electronic transport in nanowires (NW) is mandatory. Furthermore, the transport properties can be improved by changing the channel materials. For instance, replacing silicon with Ge or SiGe allows obtaining higher mobility for the holes.

Germanium nanowires for multi-channel technology appear as promising candidates for device performance improvement (3). One of the issues about these nanostructures is the low level of drain current that can be supported. Bundling more NWs per device and vertically stacking them improves the on-state current (4). Our devices feature a nanowire channel, made of germanium-enriched SiGe or SiGeC, in a vertically 3D stacked architecture fabricated by an innovating top-down technique. Once the NWs have been grown and isolated, it is important to probe their electrical properties before completing the long CMOS process (gate stack, source/drain implants, back end). This task is challenging because undoped NWs are fully depleted, preventing current flow. The only subsisting solution for inducing a mobile charge is to bias the substrate as a back-gate. This is exactly the principle of the Pseudo-MOSFET (5) which, in this paper, is adapted to investigate the electrical transport properties of NWs. We will show the merits and limitations of this method. The impacts of device geometry and germanium concentration on the current flowing through the NWs will be discussed.

Fabrication

The innovative fabrication process is based on the germanium condensation technique on silicon on insulator (SOI) (6) and is detailed elsewhere (7). Fabricated nanowire channels are suspended and surrounded by silicon dioxide. Two levels stacked SiGe (Figure 1a) and three levels stacked SiGeC wires were formed. From the fabrication point of view, the interest in adding carbon in the super-lattice structure is to obtain a good agreement between the lattice parameter of the SiGeC and the SOI substrate. In this case, the selective etching used to free the nanowires is more stable and avoids bowing of the structures.

When referring to the geometry of the structure, one should define two kinds of widths. The designed width corresponds to the designed pattern on the mask. During the condensation process, the Si is oxidized selectively with respect to Ge, so a germanium enriched channel surrounded by silicon oxide is obtained (see Figure 1b). The effective width ($w\_effective$) of the Ge enriched channel is smaller than the designed width ($w\_designed$). Figure 1c shows a scanning electron microscopy (SEM) picture of multiple parallel nanowires after the selective etching of Si versus SiGe. An important consequence is that starting from the same material but with two different designed widths it is possible to obtain, after the same oxidation time (meaning the same silicon consumption), wires with different germanium concentrations. Table I is showing the designed and the effective geometry, as well as the germanium concentration for the nanowires that were electrically characterized.

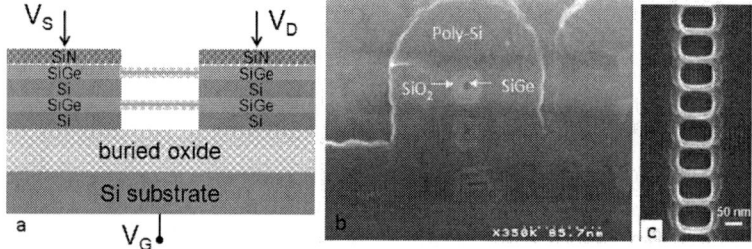

Figure 1. (a) Schematic view of two suspended germanium enriched nanowires fabricated from SiGe layers. (b) Scanning electron microscopy image of nanowires after condensation of germanium. A Ge-enriched wire surrounded by silicon dioxide is obtained. (c) Plan view SEM picture of multiple parallel nanowires with 9nm effective width and 100nm length.

Results and discussion

We will focus here on the results obtained on SiGeC-based nanowires. The behavior is similar to that obtained for the SiGe-based structures. The electrical set-up is shown in Figure 1a: the substrate is biased and used as a gate. The buried silicon oxide (145nm thickness) serves as a gate dielectric. Two probes placed on the stack of SiN/SiGe/Si/SiGe/Si are used as source and drain.

The nanowires have an ohmic behavior (see Figure 2). Moreover, the drain current I_D is obviously depending on the voltage applied on the substrate V_G, which opens the way to a MOSFET-like behavior. Without substrate biasing, the current is zero (Figure 3) because the NWs are fully depleted. The peculiarity of the two graphics in Figure 2 is coming from the fact that it is possible to activate a drain current with both positive and negative back-gate voltage. As the chemical content of the nanowires is mainly germanium and in germanium the mobility of holes is much higher than the mobility of electrons, one would expect to primarily observe a channel of holes. This means that only a negative back-gate voltage should give rise to a drain current.

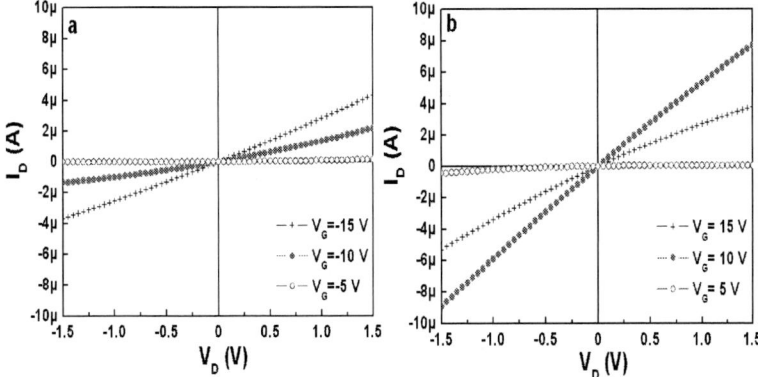

Figure 2. Drain current I_D versus drain voltage V_D for various back-gate voltages (negatives in (a) and positives in (b)). The designed geometry under test is 100nm long and 100nm wide. Ohmic behavior is evidenced for hole channel in (a) and for electron channel in (b).

In order to study this atypical phenomenon, Figure 3 is showing I_D-V_G curves obtained for two nanowires of 100nm length and of designed widths of 100nm (a) and of 50nm (b). The curve obtained for a width of 100nm confirms the results in Figure 2: transport shows both electron and hole channels. Surprisingly, when the same electrical characteristic is traced for a narrower nanowire ($w\_designed$ = 50nm), only a channel of holes is created. Moreover, the study of a 100nm wide and 500nm long structure also shows a loss of the electron channel (inset of Figure 3b). Therefore, the aspect-ration w/L seams to play an important role. Two possible phenomena could explain this behavior:

- during the condensation process, a parasitic wire of silicon remains under the wires and it allows conduction for both electrons and holes;
- bowing and sticking of the nanowires can occur during the selective etching process.

These two phenomena are geometry-dependent. The following discussions will concentrate on the hole transport.

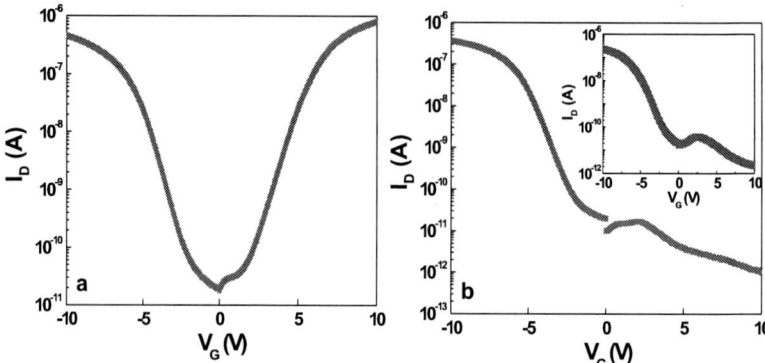

Figure 3. Drain current I_D versus back-gate voltage V_G for a SiGeC-based NW of 100nm length with w_design = 100nm (a) and w_design = 50nm (b). Inset: I_D versus gate voltage V_G for a nanowire with w_design = 100nm and L = 500nm. The applied drain voltage is of 0.2V.

For more quantitative studies, the simple drain current model of a MOS field effect transistor is used. The drain current, I_D, and the transconductance, g_m, are given by:

$$I_D = \frac{w}{L} C_{ox} V_D \frac{\mu_0}{1 + \theta(V_G - V_T)} (V_G - V_T) \qquad [1]$$

$$g_m = \frac{dI_D}{dV_G} = \frac{w}{L} C_{ox} V_D \frac{\mu_0}{[1 + \theta(V_G - V_T)]^2} \qquad [2]$$

w and L represent the width and length of the nanowire. C_{OX} is the buried oxide capacitance, μ_0 is the low-field mobility, V_D the drain bias, V_G the back-gate bias and θ is the mobility reduction factor due to the vertical field. Behind V_T, one can find either the threshold voltage V_T (for electrons channel) or the flat-band voltage V_{FB} (for holes channel).

The extraction of parameters (μ, V_T) is done using the Y-function (8):

$$Y = \frac{I_D}{\sqrt{g_m}} = \sqrt{f_g C_{ox} \mu_0 V_D} (V_G - V_T) \qquad [3]$$

which has a linear behavior for $|V_G| \gg V_T$. The intercept of Y(V_G) with the V_G axis yields V_T, whereas the low-field mobility is extracted from the slope.

Table I is showing the geometry, the Ge concentration and the extracted mobility of holes. The mobility was calculated by accounting for the effective width and actual number (three) of nanowires. Superior hole mobility values were obtained in the nanowire with a higher Ge concentration, i.e. narrow width. This result is in agreement with previous data obtained for SiGe thin films on insulator (5). It suggests that hole transport is primarily dominated by the material nature rather than by the interface traps.

TABLE I. Geometry and electrical properties of SiGeC vertically stacked 3-level nanowires in a single row.

Length (nm)	Designed w (nm)	Effective w (nm) per nanowire	Ge concentration	μ_{HOLES} (cm²/Vs) 3-level nanowires	I_D (nA) (for V_G=-10V)
100	50	9	100%	80	390
100	100	60	40%	14	450
500	100	60	40%	34	230

The relatively low values of extracted mobility are explained by the difficulty to determine the precise effective width and also by the fact that the series resistance in these samples is very high (due to undoped source and drain). The series resistance was explored by conducting experiments on the source stack only (Figure 4a). Figure 4b shows that the current through the source stack is depending on the gate voltage. The contribution of the gate-dependent series resistance has not been taken into account in the extracted values of mobility presented here.

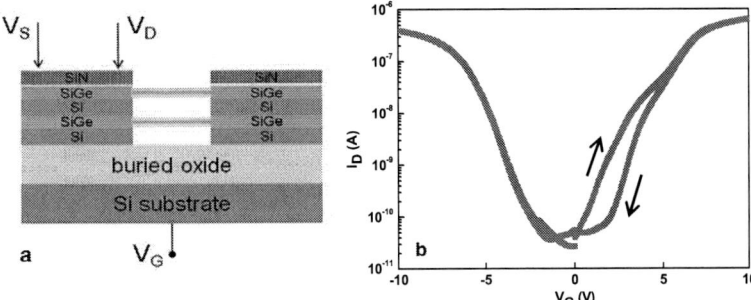

Figure 4. (a) Schematic view of the measurement on the contact stack: probes are on the same contact pad. (b) Drain current versus back gate voltage in double sweep mode (forward and backward) for the SiGeC-Si super-lattice.

Another way to validate our experimental results is suggested by the equation [1]. The drain current should be proportional to the aspect ratio multiplied by the mobility. We know that the mobility is Ge-concentration dependent, so the drain current should be roughly proportional to Ge concentration multiplied by w/L. Figure 5 shows that this proportionality is qualitatively verified. We used here experimental values of drain current measured for -10V on the back gate (strong inversion).

The results presented above demonstrate the adaptability of the Pseudo-MOSFET method for NW characterization. However, two pending aspects deserve further attention. First, the series resistance effect in the terminals is complicated due to the contributions of the *vertical* current flow (through the Si/SiGe superlattice) and *horizontal* flow (along the SiGe layers contacting the NWs). Second, the comparison of structures composed of a single row or 50 parallel rows of 3-stacked NWs is intriguing. The latter structure does not yield 50 times more current. This implies that the current is defined either by the contact resistance or by the most conductive nanowire out of 150 NWs. Multiple threshold voltages could not be evidenced: when plotting the second derivative of current with respect to gate voltage, we obtained one peak, not multiple peaks.

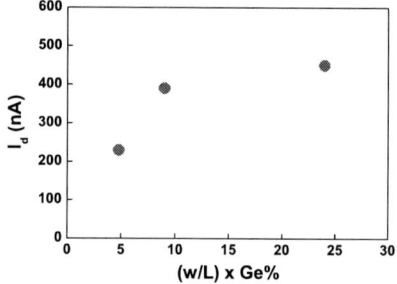

Figure 5. Drain current at V_G = -10 V function of (w/L) multiplied by Ge-concentration.

Conclusion

3D vertically stacked suspended gateless nanowires behave as MOS transistors controlled by the back gate. An original characterization method, inspired by the Pseudo-MOSFET, has been demonstrated. The hole mobility is improved by the Ge concentration. Another key parameter is the series resistance which is higher for SiGe than for SiGeC. The doping of the contacts could clarify this problem. The benefit of integrating multiple parallel nanowires (with gate-all-around configuration) for increasing on-current is obvious. Co-integration of SiGe nanowires with Si structures is very promising.

References

1. C. Dupré, T. Ernst, V. Maffini-Alvaro, V. Delaye, J.-M. Hartmann, S. Borel, C. Vizioz, O. Faynot, G. Ghibaudo and S. Deleonibus, *Solid-State Elect.*, **52**, 512 (2008).
2. N. Singh, A. Agarwal, L. K. Bera, T. Y. Liow, R. Yang, S. C. Rustagi, C. H. Tung, R. Kumar, G. Q. Lo, N. Balasubramanian and D.-L. Kwong, *IEEE Electron Device Lett.*, **27**, 383 (2006).
3. Y. Jiang, N. Singh, T. Y. Liow, W. Y. Loh, S. Balakumar, K. M. Hoe, C. H. Tung, V. Bliznetsov, S. C. Rustagi, G. Q. Lo, D. S. H. Chan and D. L. Kwong, *IEEE Electron Device Lett.*, **29**, 595 (2008).
4. W. W. Fang, N. Singh, L. K. Bera, H. S. Nguyen, S. C. Rustagi, G. Q. Lo, N. Balasubramanian and D.-L. Kwong, *IEEE Electron Device Lett.*, **28**, 211 (2007).
5. Q.T. Nguyen, J.F. Damlencourt, B. Vincent, L. Clavelier, Y. Morand, P. Gentil and S. Cristoloveanu, *Solid-State Elect.*, **51**, 1172 (2007).
6. J.F. Damlencourt, Patent No FR2905197.
7. E. D. Saracco, J.F. Damlencourt, D. Lafond, S. Bernasconi,V. Benevent, P. Rivallin, Y. Morand, J.M. Hartmann, P. Gautier, C. Vizioz, T. Ernst, C. Bonafos and P. Fazzini, *ECS Trans.*, **19**, 207 (2009).
8. G. Ghibaudo, *Electronics Lett.*, **24**, 543 (1988).

ECS Transactions, 35 (5) 163-168 (2011)
10.1149/1.3570792 ©The Electrochemical Society

FISH SOI MOSFET: An Evolution of the Diamond SOI Transistor for Digital ICs Applications

Salvador Pinillos Gimenez and Daniel Manha Alati

Electrical Engineering Department, FEI University Center, São Paulo, Brazil

This work introduces and studies a new transistor layout style, called Fish SOI MOSFET using 3D numerical simulations, where two trapezes compose the transistor gate area, generating a "smaller than (<)" mathematical signal shape. This innovative layout structure is an evolution of the Diamond SOI MOSFET. The Fish structure was carefully designed to be used in the digital integrated circuits applications, because now its channel length can be implemented with the minimum dimension allowed by the manufacture process, in contrast to Diamond transistor in which this is not possible. The Fish transistor also uses the Longitudinal Corner Effect (LCE) to increase the resultant longitudinal electric field along to the channel length, that results in an improvement in the average carriers drift velocity in the channel, in the drain current, in the transconductance and in the on-state series resistance parameters.

Introduction

Silicon-On-Insulator (SOI) Complementary Metal-Oxide-Semiconductor (CMOS) Field Effect Transistor (FET) is the most recommended manufacture process technology for low power low voltage (LPLV) integrated Circuits (ICs) applications (1).

Several efforts have been done to improve the electrical performance of the planar and the three-dimensional (3D) transistors and consequently increase the performance of the ICs and several new planar and 3D devices have been introduced with this intention. Some examples of these transistors are: (a) Planar: Circular Gate Transistors (2), Wave Transistor (3), Overlaping-Circular Gate Transistor (4), Diamond Transistor (5-7) and OCTO Transistor (8); (b) 3D Devices: Dual-Gate Transistor (1), FinFET (1), Three-Gate Transistor (1), Four-Gate Transistor (1) and Gate-All-Around Transistor (Pillar and Cynthia) (9). Recently, a new approach of the channel engineering was created to improve the performance of the transistors, named Longitudinal Corner Effect (LCE), and applied in the Diamond SOI MOSFET (DSM), where the resultant longitudinal electric field in the channel direction is enhanced as compared to the Conventional SOI MOSFET (CSM), resulting in an improvement in the average carriers drift velocity ($\overrightarrow{v_{//}}$), the drain current (I_{DS}), the transconductance (g_m), transconductance over drain current (g_m/I_{DS}) and on-state series resistance (R_{DS_on}) (5-7). The way we reached the implementation of the LCE in the DSM was by changing the gate geometry from rectangular to hexagonal (5-7). Diamond layout style presents an effective channel length (L_{eff}) higher than the minimal dimension allowed by the manufacture process and therefore can limit its use for the digital ICs applications (5-7). So, in order to overcome the limitation of the Diamond transistor for the digital ICs applications, we carefully designed the Fish layout style to use the LCE for improving the resultant longitudinal

163

electrical field that results in an increase of the I_{DS}, g_m, g_m/I_{DS} and R_{DS_on}, when compared to the CSM counterpart when we consider the same bias conditions. Figure 1 presents an example of the Fish SOI MOSFET (FSM) layout and can be implemented with any manufacture process technology of MOSFETs (planar and 3D transistors).

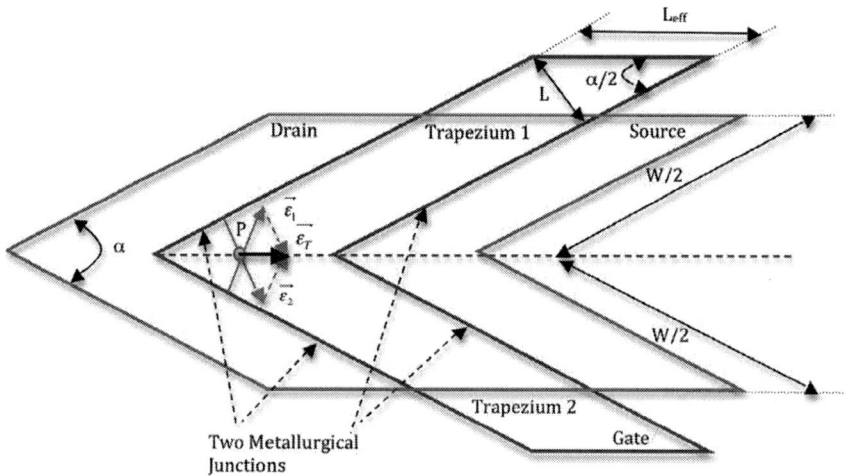

Figure 1. Example of the Fish SOI MOSFET layout.

In Figure 1, α is the angle defined by the two metallurgical junctions composed by the drain/source and channel regions, W is the effective channel width, L is the length of the gate material that can be implemented with the minimum dimension allowed by the manufacturing process technology, $\vec{\varepsilon_1}$ and $\vec{\varepsilon_2}$ are the two electrical field components in the P point as the result of the application of the drain bias (perpendiculars to the two metallurgical junctions between the drain and channel regions) and $\vec{\varepsilon_T}$ is the resultant electric field $(=\vec{\varepsilon_1}+\vec{\varepsilon_2})$ that is higher than the one found in the transistor with rectangular gate shape (CSM).

The Fish I_{DS} (I_{DS_FSM}) presents the same direction of the $\vec{\varepsilon_T}$ and consequently the effective channel length (L_{eff}) of the FSM is given by $L/\sin(\alpha/2)$, as indicated in the Figure 1. Therefore, the Fish layout style has the ability to increase the transistor L_{eff} and consequently increase the figure of merit defined as $I_{DS}/(W/L_{eff})$ that represents the transistor capability to drive drain current (I_{DS}) for a specific W/L_{eff} ratio. Therefore, the Fish layout style brings an innovative and extraordinary possibility in the lengthening of the effective channel length of the transistor, keeping the channel length with the minimum dimension allowed by manufacturing process technology used in digital ICs applications. This happens mainly in the center of the Fish structure as shown in the Figure 1 above (P point) and reduces its strength as P point reaches the extreme sides due to the smaller interaction between the two electrical fields ($\vec{\varepsilon_1}$ and $\vec{\varepsilon_2}$).

Equation [1] shows us the ratio of the $I_{DS}/(W/L_{eff})$ between the FSM [$I_{DS\_FSM}/(W/L_{eff})$] and CSM counterpart [$I_{DS\_CSM}/(W/L_{eff})$].

$$\frac{\dfrac{I_{DS\_FSM}}{(W/L_{eff})}}{\dfrac{I_{DS\_CSM}}{(W/L)}} = \frac{I_{DS\_FSM}}{I_{DS\_CSM}}\left[\frac{1}{\sin(\alpha/2)}\right] \geq 1, \text{ for } 0<\alpha\leq180^\circ. \qquad [1]$$

The equation [1] shows us that FSM can produce higher $I_{DS}/(W/L_{eff})$ than the one found in the CSM for two reasons, i.e., because $I_{DS\_FSM}>I_{DS\_CSM}$ due to LCE and $1/\sin(\alpha/2)>1$ due to Fish structure, if we consider the same L (Figure 1), the same W (Figure 1), the same gate area (A_G) and bias conditions. This is one of the benefits of the use of Fish layout style that we can use to improve the current drive capacity when we use it for current driver and buffer for digital ICs. Another benefit of the use of Fish structure is the possibility of significantly reducing the W in order to keep the drain current of the conventional SOI pMOSFET in a combinational and sequential digital circuits and consequently obtaining a die area reduction of the digital ICs applications. This is only possible because the dimensions of these devices are determined considering the ratio between the W/L ratios of the nMOSFETs and pMOSFETs, respectively. This same approach can also be largely used for the design of the current mirrors of the analog ICs for the same reason that is another benefits of the use of the FISH structure.

Three-Dimensions Simulation Results

Three PD SOI nMOSFETs were implemented in Sentaurus Structure Editor, being two PD FSM with α equal to 90° and 135°, respectively and one PD CSM counterpart that presents the same gate area (A_G). The constructive parameters of these devices are: gate-oxide (t_{ox}), silicon-film (t_{Si}) and buried-oxide (t_{BOX}) thickness are 2.5 nm, 100 nm and 400nm, respectively. The drain/source and channel doping concentrations are 1×10^{20} cm^{-3} and 5.5×10^{17} cm^{-3}, respectively. The L and W dimensions of all devices are equals to 1µm and 6 µm, respectively (Table I).

Table I. Transistors Dimensions.

Transistors	A_G (µm^2)	α (°)	W (µm)	L (µm)	L_{eff} (µm)	W/L_{eff} (µm)
CSM	6.00	180	6.00	1.00	1.00	6.00
FSM	6.00	135	6.00	1.00	1.08	5.56
FSM	6.00	90	6.00	1.00	1.41	4.26

Philips unified mobility, Enormal, Shockley-Read-Hall, Band Gap Narrowing and Avalanche Generation for Driving Force models were used in the three-dimensions numerical simulations (3DNS) (Sentaurus).

Figure 2 presents the curves of $I_{DS}/(W/L_{eff})$ as a function of V_{GS} of the FSMs and CSM, respectively. The threshold voltages (V_{TH}) of the devices were extracted and all presented the same value (230 mV). Additionally, Table II shows the values of $I_{DS}/(W/L_{eff})$ of the FSM and CSM with the same A_G, considering V_{DS} and V_{GS} equals to 10 mV and 1V, respectively.

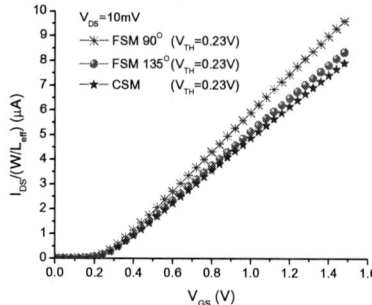

Figure 2. $I_{DS}/(W/L_{eff})$ as a function of V_{GS}.

Table II. $I_{DS}/(W/L_{eff})$ of the FSM and CSM with the same A_G, considering V_{DS}=10mV and V_{GS}=1V (Triode Region).

Transistors	α (°)	$I_{DS}/(W/L_{eff})$ (μA)	$I_{DS}/(W/L_{eff})$ improvement (+) / worsening (-)
CSM	180	4.87	-
FSM	135	5.14	+5.54%
FSM	90	5.91	+21.4%

Observing the 3DNS results, all the FSMs I_{DS} found were higher than those found in the CSM counterpart, considering the same gate area. We got better gain in the FSM $I_{DS}/(W/L_{eff})$ for α equal to 90° (21.4%) due to a higher interaction between the electric field components in the channel and a higher L_{eff} FSM value than the one found in the CSM counterpart. This is a remarkable result of the Fish layout style and therefore it is an important alternative of device to be used as current driver/buffer in the digital ICs applications.

Figure 3 presents the curves $g_m/(W/L_{eff})$ as a function of V_{GS} of the devices, considering the same A_G and Table III shows the values of the maximum transconductance normalized as the function of the geometric factor [$g_{m\_max}/(W/L_{eff})$] of the FSMs and CSM counterpart, considering V_{DS} equal to10 mV.

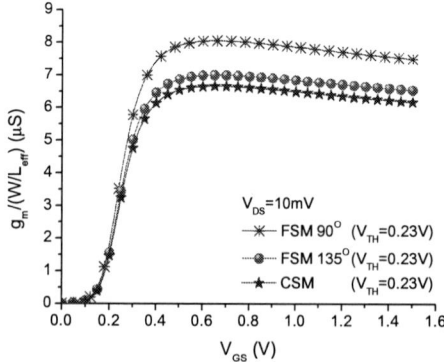

Figure 3. Curves of the $g_m/(W/L_{eff})$ as a function of V_{GS} of the FSMs and CSM counterpart, considering V_{DS}=10 mV.

Table III. $g_{m\_max}/(W/L_{eff})$ of the FSMs and CSM counterpart for V_{DS}=10 mV.

Transistors	α (°)	$g_{m\_max}/(W/L_{eff})$ (µS)	$g_{m\_max}/(W/L_{eff})$ improvement (+) / worsening (-)
CSM	180	6.66	-
FSM 135°	135	7.01	+5.26%
FSM 90°	90	8.06	+21.0%

Observe that FSMs $g_m/(W/L_{eff})$ present higher values than the one found in the CSM counterpart, considering the same A_G and V_{DS}=10 mV. We got better gain in the FSM $g_m/(W/L_{eff})$ for α equal to 90° (21%) due to the same reason present for the FSM $I_{DS}/(W/L_{eff})$. Therefore, we can also use the Fish structure to increase significantly the unit voltage gain frequency (f_T) of the amplifiers circuits (1) in the analog ICs, depending the α angle used.

The Figure 4 shows $I_{DS}/(W/L_{eff})$ as a function of V_{DS} curves of the FSMs and CSM counterpart for V_{GS} equal to 400 mV and Table IV presents the values of saturation drain current normalized as a function of geometric factor [$I_{DS\_sat}/(W/L_{eff})$] of the FSMs and CSM counterpart, considering V_{GS} and V_{DS} equals to 400 mV and 1 V, respectively.

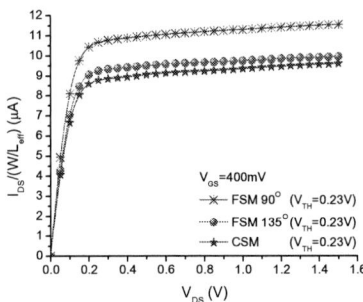

Figure 4. $I_{DS}/(W/L_{eff})$ as a function of V_{DS}.

Table IV. $I_{DS\_sat}/(W/L_{eff})$ of the FSMs and CSM counterpart for V_{DS}=1 V and V_{GS}=400 mV.

Transistors	α (°)	$I_{DS\_sat}/(W/L_{eff})$ (µA)	$I_{DS\_sat}/(W/L_{eff})$ improvement (+) / worsening (-)
CSM	180	9.35	-
FSM	135	9.75	+4.28%
FSM	90	11.3	+20.9%

Notice that I_{DS_sat} improvements can be reached when we use Fish layout instead conventional transistors. For example, I_{DS_sat} improvements of approximately 21% can be obtained if we use FSM with α angle of 90° instead CSM counterpart, considering the same A_G and bias conditions. This means to say also that, considering the same I_{DS_sat}, we can replace the CSMs by FSMs counterparts in order to reduce significantly the die area of analog and digital ICs, depending the α used. Besides that, Table V shows the R_{DS_on} values of the FSMs and CSM counterpart.

Table V. The values of R_{DS_on} of the FSMs and CSM counterpart for V_{GS}=400 mV.

Transistors	α (°)	R_{DS_on} (kΩ)	R_{DS_on} improvement (+) / worsening (-)
CSM	180	12.8	-
FSM	135	12.1	+5.47%
FSM	90	10.5	+18.0%

Observe that when we perform the comparison between the R_{DS_on} of the FSMs and CSM counterpart, better values of R_{DS_on} are found for smaller FSM α angles, due to the higher $\vec{\varepsilon_T}$ and L_{eff}. This is a remarkable result of the Fish structure and can significantly improve the velocity of the digital ICs when FSM is used as pass transistors and switches.

Conclusion

The Fish layout style as an evolution of the Diamond structure and it was specially designed for digital integrated circuits applications (it can be implemented with the minimum dimension of manufacture process in contrast of the Diamond transistor) and uses the LCE to improve its electrical parameters when we compare to the conventional counterpart. For the first time, the Fish layout style is introduced, studied and compared to the conventional counterpart by three-dimensional numerical simulations. It is verified that the Fish SOI nMOSFET present better values of the drain current normalized as a function of geometric factor in triode (5.54% and 21.4% for α equal to 135° and 90°, respectively) and saturation (4.28% and 20.9% for α equal to 135° and 90°, respectively) regions, maximum transconductance as a function of geometric factor (5.26% and 21.0% for α equal to 135° and 90°, respectively) and on-state series resistance (5.47% and 18.0% for α equal to 135° and 90°, respectively) in comparison to the conventional counterpart. In conclusion, the Fish layout style is a device alternative to improve the performance (current driver and velocity) of the integrated circuits applications. Additionally, Fish structure can be used to improve the unit voltage gain frequency in analog integrated circuits due to present better values of the transconductance than the one observed in the conventional counterparts.

Acknowledgement

The authors would like to thank the financial support provided by CNPq (Grant 311149/2009-0 and 556756/2009-6, respectively) and Salvador Pinillos Gimenez acknowledges Vera Lucia Cardoso Berk for the English support.

References

1. J. P. Colinge, *Silicon-On-Insulator Technology: Materials to VLSI*, p. 156, Kluwer Academic Publishers, Boston (2004).
2. S. P. Gimenez et. al., in *SBMicro 2006*, J. Diniz et. al. Editors, ECS Trans. 4 (1), p. 309, Ouro Preto, Brazil (2007).
3. S. P. Gimenez, in *215th ECS Meeting*, Y. Omura et. al. Editors, ECS Trans. 19 (4), p. 153, San Francisco, CA (2009).
4. J. A. De Lima and S. P. Gimenez, in SBMicro 2009, ECS Trans. 23 (1), p. 361, Natal, Brazil (2009).
5. S. P. Gimenez et. al., in *EUROSOI 2009*, PV 1, p. 87, Goteborg, Sweden (2009).
6. Salvador Pinillos Gimenez, *Solid-State Electronics*, **54** (12), 1690 (2010).
7. S. P. Gimenez et. al., in *218th ECS Meeting*, ECS Trans. 33, p. 121, Las Vegas, NV (2010).
8. S. P. Gimenez et. al., in *EUROSOI 2011*, PV 1, pp. 91, Granada, Spain (2011).
9. D. R. Oliveira et. al., in *SBMicro 2009*, D. De Lima Monteiro et. al. Editors, ECS Trans. 23 (1), p. 381 (2009).

ECS Transactions, 35 (5) 169-176 (2011)
10.1149/1.3570793 ©The Electrochemical Society

Nonlinear Properties of Si-based Substrates for Wireless Systems and SoC Integration

C. Roda Neve and J.-P. Raskin

Microwave Laboratory, Université catholique de Louvain, Place du Levant, 3,
B-1348 Louvain-la-Neuve, Belgium
cesar.rodaneve@uclouvain.be, Tel: +32 (0)10 47 80 96

The nonlinear behaviour of silicon substrates with different resistivities is analyzed using coplanar structures. In order to compare the nonlinear performance for different substrates and technologies, the harmonic distortion of crosstalk test structures is investigated, as well as the dependence on the distance. The generated harmonic components due to a large signal at 900 MHz are measured using a one-tone network analyzer based setup. Below the crosstalk tap, harmonic levels as high as -43 and -54 dBc for 15 dBm are generated for standard and high-resistivity Si substrate, respectively. The introduction of a trap-rich layer at the interface between the BOX and the high-resistivity Si (HR-Si) provides a reduction of at least 45 dB in the harmonic distortion generated into the substrate. It has been proven that these results can be easily extrapolated to crosstalk structures with different dimensions.

Introduction

During the last few years, there has been a growing interest in silicon-based substrates, and more precisely in Silicon-on-Insulator (SOI), to be used as handle substrates for the fabrication of wireless systems and Systems-on-Chip (SoC) integration (1)-(3). For RF and SoC applications, SOI presents the major advantage of providing high resistivity (HR) substrate capabilities, leading to substantial reduced substrate losses, in addition to CMOS compatibility and cost effective solution for the fabrication of microsystems.

However, HR-Si substrates, as well as HR SOI, suffer from parasitic surface conduction (PSC) at the Si/SiO2 interface, due to the presence of fixed charges in the oxide (4), and other wafer inhomogeneities, thereby reducing the effective resistivity of the substrate by more than one order of magnitude (5). It has been already proven that this effect can be overcome by introducing a trap-rich layer at the interface between the oxide layer and the Si substrate (4). The trap-rich layer can been achieved by using different techniques such as ion implantation (7), and the deposition of amorphous silicon (4) or polycrystalline silicon layer (PSi) (8).

It is only recently (6), due to the stringent linear requirement of new 3G standards, that the HR-Si non-linear behaviour has been pointed out as a major limitation for the design of new communications and high-power systems on HR SOI. The degree of linearity, or nonlinearity, of a substrate can be measured using a one-tone characterization setup to determine the relation between the fundamental tone and its harmonics. This is typically done with a tone at 900 MHz and measuring a CPW line lying on the investigated substrate (3), (6). However, this type of characterization is difficult to compare from a technology to another as the harmonic level does not only depend on the substrate non-linear properties, but also on the size and length of the CPW line. In this work, we propose to analyze the harmonic distortion by means of a crosstalk test structure. This approach allows us to separate the possible harmonic contribution of the metallic lines, as well as to clearly identify the origin of the non-linearity from the induced changes in the electrons and holes distribution below the metallic pads as a function of the applied large signal.

Nonlinear Properties of Si-Based Substrates

The bias voltage dependence is an inherent property of all semiconductor substrates, at least at quasi static operation and low frequencies. At very high frequencies, e. g. microwaves frequencies and above, silicon has been typically considered as a lossy dielectric with a constant permittivity and conductivity. The introduction of HR-Si substrates, the increasing frequency of operation and handled power, in addition to the use of coplanar devices for RF circuits, which are more sensitive to undesired effects located close to the top of the handle substrate, have brought to light the nonlinear properties of Si when large signals are applied at microwaves frequencies (3), (6).

This phenomenon can be roughly explained by looking into the steady state of the carrier distribution of a coplanar device on top of an oxidized p-type HR-Si under various bias conditions. In Fig. 1(a), for zero voltage (Vs = 0 V), the metallic work function, fixed charges inside the oxide and other wafer inhomogeneities induce a shallow inversion layer at the Si-SiO2 interface where the carriers can laterally move, hence creating a conductive layer, also know as parasitic surface conduction (PSC) and increasing RF losses (i.e. reducing the effective substrate resistivity). In Figs. 1(b) and 1(c) we can see that when applying a positive or negative voltage, a larger inversion or an accumulation layer is created, varying the conductive characteristic at the interface and having a bias-dependent conductive layer.

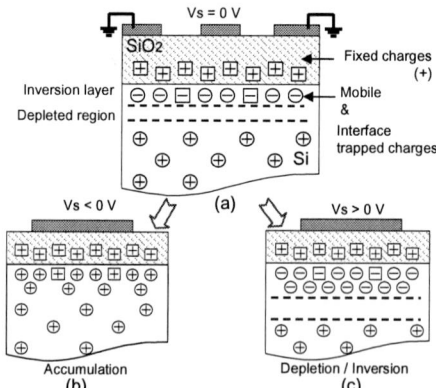

Figure 1. Cross section of a coplanar MOS p-type structure under zero, (a) negative (accumulation) (b) and positive bias voltage (depletion/inversion).

For small-signals this bias dependence translates only into a different attenuation level and effective resistivity at different bias points (5). However, in the case of large signals the amplitude of the signal may become large enough to vary the conductive characteristic of the substrate. The mobile carriers at the interface have a response time fast enough to follow signals at high frequencies (> 100 MHz).

Impact of Substrate Nonlinearities in Coplanar Structures

Coplanar structures could be considered as the most suitable structures for the design of RF circuits and high-frequency on-wafer interconnections. Added to their simple fabrication process, coplanar structures enable using wider lines to achieve the same characteristic impedance than microstrip structures which leads to reduced metallic losses (9). Besides, in the case of HR SOI, they fully benefit of the high-resistivity properties of the handle substrate, allowing the use of the whole back-end-of-line of the semiconductor process (10). They also can have more relaxed design specifications, making them more suitable for being reused in future technological nodes.

Unfortunately, as mentioned above, coplanar structures are more sensitive to PSC and to the nonlinear properties of handle substrates. As it is shown in Fig. 2 (11), the lineic capacitance and conductance of the RLCG equivalent model of a CPW transmission line strongly depends on the applied bias. Consequently, the higher measured harmonic component of the CPW, see Fig. 3 (11), can be as high as -57 dBc (dB relative to carrier) at +35 dBm of input power for a substrate with a nominal resistivity of 5 kΩ-cm. It is worth to notice that for all types of Si substrates which do not include a trap-rich layer, the highest harmonic level always corresponds to the 2nd harmonic component (1), (3), (11).

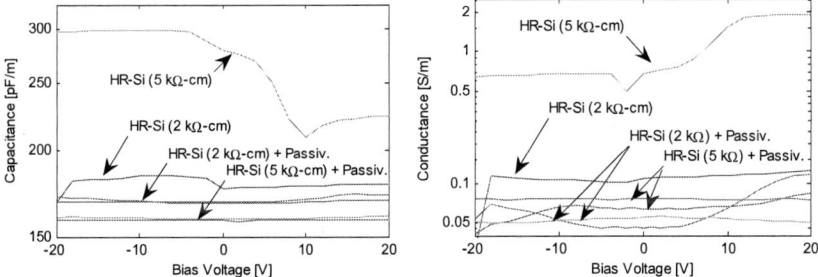

Figure 2. Extracted capacitance (left) and conductance (right) per unit length of a 50 Ω 8000 μm-long CPW line versus DC bias at 900 MHz.

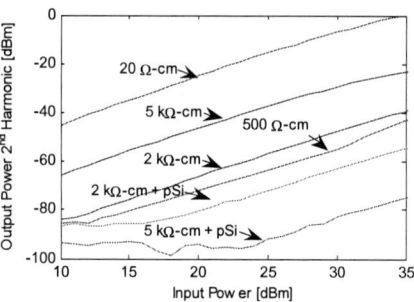

Figure 3. Measured 2nd harmonic component (carrier frequency = 900 MHz) of a 2.2 mm-long CPW line lying onto different Si-based substrates.

From the previous figure we observe that the harmonic distortion of a CPW line depends on the Si resistivity, on the quality and thickness of the oxide, and on the work function of the metal. PSC increases the harmonics levels as it reduces the effective resistivity seen by the CPW or other coplanar structures. In the case of a CPW, the dimensions of the line affect the measured distortion, as it does it for the PSC (12). The length of the line also influences the detected harmonic power at the end of the line. For long lines, compared to the wavelength of the carrier frequency and of the harmonics, distributed elements (13) of a maximum length must be considered in the harmonic generation as well as in the attenuation and losses at the different harmonic frequencies. To be able to compare the nonlinear properties of different substrates, and its possible reduction, we need to use identical CPW lines which seem difficult as dimensions and lengths vary from a technology to another, or find a simple coplanar device that will be less sensitive to its dimensions and its results could be extrapolated to other coplanar structures.

Crosstalk Test Structure

One of the simplest high frequency coplanar devices is the crosstalk structure. In Fig. 4, a classical crosstalk structure, that is used in this work, is presented. Typically, it is used to investigate the substrate crosstalk and isolation between different devices or circuits lying onto the same substrate, which are critical issues for SoC applications.

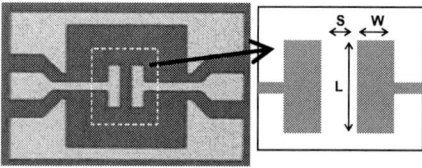

Figure 4. Top view of a crosstalk test structure. The rectangular metallic taps sizes are of LxW = 150 μm x 100 μm and the spacing is S = 50 μm.

The frequency performance of such type of structures has been previously investigated by the authors in (2) and (14). Its frequency response is synthetically shown in Fig. 5. Three different coupling mechanisms exist, each corresponding to a specific frequency range. First, at low frequencies, crosstalk is governed by the isolation provided by the oxide capacitance. It appears an intermediate zone with a flat response due to the conductive properties of the Si substrate. And finally a third zone, at higher frequencies, where the coupling is mainly capacitive. The different substrate properties and thicknesses shift the frequency limits of these three zones, while the spacing of the taps and their dimension only modify crosstalk levels by the same value along the whole frequency spectrum.

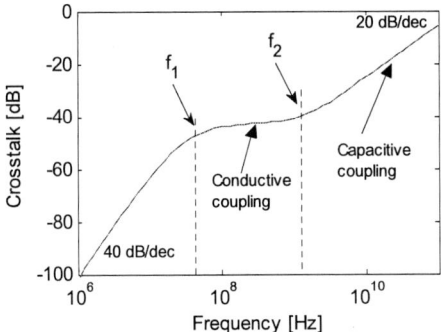

Figure 5. Frequency response of a crosstalk test structure lying onto an oxidized Si substrate.

Regarding the harmonic generation of the crosstalk structure, it is mainly located below one of the metallic taps, where the large signal is applied. The fundamental tone and all the harmonics are coupled through the substrate and detected, at the other metallic tap, without additional harmonic generation. It is in the substrate coupling where the PSC plays a major role, and not in the nonlinear behaviour. The main inconvenience for the use of this structure is the extremely weak power levels that must be detected. In addition to the low nonlinear levels in Si-based substrates, we have to add an attenuation of 20-30 dB for the frequencies of interest. A simulation of the expected harmonic distortion is shown in Fig. 6 for a crosstalk structure with the following dimensions L = 150, W = 100, and S = 50 μm. It presents harmonic levels higher than -75 dBc for bulk Si substrates (ρ = 20 Ω-cm) and -112 dBc for a HR-Si substrates (ρ > 5 kΩ-cm) in the presence of fixed charges in the oxide (Q_f), for an input power of 15 dBm..

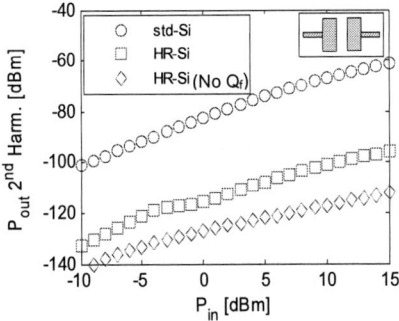

Figure 6. Simulated 2nd harmonic component of a crosstalk test structure on different Si substrates, for a sinusoidal signal at 900 MHz.

Harmonic Distortion Characterization of Crosstalk Test Structures on Si-Based Substrates

To identify the degree of nonlinearity of the substrate, we use a one-tone characterization setup to determine the relationship between the input amplitude of a fixed frequency with its output at the fundamental frequency and its harmonics (15). The characterization setup is based on a 4-port Agilent PNA-X network analyzer, which easily allows us to accurately measure not only the amplitude and phase of the fundamental tone but also the amplitude of the corresponding harmonics (16). Additionally, the device under test (DUT) can be biased at different DC voltages through the internal bias-tees of the network analyzer, and the S-parameters from 10 MHz to 26.5 GHz are obtained at the same time by switching off the band-pass filter. After proper calibration, and by carefully chosen the filters and attenuators at the source input an at the harmonics output detector, we can detect harmonic levels of a 900 MHz signal as low as -110 dBm for a maximum input power of 15 dBm.

The frequency response of a crosstalk structure with dimensions L = 150, W = 100, and S = 50 µm, is shown in Fig. 7(a) for three different Si-based substrates: a bulk-Si with ρ = 20 Ω-cm (standard resistivity – Std), a HR-Si with ρ > 5 kΩ-cm but a $\rho_{eff} \approx 200$ Ω-cm, and the same HR-Si with a 278 nm-thick intermediate layer of PSi between the oxide and the silicon substrate. All of them have a 50 nm-thick top layer of thermally grown SiO_2 and 1 µm-thick pure Al layer for the pad contact. Depending on the substrate properties, the different operation zones can be identified, being the substrate with the trap-rich layer the one that behaves better. In Fig. 7(b), we can see how an increase of 70 µm in the spacing between taps implies a decrease of approximate 10 dB in the crosstalk for the Std-Si.

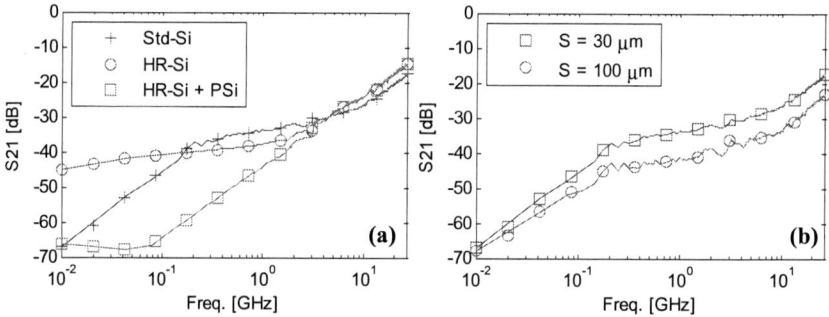

Figure 7. Measured crosstalk for (a) different oxidized Si-based substrates, and (b) a Std-Si substrate with different spacings between crosstalk taps.

The harmonic distortion of the investigated substrates is shown in Fig. 8 when a sinusoidal signal at 900 MHz from -25 to 15 dBm is applied at one of the pads. The higher output power corresponds to the 2nd harmonic component, being -67 and -90 dBc at 15 dBm input power for the Std-Si and HR-Si substrates, respectively. In both cases it is approximately 30 dB lower than a 3.8 mm-length 50 Ω CPW line lying onto the same substrates. If we subtract the attenuation due to the substrate coupling, we obtain -43 and

-54 dBc for the Std-Si and HR-Si, respectively, below the tap. As it was expected, the presence of a trap-rich layer reduces the harmonic distortion by more than 45 dB, always below the minimum detectable power.

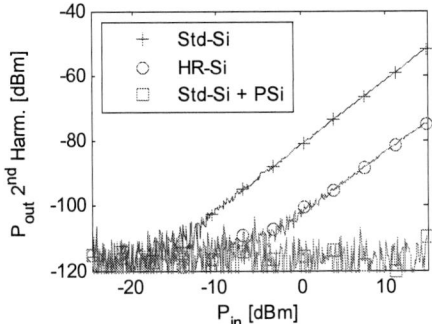

Figure 8. Measured 2nd harmonic component distortion for a sinusoidal signal at 900 MHz, for the three investigated substrates.

To verify the insensitivity of the harmonic levels to the distance between the taps, we compare the decrease with the spacing in the fundamental tone and in the 2nd harmonic for the Std-Si and HR-Si substrates, to the decrease in the crosstalk measurement, see Fig. 7(b). The decrease in the power level of the fundamental tone (H1) and of the 2nd harmonic (H2) from a spacing between taps of 30 to 100 μm can be seen in Fig. 9. For H1, the values are identical to the crosstalk difference at 900 MHz, -8 and -9.5 dB for the Std-Si and HR-Si, respectively. However, for H2, even if we can consider them constant with power, when the detected levels are high enough, the values differ slightly (~1 dB) from the crosstalk at 1.8 GHz: -6.7 and -8.8 dB for the Std-Si and HR-Si. Despite this small difference, that could be simply due to measurement inaccuracies in the crosstalk or in the harmonics, we can conclude that our assumption is still valid. The characterization of several substrates with different resistivities and layer thicknesses will help us to definitively confirm the possibility to extrapolate the detected harmonic levels in the crosstalk structures to other dimensions and coplanar devices.

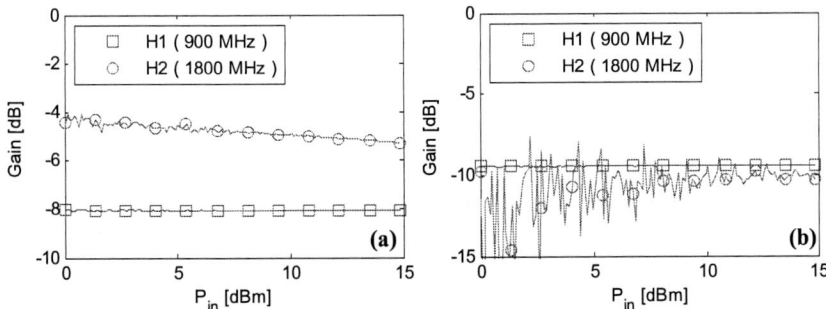

Figure 9. Difference in the measured output power levels at the fundamental tone (900 MHz) and at the 2nd harmonic component for an increment in spacing between crosstalk taps, for (a) a Std-Si, and (b) HR-Si substrate.

175

Conclusion

In this work we have successfully characterized the nonlinear properties of three different silicon substrates using coplanar crosstalk structures. The higher harmonic component correspond to the 2nd harmonic (freq. = 1.8 GHz), as it was already proved for CPW lines fabricated on similar substrates. It presents absolute power levels of -75 and -52 dBm for a Std-Si and HR-Si substrates, respectively, for an input power of 15 dBm. It has been proved that the harmonic generation is due to the substrate itself and that it is mainly located below the tap where the large signal is applied. If the spacing of the metallic taps of the crosstalk structure is known, it is possible to extrapolate the values to other dimensions by using the crosstalk measurement for the new spacing or by means of analytical models. This approach allows compare different technologies using a more standard and simpler coplanar structure.

Compared to initial simulations, the experimental harmonic distortion for crosstalk structures is 10 to 15 dB higher than expected. It shows the importance of taking into account for crosstalk consideration not only the frequency of the signal of interest in the performance of neighbour circuits in SoC application, but also the harmonics of that signal when RF power circuitry is integrated together into the same chip.

As it was previously proved in (6) and (11), the introduction at the interface between the BOX and the HR-Si provides a reduction of at least 45 dB in the harmonic distortion generated into the substrate. This technological solution in combination with HR SOI will provide a lossless harmonic-free high-resistivity silicon substrate, with low crosstalk levels for frequencies below 10 GHz.

References

1. T. G. McKay *et al.*, *IEEE Intl. SOI Conf.*, (2007).
2. J.-P. Raskin *et al.*, *IEEE Trans. on Elect. Dev.*, **44**(12), p. 2252, (1997).
3. A. Botula *et al.*, *IEEE SiRF 2009*, pp. 1, (2009).
4. H.S. Gamble *et al.*, *IEEE Microwave and Guided Wave Letters*, **9**(10), p. 395, (1999).
5. D. Lederer and J.-P. Raskin, *Solid-State Electr.*, **49**(3), p. 491, (2005).
6. D. C. Kerr *et al.*, *IEEE SiRF 2008*, p. 151, (2008).
7. Y. H. Wu *et al.*, *IEEE MTT-S Digest*, (2000).
8. D. Lederer *et al.*, *IEEE Trans. on Elect. Dev.*, **55**(7), p. 1664, (2008).
9. M. Si Moussa *et al.*, *Solid-State Electr.*, **50**(11-12), p. 1822, (2006).
10. F. Gianesello *et al.*, *IEEE MTT-S*, (2007).
11. C. Roda Neve *et al.*, *Proc. of the EuMIC Conf. 2008*, p. 36-39, (2008).
12. D. Lederer, *Ph.D. Thesis, UCL, Louvain-la-Neuve, Belgium*, (2006).
13. D. Zelenchuk *et al.*, *Proc. of the EuMC 2007*, p. 396, (2007).
14. K. Ben Ali, C. Roda Neve and J.-P. Raskin, *SiRF 2010*, p. 212, (2010).
15. J. C. Pedro, and N. B. Carvalho, *Intermodulation Distortion in Microwave and Wireless Systems*, Artech House, Boston, p. 31, (2003).
16. Agilent Technologies, *Application Note 1408-8*, (2006).

CHAPTER 8

DEVICES AND CIRCUITS

FDSOI Process Technology for Subthreshold-Operation Ultra-Low Power Electronics

S. A. Vitale, P. W. Wyatt, N. Checka, J. Kedzierski, C. L. Keast[1]

MIT Lincoln Laboratory, Lexington, Massachusetts 02420, USA

Ultralow-power electronics will expand the technological capability of handheld and wireless devices by dramatically improving battery life and portability. In addition to innovative low-power design techniques, a complementary process technology is required to enable the highest performance devices possible while maintaining extremely low power consumption. Transistors optimized for subthreshold operation at 0.3 V may achieve a 97% reduction in switching energy compared to conventional transistors. The process technology described in this article takes advantage of the capacitance and performance benefits of thin-body silicon-on-insulator devices, combined with a workfunction engineered mid-gap metal gate.

Introduction

Ultra-low-power transistors are an enabling technology for many proposed applications. Ubiquitous sensor networks, RFID tags, implanted medical devices, portable biosensors, handheld devices, and space-based applications are among those which would benefit from extremely low power circuits (1-3). Other applications include energy-harvesting devices which recharge batteries by scavenging power from motion or solar cells, such as a recently demonstrated wristwatch design requiring a maximum 50nA of on-current at 0.42 V operation (4-5). In general, low standby power (LSTP) applications require less than 100 pA/μm leakage current, while maximizing the on-current at a modest power supply voltage.

Subthreshold operation transistors hold great promise for integration into ultra-low-power designs. The most efficient way to reduce power is to reduce the operating voltage (6). With an operating voltage of 0.3 V, and an on-current of less than 1 μA/μm, subthreshold transistors use orders of magnitude less power than transistors operated in strong inversion. Subthreshold operation also provides the highest transconductance (g_m) for a given drain current (7). In subthreshold conduction is by diffusion rather than drift, which implies that it is possible to have equal NMOS and PMOS drive currents per micrometer of transistor width. This allows equal sizing of NMOS and PMOS transistors (in contrast to the typical 2x wider PMOS transistor size for conventional devices) which would then allow reduced circuit area and device capacitance.

---

[1] This work was sponsored by the Air Force under contract #FA8721-05-C-0002. Opinions, interpretations, conclusions, and recommendations are those of the author and are not necessarily endorsed by the United States Government.

Simply lowering the operating voltage of a conventional high-performance transistor will not produce very good device performance in subthreshold operation. Conventional transistors will have comparatively high off-state leakage and overlap capacitance, as well as poorer subthreshold slope. By designing a fabrication process from the substrate material through the interconnect metal, optimized for subthreshold transistor performance, it is possible to realize a device with the minimum switching energy and off-state current without significant impact to the energy-delay product. This paper explores advantages of SOI technology in ultra-low power applications, and details the processing techniques used to optimize the devices for subthreshold operation.

Bulk silicon vs. SOI

SOI-based CMOS is commonly used today in high performance applications, such as gaming consoles (5). Though double gate designs (including FinFETs) can provide improved channel control and lower leakage, (8) the high cost associated with the significant increase in process integration complexity makes them less desirable for ultra-low-power electronics. To introduce ultra-low-power process technology in the shortest practical timeframe, it is appropriate to consider planar architectures.

The benefits and disadvantages of SOI vs. bulk silicon technology have been discussed many times (2,4,9-10). Compared to bulk silicon, SOI provides up to 90% lower junction capacitance, near-ideal subthreshold swing, reduced device cross-talk, lower junction leakage, no latch-up, increased radiation hardness, and full dielectric isolation of the transistor. The low junction capacitance is extremely valuable to ultra-low-power devices, as it allows reduction of the CV^2 switching energy of the transistor. Another significant advantage for low-power operation is that SOI devices do not suffer from substrate reverse bias effects, in that the depletion charge does not increase when a source potential is applied. Thin-body SOI also provides better electrostatic channel control, leading to reduced source-to-drain leakage and reduced short channel effects (SCE) (8). In contrast, it has been suggested that bulk silicon is now facing GIDL limits with device scaling, making it inappropriate for ultra-low-power applications (5).

SOI technology can be fully depleted (FDSOI) or partially depleted (PDSOI). The depletion depth is given by:

$$T_{dep} = \sqrt{\frac{4\varepsilon\Phi_f}{qN_{ch}}} \tag{1}$$

where $\Phi_f = kT\ln(N_{ch}/n_i)/q$ is the Fermi potential, N_{ch} is the channel doping concentration, and n_i is the intrinsic carrier concentration (6). When the depletion depth is larger than the physical silicon thickness, a neutral region no longer exists between the source and drain, and the silicon becomes fully depleted (4). For a highly doped channel where $N_{ch} = 1\times10^{18}$ cm^{-3}, $T_{dep} = 32$nm by Equation (1).

FDSOI is more difficult to fabricate than PDSOI, as the silicon channel must be reduced to a very small and well-controlled thickness. Another disadvantage is that series resistance in thin FDSOI can be a significant issue (11). Further, unlike a PDSOI device, the FDSOI device is very susceptible to charge in the BOX layer, which can capacitively couple through the depleted silicon of the body, changing the front channel threshold voltage. However, FDSOI also has many important advantages over PDSOI, such as higher g_m and reduction of floating body effects including transient V_t shifts and the kink effect (4,7,11).

For subthreshold transistors, the most important FDSOI advantage is the near-ideal subthreshold swing. The drive current in the subthreshold regime is given by:

$$I_{sub} = I_o x10^{([Vgs+\eta Vds]/S)} x(1 - e^{-Vds/Uth})$$

(2)

where I_o is a function of the transistor L and W, η is the DIBL factor, S is the subthreshold swing, and U_{th} is the thermal voltage (1). Note that subthreshold transistor performance is a strong function of η and S in Equation (2), and will be very sensitive to short channel effects (SCE) since both η and S increase as gate length decreases. Buried oxide (BOX) thickness can be varied to achieve a tradeoff between on-current and off-current; thinner BOX improves DIBL but degrades S (1). Simulations using the Atlas device simulator predict that at low operating voltages ($V_{dd} = 0.3$ V), FDSOI devices still provide a superior subthreshold slope to bulk silicon devices, as shown in Figure 1. Therefore an optimized ultra-low-power process technology will greatly benefit from the lower subthreshold swing and capacitances provided by FDSOI.

Channel Doping

The threshold voltage is a strong function of silicon film thickness when using FDSOI with highly doped channels. V_t changes by approximately 4 mV per 1 nm silicon thickness when the doping level is $1x10^{17}$ cm^{-3} (12). By comparison, when the channel doping is extremely low, below ~$5x10^{15}$ cm^{-3}, V_t is effectively independent of silicon film thickness. Since V_t control is critical for subthreshold transistors, it is highly desirable to use undoped (or more precisely, lightly doped) FDSOI, particularly for highly scaled designs (11).

An undoped channel has additional benefits, including no V_t variation due to random dopant fluctuations, as well as higher carrier mobility. Unlike bulk silicon which requires higher channel doping to control the SCEs as gate length scales, thin-body FDSOI is less sensitive to SCEs, and enables the use of undoped channels (8). In addition, the depletion thickness given by Equation (1) is large when the channel is undoped, which allows a more manufacturable silicon thickness to be used while maintaining the benefits of a FDSOI as opposed to a PDSOI device. Furthermore, low channel doping will reduce band-to-band tunneling and increase the S/D breakdown voltage, which could be important when integrating 3.3V I/O transistors on the same chip as the subthreshold transistors (2,11).

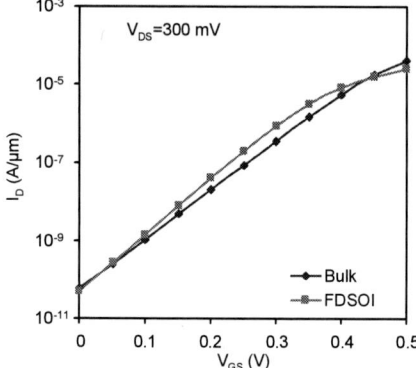

Figure 1: Simulation of ultra low power transistors, for 65nm gate length. FDSOI exhibits improved subthreshold slope and thus a 2.5x improvement in I_{on}/I_{off} ratio at 0.3V operating voltage compared to bulk silicon.

Gate Materials

When simplified to ignore back-channel and short channel effects, the threshold voltage of an SOI transistor is given by:

$$V_t = \phi_{ms} + 2\Phi_f - \frac{1}{C_{ox}}\left(Q_{it} + q\int_o^t \rho(x)dx - qN_{ch}t_{soi}\right) \qquad (3)$$

where Φ_{ms} is the gate-to-semiconductor workfunction difference, C_{ox} is the gate dielectric capacitance, Q_{it} is the dielectric interface charge, $\rho(x)$ is the charge density in the dielectric, and t_{soi} is the silicon thickness. When N_{ch} is high, V_t is a sensitive function of t_{soi}, which is a drawback for thin FDSOI. However, when the channel is undoped, the Fermi potential and the depletion charge are approximately zero, and the expression for the threshold voltage becomes:

$$V_t = \phi_{ms} - \frac{Q_f}{C_{ox}} \qquad (4)$$

where Q_f is the total charge in the gate dielectric. Thus the threshold voltage of the transistor is essentially set by the gate workfunction.

Figure 2 shows a graphical representation of Equation (3), illustrating V_t as a function of N_{ch} for N^+ poly, P^+ poly, and mid-gap gates (10). There are two solutions for achieving a threshold voltage of ~0.35 V, band-edge gate materials with high channel doping ($1x10^{18}$ cm^{-3}), or a mid-gap gate material with very low channel doping ($1x10^{15}$ cm^{-3}). As described in Section III, the undoped channel solution is preferred for subthreshold optimized transistors.

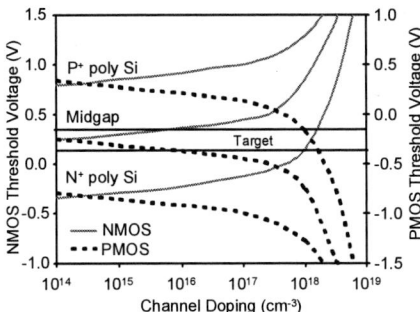

Figure 2: Threshold voltages of FDSOI NMOS and PMOS with mid-gap, N^+ poly, and P^+ poly gates as a function of SOI doping concentration. 10-nm thick SOI and 5-nm gate oxide are assumed. From Shimada (10).

Therefore, a workfunction-engineered mid-gap metal gate material should be used to provide symmetric threshold voltages for NMOS and PMOS. There are several literature examples of successful integration of mid-gap metal gate transistors, including SiGe, Ta , and TiN. (12-17) A metal gate stack typically consists of a thin metal layer sandwiched between a thicker polysilicon layer above, and the gate dielectric below. The gate dielectric may be a conventional SiO_2 or SiON gate oxide, or a high-k gate dielectric such as HfSiON. To prevent GOI issues, it is important that there is little diffusion of metal into the gate dielectric, or reaction between the metal and the dielectric. It is also important to minimize trapped charges in the gate dielectric during plasma sputtering of the metal gate material, which can shift the V_t of the transistor according to Equation 4. Such charging effects have been noted after Ta gate deposition by Ar^+ sputtering, where an oxide charge density of 5×10^{11} cm^{-2} caused a 0.1 V shift in V_t (10).

TiN is a suitable mid-gap metal gate material, with the advantage of being already commonplace in the backend of fabrication flows. A gate stack was fabricated using 4nm-thick SiO_2, 20nm-thick TiN, and 200nm-thick polysilicon. Large (1 mm^2) capacitors were then phosphorous doped and capped with 250nm Al. Figure 3 shows C-V curves comparing poly and two TiN gates with different N_2 flow during TiN PVD. The curves are fit with a quantum-corrected capacitor model from NCSU (18) to extract workfunction, equivalent oxide thickness (EOT), and other parameters. The workfunction of the TiN gates increases toward mid-gap compared to the N^+ poly gates, with Φ_m = 4.45eV and 4.60 eV for 100% N_2 flow and 66% N_2 flow (balance Ar) during TiN deposition. To increase Φ_m further, several post-deposition TiN anneal experiments were performed. A sub-atmospheric 626°C N_2 anneal after TiN deposition increases Φ_m by 0.10-0.15eV, enabling tuning of the effective workfunction across the mid-gap range.

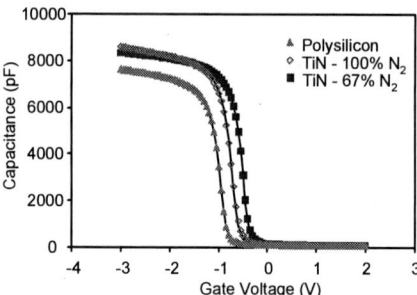

Figure 3: MOS capacitors with TiN metal gates under two different TiN deposition conditions, and a polysilicon gate. TiN gates show an increase in flatband voltage, and thus Φ_{ms}, toward the silicon mid-gap, as well as increased capacitance due to elimination of poly depletion. Lines are quantum-corrected model fits to the data.

Etching the TiN gate metal without damaging the polysilicon above the TiN or the gate dielectric below requires a delicate balance of plasma processing conditions. Lateral etching or notching of the overlying polysilicon could lead to undesirable penetration of implant species into the active channel beneath the gate, causing severe SCE's. Notching of the metal gate may lead to similar implant issues, as well as delamination of the narrower gates. A large foot on the metal must be avoided, as this will cause an undesirable increase in C_{gd}. Plasma etching selectivity to the underlying gate dielectric material is crucial, as punch-through of the thin gate dielectric will cause severe leakage or complete failure of the thin FDSOI device. Microloading effects must be minimized, to ensure that both dense and isolated gates have similar critical dimension (CD) and profile, in order to reduce variation in the transistor parametrics across the chip.

Source/ Drain Underlap Optimization

Eliminating the S/D extension implants (LDD implants) and increasing the spacer thickness results in a gate-to-S/D underlap which provides several benefits for ultra low power operation. Most importantly, reduced overlap capacitance will allow lower CV^2 switching energy. In addition, increased spacer thickness will reduce subthreshold leakage, gate leakage, and DIBL. Simulations have shown that an optimized underlap can yield a 70% reduction in SRAM cell leakage and 200x lower cell read failure probability (8). Since channel hot carrier (CHC) effects are not significant at low V_{dd}, the LDD implants are not required to mitigate gate oxide integrity (GOI) issues.

Simulations of NMOS and PMOS I_d-V_g curves for an xLP device at various gate-to-S/D underlap distances demonstrate that an optimized subthreshold slope and off-current occur for a 60 nm underlap. Figure 5 compares experimental C-V curves from conventional devices, and subthreshold-optimized devices with a 60 nm gate-to-S/D underlap. The capacitance for the underlapped devices is reduced by 71%. Connelly has also described the use of an underlapped S/D technology for ultra-low-power FDSOI, though in that case the underlap was only 4 – 9 nm (19). That work also proposed the use of Schottky S/D to reduce the parasitic resistance of thin-body SOI.

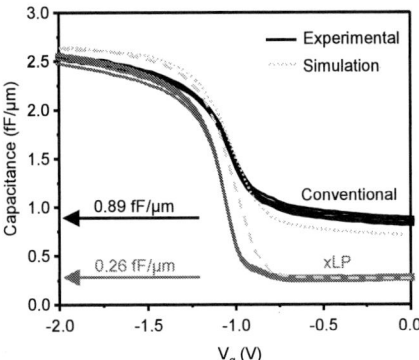

Fig. 5. Comparison of total gate-to-S/D overlap capacitance of conventional and ultra-low power (xLP) PMOS transistors. $L_g = 180$ nm, $W = 100$ μm. Solid lines: Measured data, dashed lines: Atlas model simulation. xLP underlap design reduces capacitance by 71% compared to the standard transistor design

Because of the gate-to-S/D underlap, carrier injection into the channel relies on diffusion from the source. Simulations provided in Figure 6 predict that heavy channel doping causes the drive current to collapse under these conditions, so the channel doping must be kept very light, which is consistent with the requirements for threshold voltage.

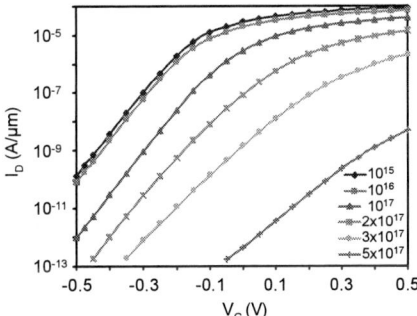

Figure 6: Model NMOS transistor I-V characteristics for 150nm polysilicon gate xLP FDSOI device as a function of body doping (atoms/cm^3). Increasing body doping causes a strong decrease in drive current.

To summarize, a schematic comparison between conventional and subthreshold-optimized ultra-low-power transistors is shown in Figure 7, illustrating the undoped body, elimination of S/D extension implants, and wide spacers.

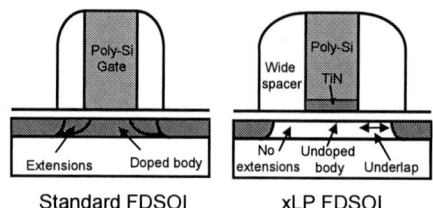

Figure 7: Schematic of standard FDSOI and subthreshold-optimized (xLP) ultra-low-power FDSOI transistors.

Interconnect Optimization

The current through the interconnect routing will be relatively small for ultra-low-power circuits, and the standard interconnect metallization will be significantly oversized. At low current densities, electromigration is not a serious issue, nor is ohmic heating. It is therefore possible to reduce the capacitance of the circuit further by reducing the thickness of the interconnect metal. The increase in interconnect resistance is not significant, since the resistance of the transistor will be much larger than that of the interconnects due to the subthreshold operation.

Though the maximum possible reduction in interconnect thickness will depend on the details of the individual circuit design, a 50% reduction is conservative for most cases. Device simulations have been performed on a transistor driving a given length of interconnect to calculate the CV^2 switching energy and the characteristic switching time of the device, given by $t_c = (R_d + R_m)(C_d + C_m)$, where d and m indicate the device and interconnect metal, respectively. A 50% reduction in interconnect metal thickness reduces t_c by 40%. The simulation predicts that the total CV^2 switching energy of the optimized FDSOI ultra-low-power device is reduced by 97% compared to a traditional 1.2V transistor.

Subthreshold Optimized Transistor Performance

FDSOI ultra-low-power transistors were fabricated with the integration optimized for subthreshold operation as outlined above. NMOS and PMOS I_d-V_g characteristics are shown in Figure 8, for mid-gap metal gate transistors with $L_g = 150$ nm and $W = 8$ μm. Nitride spacer thickness is 90 nm, yielding a 60 nm gate-to-S/D underlap after a 10 s, 1000°C activation anneal. The TiN gate workfunction tuning provides closely matched NMOS and PMOS I_{on} ($V_g = +/- 0.3$ V) and I_{off} ($V_g = 0$ V) performance. It has been proposed that a suitable leakage limit for ultra-low-power handheld electronics is 20-50 pA/μm, (5) which is well above the 4 pA/μm off-current of these FDSOI subthreshold-optimized transistors.

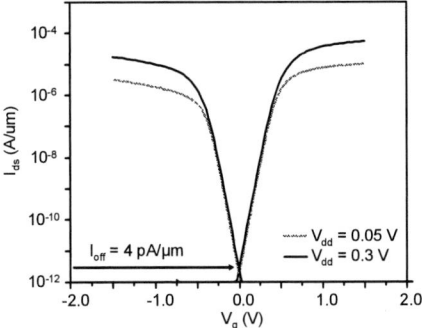

Figure 8. Representative xLP transistor I-V curves with a 150 nm TiN metal gate (W = 8 μm), showing good subthreshold performance, and nearly ideal workfunction tuning.

The subthreshold swing (S) for long-channel mid-gap gate ultra-low-power transistors (L_g = 500 nm) nearly ideal at 64 mV/decade. As the gate length decreases, S increases to 80 mV/decade due to SCE. The subthreshold swing is smaller for PMOS than for NMOS due to the difference in effective channel length for these underlapped devices; under the current process conditions, the As implant used for NMOS apparently diffuses farther than the B PMOS implant resulting in a shorter NMOS channel for a given gate length.

Conclusions and Future Directions

Scaling gate length while maintaining reliable device performance is as challenging for ultra-low-power transistor design as it is for high-performance transistors. For gate lengths below 100nm, SCE's will increase DIBL and subthreshold swing unless the silicon channel thickness is also scaled reduced to ultra-thin values below 15 nm (11). At very thin silicon thicknesses, quantization effects have significant effects and V_t is again a function of t_{si}. Mobility is also degraded, due to higher phonon and surface scattering. If a practical silicon thickness limit of 5 nm is assumed, the minimum gate length which allows acceptable ultra-low-power performance is 25 – 30 nm (11). Beyond this, non-planar devices with enhanced channel control such as FinFETs may be required. Dual- or tri- gate designs will provide better electrostatic control of the channel, minimizing SCE's. Looking forward to these double gate designs, it has been shown that a mid-gap double gate will have lower subthreshold leakage and gate leakage than designs with band-edge gates (8).

The performance of subthreshold-optimized transistors has been demonstrated to meet ultra-low-power performance requirements. By designing the transistor from the substrate through the interconnect levels for subthreshold operation, the switching energy of the device is decreased by 97% with modest impact to the energy-delay product. SOI-based devices have significant advantages over bulk silicon devices for sub-threshold operation. Widespread adoption of FDSOI technology will require significant performance advantage over bulk silicon to justify the higher cost of the SOI substrate and the additional processing steps associated with thinning and control of the active silicon thickness. The processing technology for subthreshold-optimized transistors described in this paper has enabled the verification of some of the performance advantage of FDSOI devices necessary for ultra-low power designs.

References

1. D. Bol, R. Ambroise, D. Flandre, and J.-D. Legar, *2008 IEEE International SOI Conference Proceedings*, p. 57, (2008).
2. A. Uchiyama, S. Baba, Y. Nagatomo, and J. Ida, *2006 IEEE International SOI Conference Proceedings*, p. 15, (2006).
3. B. H. Calhoun, D. C. Daly, N. Verma, D. F. Finchelstein, D. D. Wentzloff, A. Wang, S.-H. Cho, A. and P. Chandrakasan, *IEEE Transactions on Computers*, **54**, 727, (2005).
4. A. Ebina, T. Kadowaki, Y. Sato, and M. Yamaguchi, *Proceedings of the Custom Integrated Circuits Conference*, p. 57, (2000).
5. J. Cai, Z. Ren, A. Majumdar, T. H. Ning, H. Yin, D.-G. Park, and W. E. Haensch, *2008 IEEE International SOI Conference Proceedings*, p. 15, (2008).
6. J.-L. Pelloie, *Microelectronic Engineering*, **39**, 155, (1997).
7. E. Vittoz, "Micropower techniques," in *Design of Analog-Digital VLSI Circuits for Telecommunications and Signal Processing*, J. Franca and Y. Tsividis, Eds. Englewood Cliffs, NJ: Prentice-Hall, 1994.
8. K. Roy, H. Mahmoodi, S. Mukhopadhyay, H. Ananthan, A. Bansai, and T. Cakici, *Proceedings of the 19th International Conference on VLSI Design*, p. 445, (2006).
9. J. L. Pelloie, C. Raynaud, O. Faynot, A. Grouillet, and J. Du Port de Pntcharra, **48**, 327, (1999).
10. H. Shimada, Y. Hirano, T. Ushiki, K. Ino, and T. Ohmi, *IEEE Transactions on Electron Devices*, **44**, 1903 (1997).
11. V. P. Trivedi and J. G. Fossum, *IEEE Transactions on Electron Devices*, **50**, 2095, (2003).
12. T. C. Hsiao, A. W. Wang, K. Saraswat, and J. C. S. Woo, *Proceedings 1997 IEEE International SOI Conference*, p. 20, (1997).
13. H. Shang, M. H. White, *Solid State Electronics*, **44**, 1621 (2000).
14. J. Chen, B. Maiti, D. Connelly, M. Mendicino, F. Huang, O. Adetutu, Y. Yu, D. Weddington, W. Wu, J. Canelaria, D. Dow, P. Tobin, and J. Mogab, *1999 Symposium on VLSI Technology Digest of Technical Papers*, p. 25-26, (1999).
15. H. Shimada, T. Ushiki, Y. Hirano, and T. Ohmi, *Proceedings 1995 IEEE International SOI Conference*, p. 96, (1995).
16. Y. Liu, S. Kijima, E. Sugimata, M. Masahara, K. Endo, T. Matsukawa, K. Ishii, K. Sakamoto, T. Sekigawa, H. Yamauchi, Y. Takanashi, and E. Suzuki, *IEEE Transactions on Nanotechnology*, **5**, 723, (2006).
17. R. Singanamalla, J. Lisoni, I. Ferain, O. Richard, L. Carbonell, T. Schram, H. Y. Yu, S. Kubicek, S. De Gendt, M. Jurczak, and K. De Meyer, *Materials Research Society Symposium Proceedings*, **917**, 174, (2006).
18. N. Yang, W. K. Henson, J. R. Hauser, J. J. Wortman, *IEEE Transactions on Electron Devices*, **46**, 1464, (1999).
19. D. Connelly, P. Clifton, C. Faulker, and D. E. Grupp, *IEEE International Electron Devices Meeting, 2005 IEDM - Technical Digest*, p. 972, (2005).

ECS Transactions, 35 (5) 189-194 (2011)
10.1149/1.3570795 ©The Electrochemical Society

An Analytical Model for the Non-Linearity of Triple Gate SOI MOSFETs

R. T. Doria[a,*], J. A. Martino[a], E. Simoen[b], C. Claeys[b,c], M. A. Pavancello[a,d]

[a] LSI/PSI/USP, University of Sao Paulo, Sao Paulo, Brazil
* e-mail: rdoria@lsi.usp.br
[b] IMEC, Kapeldreef 75, B-3001 Leuven, Belgium
[c] E.E. Dept., KU Leuven, Kasteelpark Arenberg 10, B-3001 Leuven, Belgium
[d] Department of Electrical Engineering, Centro Universitário da FEI, São Bernardo do Campo, Brazil

This work proposes a physically-based analytical model for the non-linearity of Triple-Gate MOSFETs. The model describes the second order harmonic distortion (HD2), usually the major non-linearity source, as a function of the device dimensions, the series resistance, the low field mobility and the mobility degradation factor (θ). The model was applied to transistors of different channel lengths and fin widths and allowed to conclude that θ is the parameter which most contributes for the increase of HD2. The model was validated for both unstrained and strained FinFETs.

I - Introduction

Several recent studies (1) consider multiple gate transistors extremely promising to continue the scaling of the devices beyond the limits established for the planar transistors (2). Narrow triple gate fin shaped devices have particularly shown to be promising. In such devices, the presence of gate electrodes both in the top and sidewall surfaces improves the gate control on the channel charge with respect to the planar ones leading to higher immunity to short-channel effects (1). Figure 1 shows the scheme view of a triple gate or Trigate SOI MOSFET.

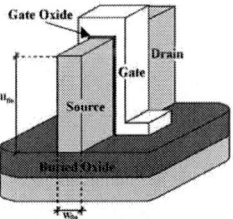

Figure 1: Schematic view of a triple gate FinFET device.

Besides the reduced short channel effects, triple-gate FETs have also shown satisfactory analog properties presenting lower drain output conductance and larger open-loop voltage gain than planar transistors (2-3). Recently, a study of the non-linearities or harmonic distortion exhibited by Trigate FETs operating as single transistor amplifiers demonstrated that in narrower devices the second order distortion (HD2), usually the most important non-linearities source, is better than in wider or planar ones (4).

189

This paper presents the development of a compact analytical model correlating the harmonic distortion to the physical parameters of the triple-gate FETs, such as, low field mobility (μ_0), mobility degradation (θ) and series resistance (R_S). The model was validated for unstrained and strained devices of different dimensions operating in saturation as single transistor amplifiers through the comparison to experimental results.

II - Model Development

The non-linearities exhibited for a system can be modeled through its Taylor series, in which a power series describes the correlation between the input and the output signals. According to (5), the Taylor series for a single MOSFET allows for the correlation of HD2 with the transconductance (g_m) of the device and it first order derivative as a function of the gate voltage overdrive ($V_{GT} = V_{GS} - V_{TH}$, where V_{GS} and V_{TH} are the gate and the threshold voltages, respectively) as expressed in [1].

$$HD2 = \frac{1}{2} Va \frac{\dfrac{dg_m}{dV_{GT}}}{2g_m} \qquad [1]$$

Aiming to clarify the roles of the different physical parameters associated to g_m, a compact model for the drain current (I_{DS}) of triple gate nFinFETs accounting for the series resistance (6) was applied to expression [1]. This model, described in eqn. [2], correlates I_{DS} when the device is operating in saturation to the effective mobility (μ_{eff}), the channel length (L) and the total device width (W = $2H_{fin} + W_{fin}$, where H_{fin} is the fin height and W_{fin} is the fin width) through the parameter K_{eff} ($K_{eff} = \mu_{eff}(\varepsilon_{ox}/t_{ox})$W/L, being ε_{ox} the oxide permittivity and t_{ox} the oxide thickness). The basic model for the effective mobility described in eqn. [3] was applied in order to consider its dependence on the gate bias and θ factor. Considering that $K_0 = \mu_0(\varepsilon_{ox}/t_{ox})$W/L, K_{eff} can be described as $K_0/[1+\theta(V_{GS}-V_{TH})]$. Thus, by applying expressions [2] and [3] in [1], HD2 can be rewritten as in [4].

$$I_{DS,sat} = \frac{K_{eff}(V_{GS} - V_{TH})^2}{2[1 + R_S K_{eff}(V_{GS} - V_{TH})]} \qquad [2]$$

$$\mu_{eff} = \frac{\mu_0}{[1 + \theta(V_{GS} - V_{TH})]} \qquad [3]$$

$$HD2 \approx \frac{Va}{V_{GT}[2 + 3(\theta + R_S \cdot K_0)V_{GT} + (\theta + R_S \cdot K_0)^2 V_{GT}^2]} \qquad [4]$$

III - Model Validation

The HD2 obtained through the model proposed in expression [4] was compared to experimental data from devices of different W_{fin} and L biased at a similar input sinusoidal amplitude (V_a) in Figure 2 considering the input bias of the devices as $V_{GS} = V_{GT} + V_a$ sin(ωt) with ωt varying between 0 and 2π. Triple gate SOI nMOSFETs analyzed were fabricated according to the process described in (7) with channel doping concentration of 10^{15} cm^{-3}, effective gate oxide thickness of 2 nm (2-nm HfO$_2$/1-nm SiO$_2$), L = 10 μm and H_{fin} = 60 nm.

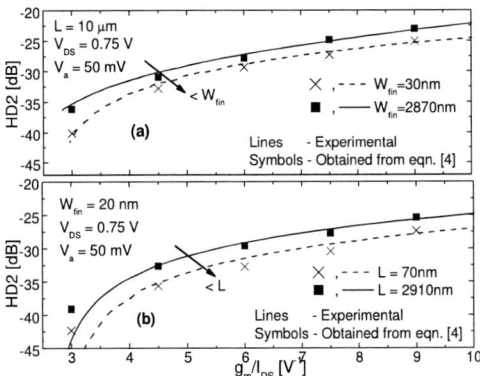

Figure 2: HD2 vs. g_m/I_{DS} obtained from IFM (Experimental) and eqn. [4] for devices of (a) different W_{fin} and (b) different L at $V_a = 50$ mV.

The values applied for $R_S.W$, θ and μ_0 in expression [4] were extracted from the measured devices through the method described in (8) and are presented in Table 1. Measured HD2 was determined through the Integral Function Method (IFM) (9), which allows the non-linearity extraction directly from the DC characteristics of the devices (I_{DS}-V_{GT}) without the need of an AC characterization as in Fourier based methods.

Table 1: Extracted $R_S.W$, θ, μ_0 R_S. K_0 and for devices of several W_{fin} and L.

[nm]	$R_S.W$ [$\Omega.\mu m$]	θ [V^{-1}]	μ_0 [$cm^2/V.s$]	$R_S.K_0$ [V^{-1}]
$W_{fin} = 30$	810	2.87	237	0.033
$W_{fin} = 2870$	338	0.95	304	0.018
$L = 70$	420	5.40	101	1.045
$L = 2910$	420	2.39	240	0.060

As one can see in Figure 2, the compact model describes satisfactorily the HD2 behavior of devices with different dimensions. Data from Table 1 shows that for all the devices evaluated θ is at least five times higher than $R_S.K_0$. Thus, through expression [4] one can conclude that the mobility degradation factor is the physical parameter which most contributes for the HD2 reduction in either narrow or short devices with respect to the wide/long ones. However, as the channel length of the devices is reduced the series resistance becomes more important raising the value of $R_S.K_0$ as can be observed in the table for the transistor of $L = 70$ nm where $R_S.K_0$ results 15 times larger than in the device of $L = 2910$ nm whereas θ increases by a factor of 2 when the channel length is reduced from 2910 nm to 70 nm.

As devices of different dimensions present different open-loop voltage gains (A_V), the model was also validated when the devices are biased aiming at a similar output voltage amplitude (V_{out}), i.e. due to the higher gain narrower devices need lower V_a than the wider ones to attain a similar V_{out}. In Figure 3, the model showed to agree perfectly to the measured results, in which lower HD2 is attained in narrower or longer devices. When an output voltage amplitude is targeted, narrower and longer devices are preferable since they can provide improved values of HD2.

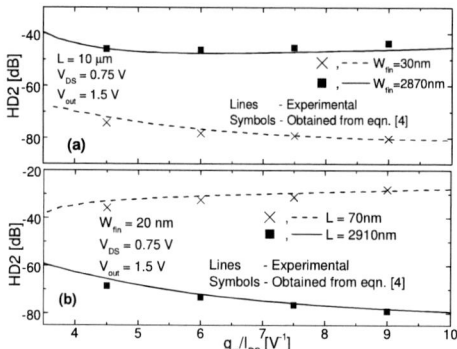

Figure 3: HD2 vs. g_m/I_{DS} obtained from IFM (Experimental) and eqn. [4] for devices of (a) different W_{fin} and (b) different L at $V_{out} = 1.5$ V.

It is known that the different conduction planes usually observed in the top and sidewall surfaces (<100> and <110>, respectively) of a FinFET present different mobility degradation coefficients. The electron's mobility in the plane <110> is severely reduced with respect to the one in the plane <100> by a factor up to 2 (6). In order to boost the carriers' mobility, strain engineering has successfully been applied in the MOS technology. Compressive mechanical stress has shown to increase the mobility of pMOS devices whereas tensile stress improves the mobility of nMOS transistors (10).

Considering that the strain engineering has vastly been used and that the strain influences on the harmonic distortion of FinFETs as demonstrated in (4), the HD2 model proposed in the current work was verified for biaxially strained nMOS FinFETs. The evaluated strained devices have similar dimensions to the ones of unstrained transistors and present tensile stress in both the channel length and width directions. Although produced in sSOI (strained SOI) wafers, the strained devices have also been fabricated according to the process described in (7).

The second order harmonic distortion was obtained for strained FinFETs of different dimensions through both the IFM applied to the experimental I-V curves and the proposed analytic model and is presented in Figure 4 as function of g_m/I_{DS}. As in the case of unstrained devices, in strained transistors the values applied for $R_S.W$, θ and μ_0 of in expression [4] were also extracted from the measured devices through the method described in (8).

As it can be observed through the comparison between Figures 2 and 4, when a similar input voltage amplitude is applied in unstrained and strained devices of similar dimensions, HD2 presents almost no variation indicating that the strain does not influence significantly the distortion. According to Figure 4, HD2 obtained from IFM and determined through eqn. [4] resulted in similar values showing that the proposed model can also be applied to calculate the second order harmonic distortion of biaxially strained FinFETs.

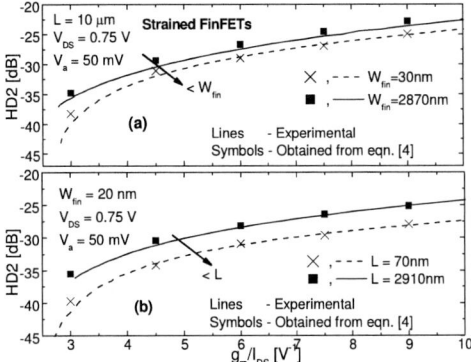

Figure 4: HD2 vs. g_m/I_{DS} obtained from IFM (Experimental) and eqn. [4] for strained FinFETs of (a) different W_{fin} and (b) different L at $V_a = 50$ mV.

The HD2 vs. g_m/I_{DS} characteristic of the strained devices were also observed at a targeted output swing in Figure 5. Thus, the strained FinFETs evaluated were biased aiming to attain an output voltage amplitude $V_{out} = 1.5$ V. According to (3), strained devices present lower intrinsic voltage gains than strained transistors of similar dimensions due to the degradation of the intrinsic Early voltage derived from the channel modulation effect. The different A_V between the devices inherently influences HD2 as stated in (11). The lower intrinsic gain observed in strained FinFETs with respect to the unstrained ones is responsible for the reduction of the output amplitude when both devices are biased at a similar input bias. Therefore, when a similar output amplitude is required strained devices need larger input voltages leading to a higher distortion and worsening HD2. The poorer HD2 obtained in strained devices can be observed when Figures 3 and 5 are compared. However, even in strained transistors biased aiming at a similar output swing, the application of the proposed model to determine HD2 was satisfactory, since expression [4] and experimental data resulted in similar values validating the model for biaxially strained FinFETs.

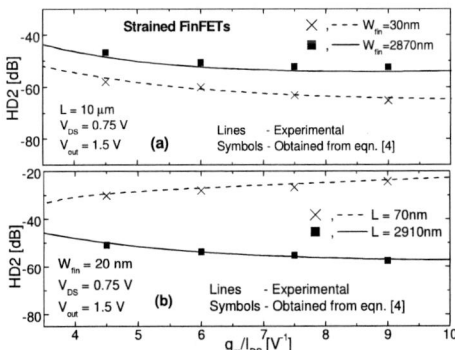

Figure 5: HD2 vs. g_m/I_{DS} obtained from IFM (Experimental) and eqn. [4] for strained FinFETs of (a) different W_{fin} and (b) different L at $V_{out} = 1.5$ V.

IV - Conclusions

This work proposed an analytical model to determine the harmonic distortion of triple-gate transistors. The model is entirely based on the physic of the devices and has associated the second order distortion to four key parameters: the devices dimensions (channel length and fin width), the series resistance, the low field mobility and the mobility degradation. When applied to devices of different dimensions, the model presented satisfactory results agreeing to the experimental curves of HD2. According to the results obtained, the mobility degradation showed to be the parameter which most influences HD2 for devices of any dimensions. The validation of the proposed model could also be observed in biaxially strained triple gate FinFETs, which have presented poorer HD2 than unstrained devices.

Acknowledgments

The authors Rodrigo T. Doria, Marcelo A. Pavanello and João A. Martino thank the Brazilian research-funding agencies FAPESP and CNPq for the financial support.

References

1. J. P. Colinge, *FinFETs and Other Multi Gate Transistors*, Springer, 2008.
2. V. Subramanian, B. Parvais, J. Borremans, A. Mercha, D. Linten, P. Wambacq, *et al.*, *IEEE Transactions on Electron Devices*, **53**, 3071(2006).
3. M. A. Pavanello, J. A. Martino, E. Simoen, R. Rooyackers, N. Collaert, and C. Claeys, *Solid State Electronics*, **51**, 285(2007).
4. R. T. Doria, A. Cerdeira, J. A. Martino, E. Simoen, C. Claeys, and M. A. Pavanello, *IEEE Transactions on Electron Devices*, **57**, 3303(2010).
5. G. Groenewold, and W. J. Lubbers, *IEEE Transactions on Circuits and Systems II.*, **41**, 569(1994).
6. V. Subramanian, A. Mercha, B. Parvais, J. Loo, C. Gustin, M. Dehan *et al.*, *Solid-State Electronics*, **51**, 551(2007).
7. N. Collaert, M. Demand, I. Ferain, J. Lisoni, R. Singanamalla, P. Zimmerman *et al. In: Symposium on VLSI technology digest of technical papers,* 108(2005).
8. P. K. McLarty, S. Cristoloveanu, O. Faynot, V. Misra, J. R. Hauser, J. J. Wortman *et al.*; *Solid-State Electronics*, **38**, 1175(1995).
9. A. Cerdeira, M. Alemán, M. Estrada, and D. Flandre, *Solid-State Electronics*, **48**, 2225(2004).
10. T. Rudenko, N. Collaert, S. De Gendt, V. Kilchytska, M. Jurczak, and D. Flandre, *Microelectronics Engineering*, **80**, 386(2005).
11. S. Sakurai and M. Ismail, *Low-Voltage CMOS Operational Amplifiers Theory, Design and Implementation*, Norwell, MA: Kluwer, 1994.

New Capacitorless Dynamic Memory Compatible with SOI and Bulk CMOS

N. Rodriguez[a], F. Gamiz[a], S. Cristoloveanu[b]

[a] Department of Electronics, University of Granada, 18071 Granada, Spain
[b] IMEP-MINATEC, 38016 Grenoble Cedex 1, France

We introduce a capacitorless DRAM cell based on a multibody transistor concept. The device features a body partitioning with dedicated regions for hole storage and electron current sensing. This separation allows aggressive scaling beyond the limits imposed by the supercoupling effect. The cell is fully compatible with both standard bulk and Silicon On Insulator substrates without changes in its architecture. Numerical simulations demonstrate its performances in terms of 1T-DRAM operation, current margins and retention time.

Introduction

The architecture and operation mechanisms of Silicon On Insulator (SOI) transistors make realistic the advent of capacitor-less single-transistor memory cells (1T-DRAMs). These cells are envisioned to replace the conventional DRAM technology, based on one transistor + one capacitor cells, where the further scaling of the storage capacitor is a blocking issue [1]. Most of the proposed 1T-DRAM cells make use of the SOI technology and the floating body (FB) effect [2], [3]: the amount of holes, stored in the body of the SOI transistor, affects the electrostatic potential inducing a shift in the threshold voltage which defines two distinct current levels ('0' and '1' states). This basic premise has been explored leading to different approaches: (i) Partially depleted (PD) transistor operated in single-gate mode [3]; (ii) Fully depleted (FD) transistors operated in double gate mode [4], and (iii) modulation of the current of the bipolar transistor intrinsic in the SOI MOSFET [5]. 1T-DRAMs are attractive as simplified memory cells where the storage capacitor is suppressed. In spite of the promising potential of the floating body memories as DRAM substitute, they are not exempt from their own issues. In particular, since the 1T-DRAM introduction has been speculated for the 22nm node, it is questioned how they will be scaled to achieve transistor bodies thinner than 30nm [6], [7].

The limitation of the body thickness in the *classical* FB-concept originates from the so-called *supercoupling effect* which inhibits the coexistence of electrons and holes in the same ultrathin body. Below a critical thickness [8], if one interface needs to be accumulated with majority carriers, it will promote the opposite interface also into accumulation. Reciprocally if the front-channel is inverted, the back-channel will also contain electrons, not holes.

The key point to break the supercoupling limitation is to magnify the potential difference between the front and back interfaces without increasing the electric field perpendicular to the channel (i.e bias). In our previous work, the use of a low-k middle insulator layer (Middle Oxide, MOX) has been proposed as an 'absorber' of the electric field moving the supercoupling thickness frontier down to the 10nm range [9]. This attractive solution, named A-RAM, requires a more complex device.

In this work, we propose a different multibody cell, A2RAM, which is able to sustain a large potential difference between the interfaces without the need of an intermediate layer. In addition this new cell can be directly exported to the still overwhelming bulk substrate because the use of the insulator layer below the device is optional. However, we will show that the performance achieved in SOI technology is superior to bulk.

The multibody A2RAM

The breaking difference between A2RAM and any other 1T-DRAMs is the use o two accumulation layers instead of one (Figure 1). One of the accumulation layers (P-type) is devoted to hole storage, whereas the bottom one is used to read the electron current (N-bridge).

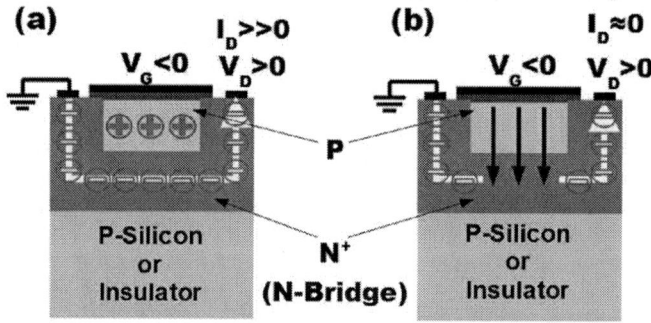

Figure 1. A2RAM schematics and principles. (a) The holes accumulated in the top P-body screen the gate electric field; the N-body (N-bridge) remains undepleted allowing the electron current to flow. (b) If the gate field is not screened due to the lack of holes in the P-body, the N-bridge becomes fully depleted blocking the electron current flow.

The operation principle of this new cell differs from the standard 1T-DRAMs. The '1' state, which is stable in time, is defined when the top P-body is charged with holes. The holes are retained at the gate interface thanks to the supply of a negative gate bias which creates the accumulation layer. In this situation, the electric field (perpendicular to the gate) is screened by the accumulation layer, having a minor effect on the electron concentration in the N-bridge. If the N-bridge is dopped enough (and the shorter the better) the current flowing through the bridge will be significant in this '1' state. Note that no current flows through the P-body from source to drain since the junctions are reverse biased. The '0' state is defined when the top P body is in deep depletion (lack of holes). In this case the electric field from the gate is not screened affecting the carrier concentration of the bridge if the top P-body is thin enough (i.e. <15nm). The central region of the bridge becomes fully depleted by the action of the electric field lines and the current flow is cut.

A2RAM operation

The transient behavior shown in the previous section can be exploited for 1T-DRAM operation. In Figure 2 we show a waveform pattern demonstrating full 1T-DRAM functionality. The top side of the figure corresponds to the voltage signals applied to gate and drain contacts (source and substrate contact are grounded), whereas the bottom side of the figure shows the output drain current of the device. The sequence is as follows: write the '1'; read twice; write the '0'; read twice.

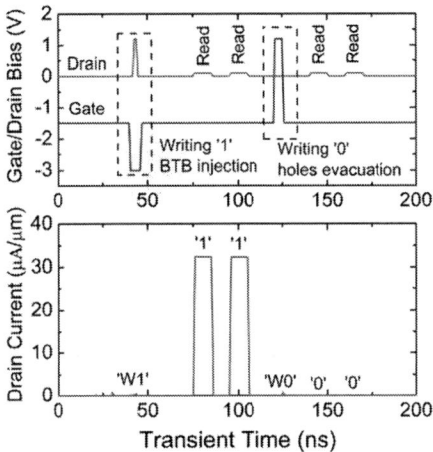

Figure 2. Waveform patterns demonstrating full 1T-DRAM functionality of the A2RAM cell. Top: Gate/Drain bias pattern; bottom: simulated drain current. L=32nm, T_{ox}=2nm, T_{Si-P}=8nm, $T_{Si-bridge}$=10nm.

The '1' state can be written either by impact ionization (I.I.) or band-to-band tunneling (BTBT) [10], [11]. The second mechanisms is best suited for low power applications since a proper synchronization of the pulse can lead two a very limited writing current (negligible in the simulation shown in Figure 2). Nevertheless the use of the BTBT requires a more negative value of the gate bias during the programing as compared with the I.I. alternative, which should be evaluated depending on the technology and particular application.

The cell state can be read simply by slightly increasing the drain voltage and testing the current flow though the source to drain resistor (N-bridge). When the gate field is screened ('1' state), this current is linear with respect to the drain voltage bias and can be modulated according to the discrimination margins between '0' and '1' states, needed for the detection circuits.

The '0' state is written by pulsing the gate voltage from the negative retention value to a positive enough value to forward bias the body-to-source and drain junctions evacuating the accumulated holes (i.e. $V_G=1V$, $V_D=0V$, $V_S=0V$). When the gate bias returns abruptly to the negative retention value (i.e. $V_G=-1.5V$), there is no immediate source to supply the holes that according to the negative gate bias should be accumulated, therefore the potential in the P-body (and the N-bridge) drops. The N-bridge becomes fully depleted increasing its resistance by several orders of magnitude. When the cell state is read again by increasing the drain bias the current flow is negligible as compared with the '1' state current.

Despite being a very thin device ($T_{Si-P}+T_{Si-Bridge}<20nm$) the use of an insulator substrate is optional, since the floating body is created in the top P-body by the action of the gate and the back channel does not need to be enhanced through back gate control. Nevertheless the SOI technology brings additional benefits due to the better confinement of the current in the N-bridge and the better electrostatic integrity. In Figure 3 we show a comparison of the initial current ratio between the states 30ns after a writing sequence (whether '1' or '0'). The ratio has been represented versus the gate retention voltage needed to create the hole accumulation. The more negative this voltage, the better the depletion of the bridge (when the P-body is in deep depletion) and therefore the better the ratio between the two states.

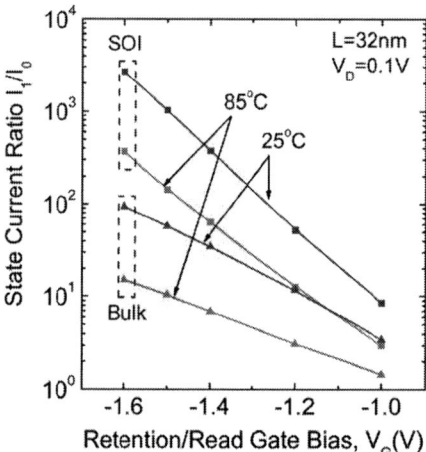

Figure 3. Simulation results of current ratio between '1' and '0' states (I_1/I_0) at 25°C and 85°C. Results on SOI and bulk substrates are compared. L=32nm, $T_{ox}=2nm$, $T_{Si-P}=8nm$, $T_{Si-bridge}=10nm$.

The use of the insulator substrate prevents any leakage current below the *N-bridge* during the '0' state reading. The current ratio between states is enhanced over a 15 factor both at room or high temperature (85°C). Note that the SOI variant of a 32nm-length, 18nm-thick

A2RAM achieves $I_1/I_0 \sim 100$ at 85°C with only a very reasonable value of retention voltage of $V_G = -1.4V$.

Since the '0' state is unstable, it defines the retention time of the cell. Thermal generation, junction leakage and gate-induced drain leakage (GIDL) contribute to supply holes to the P-body after writing the '0', eventually restoring the equilibrium concentration after the deep-depletion. For gate retention voltage below -1V ($V_G < -1V$) and low drain voltage during the reading of the cell, the simulations show that the limiting mechanisms is the GIDL. Otherwise the holes generated by impact ionization will be captured in the P-body becoming the limiting factor (for high V_D during reading).

The evolution of the '0' state current has been studied on the same device as in Figure 4. It has been observed that the source and drain junctions play a major role on the retention time (defined as the time needed to reduce the initial current margin $I_1(t=0)/I_0(t=0)$ by a factor of two). A 3nm underlap in the source and drain with respect to the gate increases the retention time by two orders of magnitude at 85°C.

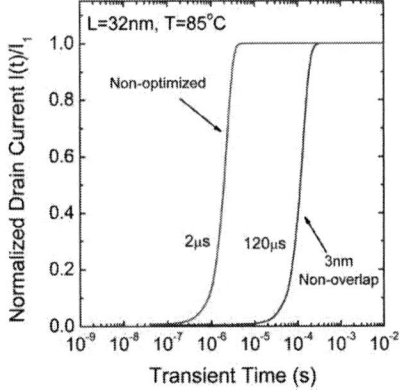

Figure 4. Evolution of the '0' state current (worst case models) under continuous reading operation ($V_D = 0.1V$, $V_G = -1.5V$, T=85°C). A 3nm gate underlap increases the retention time by two orders of magnitude.

Conclusion

The new concept of 1T-DRAM, the A2RAM cell, features N/P body partition which enables the electrical and physical separation of the hole storage and electron current. The hole concentration controls the partial or full depletion of the N-body. Its architecture can be adapted to both bulk and SOI substrates. The cell is compatible with ultimate scaling in single-gate operation and shows attractive performance (long retention, wide memory window, simple programming, non-destructive reading and low power operation) for embedded systems.

Acknowledgments

EuroSOI+, NANOSIL and WCU projects are thanked for support.

References

1. K. Kim, *Microelectronics Reliability*, 40, 11 (2000).
2. H.-J. Wann and C. Hu, *IEDM Tech. Digest* (1993).
3. S. Okhonin, M. Nagoga, J. Sallese and P. Fazan, *Proceedings IEEE International SOI Conference* (2001).
4. M. Bawedin, S. Cristoloveanu, D. Flandre, *IEEE Electron Device Lett.*, 29, 7 (2008).
5. S. Okhonin, M. Nagoga, E. Carman, R. Beffa and E. Faraoni. *IEDM Tech. Digest* (2007).
6. A. Hubert, S. Cristoloveanu, M. Bawedin and T. Ernst, *Proceedings ULIS conference* (2009).
7. T. Tanaka, E. Yoshida and T. Miyashita, *IEDM Tech. Digest* (2004).
8. S. Eminente, S. Cristoloveanu, R. Clerc, A. Ohata and G. Ghibaudo, *Solid-State Electronics*, 51, 2 (2007).
9. N. Rodriguez, F. Gámiz, S. Cristoloveanu, *IEEE Electron Device Letters*, 31, 9 (2010).
10. E. Yoshida, T. Tanaka, *IEEE Trans. Electr. Dev.*, 54, 4 (2006).
11. A. Hubert, M. Bawedin, G. Guegan, S. Cristoloveanu, T. Ernst O. Faynot, *Proceedings European Solid-State Device Research Conference* (2010).

ECS Transactions, 35 (5) 201-210 (2011)
10.1149/1.3570797 ©The Electrochemical Society

Radiation-Induced Pulse Noise in SOI CMOS Logic

Daisuke Kobayashi, Kazuyuki Hirose, Hirokazu Ikeda, and Hirobumi Saito

Institute of Space and Astronautical Science, Japan Aerospace Exploration Agency
3-1-1 Yoshinodai, Sagamihara, Chuo, Kanagawa 252-5210, Japan

A comprehensive study on radiation-induced pulse noises in SOI
CMOS logic is reviewed. The noise pulses are called single event
transients or SETs and becoming a serious source of soft errors in
logic systems. As a result of miniaturization of transistors, concern
about the soft error problems caused by the SETs are growing not
only in special applications for harsh radiation environments like
space but also in usual ones used on the ground. It is important to
reveal what the SET is in its nature. Measurement techniques and
analytical model have been developed for the purpose. They are introduced together with experimental data obtained with test circuits
fabricated by a commercial 0.2-μm fully-depleted SOI technology.
Issues to be solved for use of SOI technologies in realizing radiation
hardened devices are also described with a practical example, or a
development process of radiation hardened SOI SRAMs.

Introduction

Amongst merits of the silicon-on-insulator (SOI) technology, its potential tolerance to transient radiation effects or soft errors is noteworthy. Clear evidence for this can be seen in
its development history. The potential radiation hardness was a main driving force behind
its early development (1). The SOI technology was devotedly researched for military and
space electronics that demand the higher immunity to radiation (2). Around 50 years after this beginning stage, this radiation hard property is now drawing renewed attention.
As a result of the reduction in noise margins caused by aggressive device scaling, present
semiconductor devices suffer from radiation attacks not only in such special applications
but also in those we usually use on the ground — because of terrestrial neutrons (3,4). It
is crucial to ensure radiation hardness for today's and future semiconductor systems; see
for example a next generation Itanium processor by Intel (5). Compared with bulk counterparts, appropriate use of the SOI technology can lead to good radiation tolerance with
smaller costs of protection designs and evaluation tests additionally introduced for ensur-
ing reliability. This could redeem the demerit of the SOI technology, i.e., the higher cost in
substrate preparation.

In this paper, the authors would like to review their recent study on radiation effects on
the SOI technology. The main purpose of this study is to reveal dynamic characteristics of
pulse noises induced by irradiation. Such time-domain information of the pulse is necessary
to solve today's soft error problems. When radiation strikes a logic gate, a transient noise
pulse is generated at its output as shown in Figure 1. This noise pulse is called single
event transient or SET. The SET pulse can propagate over logic gate networks and lead
to corruption of bit information stored in sequential elements like flip-flops (FFs), thus
causing the system to malfunction. SETs are likely to become the dominant source of soft
errors for advanced CMOS logic (6–9). To deal with this SET problem, theoretical and

201

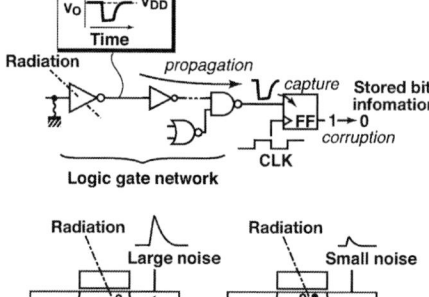

Figure 1. Single event transient (SET): a radiation-induced pulse noise in a logic gate that leads to bit data corruption in sequential elements like flip-flops (FFs).

Figure 2. Radiation effects on the bulk transistor and on the SOI transistor. The buried oxide (BOX) layer can protect drain and source terminals from noise charges created by radiation.

experimental characterizations of SETs in the time domain are essential. For this purpose, the authors have conducted a comprehensive study employing developments of a physics-based analytical model and techniques for observations and estimations.

The rest of this paper is organized as follows. First the authors would like to emphasize that the use of the SOI structure has a potential to decrease radiation tolerance, as well as to increase. Appropriate use is necessary to reap the benefit. The following part then describes a review of the authors' study on SETs in an fully-depleted SOI technology.

Radiation effects on SOI technology — merits and demerits

In this section, the authors would like to give a short review of radiation effects on the SOI technology. It is widely said that the use of the SOI structure is a good way to realize radiation hardened devices because the buried oxide (BOX) layer can prevent noise charges created by radiation from intruding into signal nodes such as drain and source terminals, as illustrated in Figure 2. This is really a merit of the SOI technology but the problem is not that simple. In some cases, the use of SOI results in a decrease in radiation hardness.

First, accumulation effects or total dose effects should be considered. A part of charges created by irradiation in the BOX layer can be trapped therein. Holes in particular remain easily because of their slow mobility. The trapped holes make the backside parasitic channel turn on, for n-type devices for example, thus resulting in an increase in the standby leakage and a shift of transistor threshold voltages. This total dose effect is more severe to the fully-depleted (FD) SOI structure than to the the partially-depleted (PD) counterpart because the latter has a neutral region at the back interface. This neutral region makes it difficult to turn on the backside channel and also mask the electrostatic effect of the trapped charge.

The second point we should take care is about the transient radiation effect. Imagine for example an OFF-state n-type SOI transistor. As illustrated in Figure 3, when a radiation particle strikes its body region, the parasitic npn-bipolar transistor, which consists of the

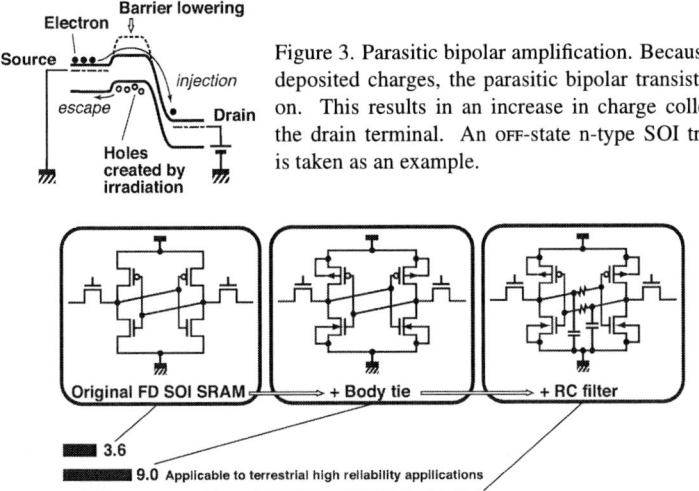

Figure 3. Parasitic bipolar amplification. Because of the deposited charges, the parasitic bipolar transistor turns on. This results in an increase in charge collected at the drain terminal. An OFF-state n-type SOI transistor is taken as an example.

Figure 4. Evolution of SOI SRAM cell structures. LET stands for linear energy transfer, which is one of the main parameters describing the radiation strength. Note that the 1-MeV·cm²/mg radiation particle generates 10-fC noise charges along its track in Si per unit length, 1 μm.

series connection of the source, body, and drain regions, turns on (10). This results in an increase in charge collected at the drain terminal. This charge amplification effect is called parasitic bipolar amplification. Typically the PD structure exhibits stronger amplification effects than the FD one (11).

The last one to be remembered is also about the transient radiation effect. Generally speaking, the main advantage of the SOI structure lies in a reduction in junction capacitance. From the view point of the transient radiation effect, however, this capacitance reduction results in a decrease in tolerance. The circuits become more sensitive to transient radiation noises.

These deterioration effects should be carefully taken into account to realize radiation hardened circuits with the SOI structure. As a good practical example on these issues, the authors would like to introduce the development process of their radiation hardened SOI SRAMs, which are based on a commercial 0.2-μm FD SOI technology. As mentioned above, FD SOI transistors are usually sensitive to the total dose effect. The first thing the authors had to do was to assess this effect. The authors fabricated an SRAM circuit and conducted an irradiation test (12). Experimental results demonstrate that the studied FD SOI technology has enough strong total dose tolerance: the SRAM operates properly up to a dose of 40 krad(SiO₂), which is higher than the typical dose requirement for space applications, or 10 krad(SiO₂). The authors also found in this test that the SRAM has low tolerance to the transient radiation effect. The memory cells lose their bit information by ion irradiation with a low energy of 3.7 MeV·cm^2/mg, which is comparable to one reported for an SRAM fabricated on a 0.25-μm PD SOI technology (13). Note that this data loss phenomenon in memories is called single event upset or SEU. This low SEU tolerance for the studied FD SOI SRAM is attributed to the parasitic bipolar amplification and the small junction capacitance. Adding an electric contact to the body region, or a body tie, increases its resistance to 9.0 MeV·cm^2/mg (12). This is because the parasitic bipolar amplification is suppressed by increasing the escape speed of noise charges generated in the body region (14). The observed tolerance is enough high for terrestrial applications, which need to deal with the natural neutrons. A neutron irradiation test clearly demonstrates that the FD SOI SRAM with the body tie structure exhibits a strong improvement of the SEU tolerance compared with bulk counterparts (15). The tolerance observed here is, however, still lower than the required value of 40 MeV·cm^2/mg for space applications. This requirement is achieved by inserting RC filters into the feedback network in the cell (16). Note that the area penalty for the RC filters, each of which consists of lightly doped polysilicon resistor and MOS capacitor, is acceptably small because the SOI structure combined with the body tie makes transient noises enough small. This evolution of the cell structure is summarized in Figure 4. As explained here, it is not so simple to develop radiation hardened circuits even with the SOI technology.

Single event transients in SOI CMOS Logic

As mentioned in the introduction section, today's logic circuits that are miniaturized by scaling become sensitive to transient noise pulses of SETs. This section reviews the authors' activities to reveal what the SET is in its nature.

Pulse-width measurements

Temporal widths of SETs are the most important parameter in evaluating its bit-distortion ability. The longer pulse can be easily captured by FFs, as explained in a theoretical formula (17). Measurements of the pulse widths are essential. A development of dedicated circuitry referred to as SNAPSHOT in this study enables efficient measurements of the width (18).

Figure 5(a) describes its basic operation principle. A chain of copies of a target logic gate (a NOR-gate chain in this example) is connected to the SNAPSHOT. An SET pulse generated at one of the gate are fed to the SNAPSHOT, and its width is digitized and recorded as a digital bit sequence. Figure 5(b) shows a circuit diagram of the SNAPSHOT. This circuitry operates conforming to a similar principle of time-to-digital converters (19, 20). Each bit recorded in this system corresponds to a delay time determined by the delay unit, a pair of inverters. The delay time is estimated through a calibration process with a built-in pulse generator, which can generate pulses with designed temporal widths. With the estimated delay time, the recorded digital bit sequence is decoded into the width in time.

One of key ideas in this system is the use of a self-triggering mechanism. The SET pulse itself works as a clock signal (CLK) to activate all storage elements (D-FFs) when it reaches at the node X, which is indicated in the figure(b). This self-triggering mechanism is essential to deal with uncertainty in time of SET generations, which are unpredictable, uncontrollable radiation-induced events.

Another important point is the concurrent use of two SNAPSHOT circuits, which are denoted by A and B in the figure(a). The purpose of this dual structure is to detect only SET

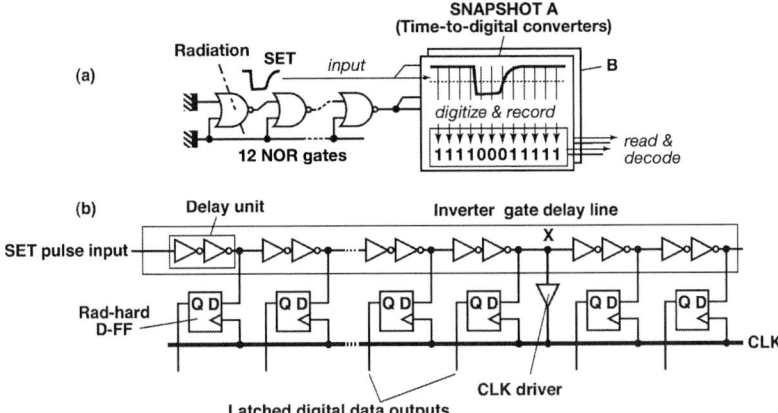

Figure 5. SET pulse-width measurements with a dedicated circuit named SNAPSHOT. (a) Basic measurement principle. (b) Circuit diagram of the SNAPSHOT circuit.

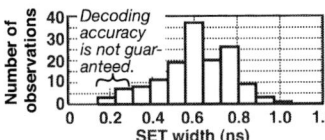

Figure 6. SET pulse-width measurements in 0.2-μm FD-SOI CMOS NOR gates under 256-MeV Ni irradiation with fluence of 2×10^8 particles/cm^2 [2].

pulses originated in the logic gates under test. When the entire region of the circuit is irradiated, SETs can be also generated at the inverter delay units in the SNAPSHOT. They should be discarded. When either SNAPSHOT A or B captures a pulse, the data correspond to these unwanted pulses. Note that the D-FFs are fabricated with the radiation-hardened design (SOI with the body ties and RC filters) in order to prevent the data loss due to a radiation attack to the D-FF. These circuit designs enable autonomous pulse-width measurements with a broad continuous radiation beam, i.e., efficient measurements.

Figure 6 shows a typical example of experimental results — a distribution histogram of observed pulse widths for a 0.2-μm FD-SOI CMOS NOR (18). The NOR gates were irradiated with high-energy Ni ions from an accelerator in the Brookhaven national laboratory. The Gaussian-like distribution is attributed to the differences in position where the ions hit. Amongst related works with similar measurement architecture (21–24), this is the first published experimental result that is taken under heavy ion irradiation.

Another experimental result is shown in Figure 7, which was obtained with an FD-SOI CMOS inverter at an irradiation facility of the Takasaki Ion Accelerators for Advanced Radiation Application (TIARA) in Japan Atomic Energy Agency (JAEA). A pulse-width distribution with a 90-nm bulk technology reported by Narasimham et al. (25) is superimposed for comparison. Compared with the bulk inverter, the FD SOI one provides lower radiation sensitivity (or lower cross sections) as well as shorter SET pulses. Figure 8 also indicates that the FD SOI technology can generate shorter pulses compared with bulk counterparts. With this measurement technique, moreover, the authors have revealed a satura-

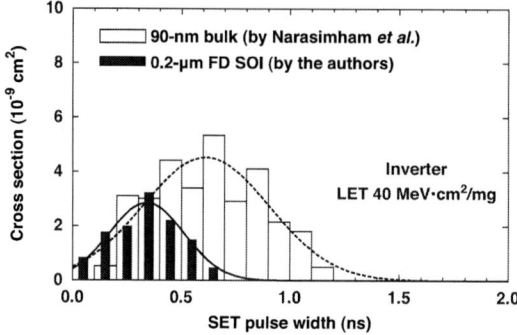

Figure 7. Histograms of SET-pulse widths for inverters irradiated by heavy ions. Experimental data for an inverter fabricated with a 90-nm bulk technology, which are reported by Narasimham *et al.* (25), are also plotted. Gaussian distributions are estimated and superimposed on the original figure reported in the authors' previous work (26) for eye guides.

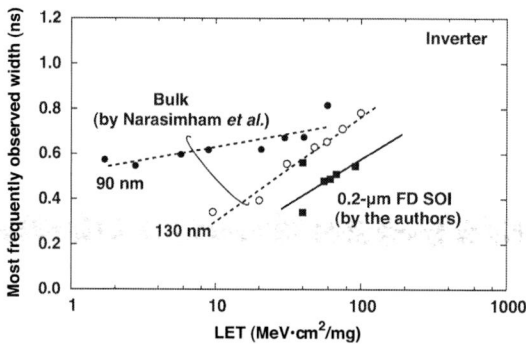

Figure 8. Comparison in SET-pulse widths experimentally observed in inverters under heavy ion irradiation. Experimental data for inverters fabricated with 90-nm and 130-nm bulk technologies, which are reported by Narasimham *et al.* (25), are compared with the authors' work with the 0.2-μm FD SOI technology. Gaussian distributions are assumed, and their mean values (pulse widths correspond to the peak locations) are plotted. Straight lines are superimposed for eye guides. The original figure was first reported in a Ph.D. presentation by Makino at the Graduate University for Advanced Studies 2008 (unpublished).

tion tendency for the pulse width (27) and validated the theoretical calculation technique for estimating bit-distortion probability (28, 29).

Full waveform observations

Although the pulse-width measurements with the SNAPSHOT circuit provide useful information efficiently, they can evaluate only pulse-width parameters and unfortunately impose a cost on circuit design. A waveform observation technique the authors developed is another approach that compensates for these disadvantages (30). Figure 9 shows its measurement principle. It has a simple and compact circuit structure. An output node of a target

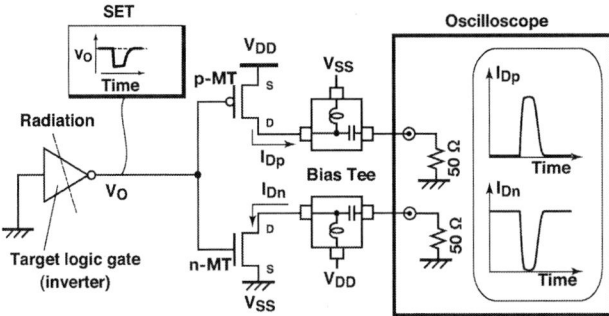

Figure 9. Monitoring transistor technique for full waveform observation of SET pulses.

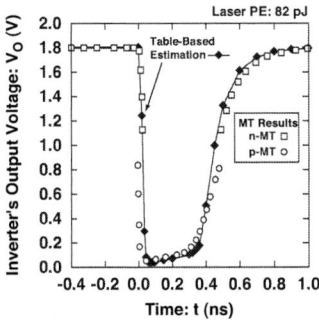

Figure 10. Full-waveform of an SET in an SOI CMOS inverter measured by the monitoring transistor technique (30). The pull-down nMOS transistor was irradiated by a pulsed laser beam. An waveform estimated from radiation responses of a single transistor, a copy of the irradiated nMOS, is superimposed.

logic gate, e.g., a low-input inverter, is connected to the gate inputs of two elementary MOS transistors, which are denoted by "monitoring transistors or MTs". Measurements are performed using a conventional 50-Ω transmission line technique, which is widely used for measuring transient characteristics of SOI transistors (31, 32). The drain terminal of each MT is connected to a 50-Ω input of an oscilloscope through a bias tee, which supplies a constant drain-source voltage. An SET generated at the target logic gate directly changes each MT's gate-source voltage and thus its drain current. This transient change in current is recorded by the oscilloscope and can be converted backwards into values in voltage through the drain current vs. gate voltage characteristics for each MT.

Figure 10 (symbols) shows an observation example. An SET waveform in an FD SOI CMOS inverter irradiated by a pulsed laser beam is plotted. Pulsed laser beams are widely used as a good approximation of ion strikes. The authors used in this study a facility in Naval Researcch Laboratory, USA (33, 34). This is the first published full waveform of SETs in a logic gate fabricated with today's miniaturized technology. The line in the figure indicates a waveform estimated with table-based modeling (35–37). This modeling enables estimation of SET waveforms in various types of logic gates from single transistor responses to radiation.

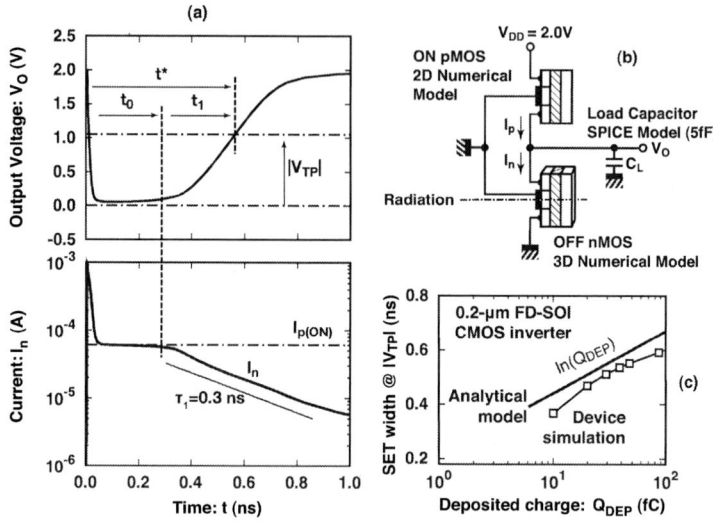

Figure 11. (a) Simulated SET in an FD SOI CMOS inverter and current transients for elementary transistors. (b) Circuit diagram. (c) Analytical model of SET width as a function of the amount of noise charges deposited in the body region. Comparison with device simulation results.

Analytical model

To gain a theoretical picture of SET pulses, we have developed an analytical model (38, 39). Figure 11(a) shows a simulated SET in an SOI CMOS inverter. Drain current transients for the nMOS and pMOS are also plotted. The simulated circuit is illustrated in Figure 11(b), where two elementary transistors are modeled with numerical device models while metal interconnections and load capacitors with theoretical SPICE models. These device- and circuit-level models are concurrently solved with a three-dimensional mixed-mode circuit-device simulator (Sentaurs TCAD from Synopsys). The SET waveform or voltage transient exhibits a typical rail-to-rail trapezoidal pattern. The output voltage levels off at around the ground for a while. As reported by Ferlet-Cavrois et al. (40), the current transient also exhibits a plateau at the level of $I_{p(ON)}$, which represents the maximum ON-state current through the pMOS. The duration of the current plateau, t_0, corresponds to the period while the SET pulse levels off at the ground level. From the simulations (38), the authors have drown a conclusion that the duration of the current plateau corresponds to the storage time observed in the turn-off behavior of saturated bipolar junction transistors (41–43). The duration t_0 is thus expressed as

$$t_0 = \tau_0 \ln\left(\frac{Q_{DEP}}{I_{p(ON)}\tau_0}\right),$$ [1]

where Q_{DEP} and τ_0 respectively represent the initial amount of noise charges deposited by irradiation and their escape speed from the body region. Combining t_0 with the charging time t_1, the authors have developed the following expression for the SET pulse width t^*, which is defined at the voltage level of $|V_{Tp}|$,

$$t^* = \tau_0 \ln\left(\frac{Q_{\text{DEP}}}{I_{p(\text{ON})}\tau_0}\right) + \tau_1 + \frac{C_L|V_{Tp}|}{I_{p(\text{ON})}}, \tag{2}$$

where V_{Tp} and τ_1 respectively represent the threshold voltage of the pMOS and the decay time of the current transient as shown in Figure 11(a).

Figure 11(c) demonstrates the validity of the model. The estimated pulse width from the analytical model (straight line) shows good agreement with the device simulation results quantitatively as well as qualitatively. The straight line relationship on the semi-logarithmic chart are also confirmed in experimental results with pulsed laser beams (39) and with heavy ions (Figure 8).

Summary

The authors reviewed in this paper their study on radiation-induced pulse noises in an FD SOI CMOS logic. The pulse noise is called single event transient or SET. Concern about the soft errors caused by the SETs are increasing as the device miniaturization progresses. It is urgent to understand what the SET is in its nature. For the purpose, the authors have developed an autonomous pulse-width measurement technique, a full waveform observation technique, and an analytical model. Various experimental results have been reported and compared with the theoretical model. In this study the authors focuses on a fully-depleted SOI technology because its potential of the higher radiation immunity but they believe that the developed measurement techniques and models could be basically applicable to the other technologies such as the partially-depleted SOI technology and also the bulk counterpart. Demerits of the SOI technology for use in radiation hardened applications are also described, as well as the merits.

Acknowledgments

This work has been supported by a lot of researchers. The authors are indebted to the following people for their kind support, in particular: S. Ishii, D. Takahashi, M. Kusano, K. Yamamoto, and Y. Kuroda of Mitsubishi Heavy Industries Ltd., Japan; S. Onoda, T. Hirao and T. Ohshima of Japan Atomic Energy Agency; V. Ferlet-Cavrois and P. Paillet of CEA, France; D. McMorrow of Naval Research Laboratory, USA; Y. Arai of the High Energy Accelerator Research Organization (KEK), Japan; M. Ohono of OKI Electric Industry, Co. Ltd., Japan. Large parts of this work were conducted by Y. Yanagawa and T. Makino in their Ph.D. activities. This work was supported in part by the Japan Society for the Promotion of Science under Grant-in-Aid for Scientific Research 18560359 and by the Japanese Ministry of Education, Culture, Sports, Science and Technology under a Grant-in-Aid for Young Scientists (B): 20760228.

References

1. J. B. Ku and K.-W. Su, *CMOS VLSI Engineering: Silicon-on-Insulator (SOI)*, Springer (1998).
2. H. Ryssel, R. Schork *et al.*, *Microelectronic Engineering*, **22**(1–4), p. 315 (1993).
3. J. F. Ziegler, *IBM Journal of Research and Development*, **40**(1), p. 19 (1996).

4　T. Nakamura, E. Yahagi *et al.*, *Terrestrial Neutron-Induced Soft Errors in Advanced Memory Devices*, World Scientific Pub. Co., Inc. (2008).

5　B. Stackhouse, S. Bhimji *et al.*, *IEEE J. Solid-State Circuits*, **44**(1), p. 18 (2009).

6　S. Buchner, M. Baze *et al.*, *IEEE Trans. Nucl. Sci.*, **44**(6), p. 2209 (1997).

7　S. P. Buchner and M. P. Baze, "Single-event transients in fast electronic circuits," in *2001 IEEE NSREC Short Course Notebook ch. V.*

8　P. Shivakumar, M. Kistler *et al.*, in *Proc. IEEE/IFP DSN*, p. 389 (2002).

9　T. Uemura, Y. Tosaka *et al.*, *Japan J. Appl. Phys.*, **45**(4B), p. 3256 (2006).

10　L. W. Massengill, J. David *et al.*, *IEEE Electron Device Lett.*, **11**(2), p. 98 (1990).

11　V. Felret-Cavrois, G. Gasiot *et al.*, *IEEE Trans. Nucl. Sci.*, **49**(6), p. 2948 (2002).

12　K. Hirose, H. Saito *et al.*, in *IEEE Radiation Effects Data Workshop Rec.*, p. 48 (2001).

13　C. Brothers, R. Pugh *et al.*, *IEEE Trans. Nucl. Sci.*, **44**(6), p. 2134 (1997).

14　K. Hirose, H. Saito *et al.*, *IEEE Trans. Nucl. Sci.*, **51**(6), p. 3349 (2004).

15　J. Baggio, V. Ferlet-Cavrois *et al.*, *IEEE Trans. Nucl. Sci.*, **52**(6), p. 2319 (2005).

16　K. Hirose, H. Saito *et al.*, *IEEE Trans. Nucl. Sci.*, **49**(6), p. 2965 (2002).

17　D. Alexandrescu, L. Anghel *et al.*, in *Proc. IEEE DFT*, p. 99 (2002).

18　Y. Yanagawa, K. Hirose *et al.*, *IEEE Trans. Nucl. Sci.*, **53**(6), p. 3575 (2006).

19　Y. Arai, T. Matsumura *et al.*, *IEEE J. Solid-State Circuits*, **27**(3), p. 359 (1992).

20　T. Watanabe, Y. Makino *et al.*, *IEICT Trans. Electron.*, **E76-C**(12), p. 1774 (1993).

21　M. Nicolaidis and R. Perez, in *Proc. IEEE IRPS*, p. 56 (2003).

22　M. Nicolaidis, U.S. Patent 7 126 320 B2, Oct. 24, 2006.

23　B. Narasimham, V. Ramachandran *et al.*, *IEEE Trans. Device Mater. Rel.*, **6**(4), p. 542 (2006).

24　P. Gouker, J. Brandt *et al.*, *IEEE Trans. Nucl. Sci.*, **53**(6), p. 2854 (2008).

25　B. Narasimham, B. L. Bhuva *et al.*, *IEEE Trans. Nucl. Sci.*, **54**(6), p. 2506 (2007).

26　T. Makino, D. Kobayashi *et al.*, *IEEE Trans. Nucl. Sci.*, **56**(6), p. 3180 (2009).

27　T. Makino, D. Kobayashi *et al.*, *IEEE Trans. Nucl. Sci.*, **56**(1), p. 202 (2009).

28　Y. Yanagawa, D. Kobayashi *et al.*, *IEEE Trans. Nucl. Sci.*, **56**(4), p. 1958 (2009).

29　T. Makino, D. Kobayashi *et al.*, *IEEE Trans. Nucl. Sci.*, **56**(6), p. 3180 (2009).

30　D. Kobayashi, K. Hirose *et al.*, *IEEE Trans. Nucl. Sci.*, **55**(6), p. 2872 (2008).

31　K. A. Jenkins and J. Y.-C. Sun, *IEEE Electron Device Lett.*, **16**(4), p. 145 (1995).

32　T. Saraya and T. Hiramoto, in *Proc. IEEE Int. SOI Conf.*, p. 84 (1999).

33　J. S. Melinger, S. Buchner *et al.*, *IEEE Trans. Nucl. Sci.*, **41**(6), p. 2574 (1994).

34　D. McMorrow, J. S. Melinger *et al.*, *IEEE Trans. Nucl. Sci.*, **52**(6), p. 2104 (2005).

35　D. Kobayashi, H. Saito *et al.*, *IEEE Trans. Nucl. Sci.*, **54**(4), p. 1037 (2007).

36　D. Kobayashi, K. Hirose *et al.*, in *Proc. IEEE SELSE* (2007).

37　D. Kobayashi, K. Hirose *et al.*, *IEEE Trans. Nucl. Sci.*, **54**(6), p. 2347 (2007).

38　D. Kobayashi, T. Makino *et al.*, in *Proc. IEEE IRPS*, p. 165 (2009).

39　D. Kobayashi, K. Hirose *et al.*, *IEEE Trans. Nucl. Sci.*, **56**(6), p. 3043 (2009).

40　V. Ferlet-Cavrois, P. Paillet *et al.*, *IEEE Trans. Nucl. Sci.*, **53**(6), p. 3242 (2006).

41　J. L. Moll, *Proc. IRE*, **42**(12), p. 1773 (1954).

42　R. Beaufoy and J. J. Sparkes, *ATE J.*, **13**(4), p. 310 (1957).

43　S. M. Sze and K. K. Ng, *Physics of Semiconductor Devices*, 3rd ed., John Wiley & Sons, Inc. (2007).

CHAPTER 9

MEMS AND PHOTONICS-1

212

A Tunable Color Filter Using Sub-micron Grating Integrated
with Electrostatic Actuator Mechanism

H. Miyao, K. Takahashi, M. Ishida, and K. Sawada,

[a] Department of Electrical and Electronic Information Engineering, Toyohashi University
of Technology, Toyohashi, Aichi 441-8580, JAPAN

We propose a NEMS (Nano Electro Mechanical Systems) tunable
color filter using sub-micron grating, which can modulate the
reflected light by changing the period of the sub-micron grating
pixel by means of electrostatic actuation. The NEMS electrostatic
actuator could be designed a high mechanical resonance frequency
with a relatively low drive voltage. The sub-micron grating was
made in a top layer of an SOI (Silicon on Insulator) wafer. The tiny
anchors were covered by parylene N during the sacrificial release
process using buffered hydrofluoric acid. The color tuning from
yellow to green was demonstrated at 20 V operation.

Introduction

MEMS (Micro Electro Mechanical Systems) has good compatibility with micro optics
in a sense that even a small motion can deliver significant optical effect. In particular,
sub-wavelength gratings are expected to have many advantages such as a high selectivity of
the wavelength and refractive index change in small region. Color filters using the sub-
wavelength grating have been fabricated using EB lithography [1] and nanoimprint
technology [2] which can be achieved to show three primary colors with high reflectivity
and transmissivity.

On the other hand, tunable filters using pitch variable sub-micron grating are expected
to deliver superior performances in terms of high diffraction efficiency and large tuning
range. Previously reported tunable filters include an electrothermal type [3] and an
electrostatic comb-drive type [4]. However, an actuator footprint was relatively large
compared to the filter area. We newly developed a pitch variable sub-micron grating
integrated with an electrostatic parallel-plate mechanism. Compared with the previous
other report, our design can accommodate the actuator mechanism and sub-micron
grating in small area, which can be used for display application. In this paper, a design,
fabrication and experimental result of the tunable color filter using sub-micron grating
with electrostatic actuator are reported.

Design and Fabrication

Figure 1 schematically illustrates a proposed structure and an operation principle of
tunable color filters using a pitch controllable sub-micron grating. The wavelength of the
reflected light is decided by the grating width and gap. The grating gap is changed by the
electrostatic force, which results in changing the wavelength of the reflected light. The
thicker part in the middle is the grating reflector, and the thinner parts near the ends are

suspensions. The grating slips are suspended over the substrate with both end anchored in a bridge style, and electrostatic force is used to tune the mutual distance.

The filter architecture was constructed by sub-micron grating integrated with parallel plate actuator mechanism. Therefore, the effective filter area is large compared with conventional MEMS actuators. In addition to this, the proposed actuator could be obtained a high mechanical resonance and a low drive voltage thanks to the small mass and electrostatic gap.

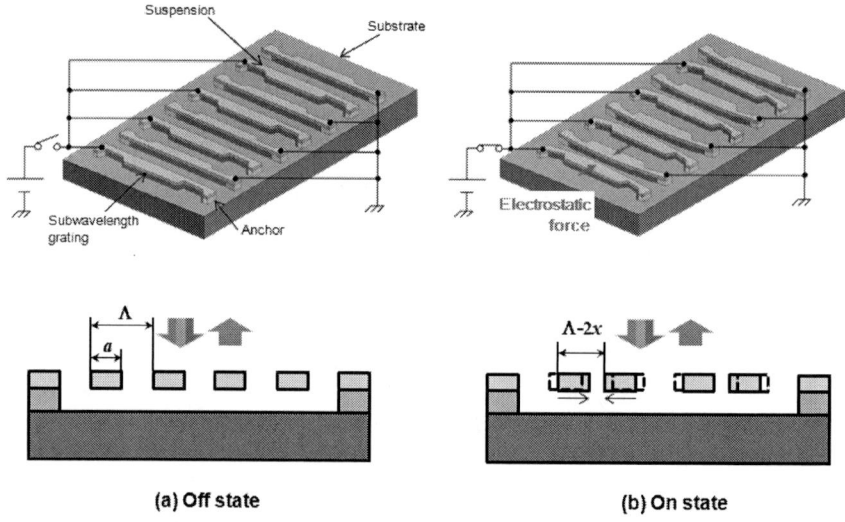

Figure 1. Schematic operation principle image of tunable color filter using a pitch controllable sub-micron grating (a) Initial state (b) Operated state.

An electrostatic actuator to drive sub-micron gratings was simulated by FEM. Grating dimensions are 20 μm long, 200 nm wide and 150 nm thick with 200 nm gap. Suspensions are designed 10 μm long and 150 nm wide. The grating slip with two fixed end was found to move without bending, as shown in Figure 3 (b). Typical displacement curve as a function of applied voltage is shown in Figure 3 (c). An electrostatic gap-closing actuator based on parallel plate has nonlinear pull-in behavior; a movable electrode is brought into contact to a counter electrode by one third displacement of initial gap. Therefore, the tuning range of the electrostatic gap was 66 nm. The proposed NEMS actuator was operated below 5 volt. The color filter could achieve an ultralow-power device. The resonant frequency of the actuator was 775 kHz. It's corresponding to sub-micron sec response speed, which is enough for display application.

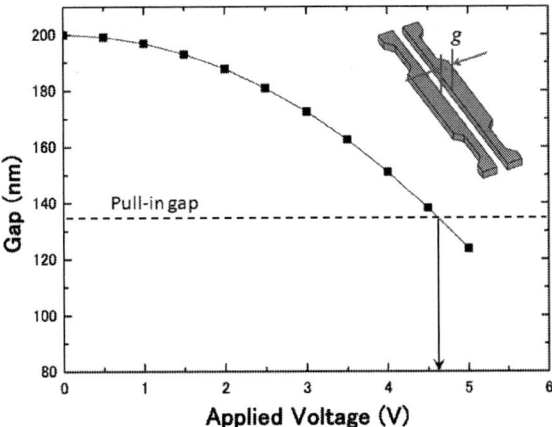

Figure 2. Schematic operation principle image of tunable color filter using a pitch controllable sub-micron grating (a) Initial state (b) Operated state.

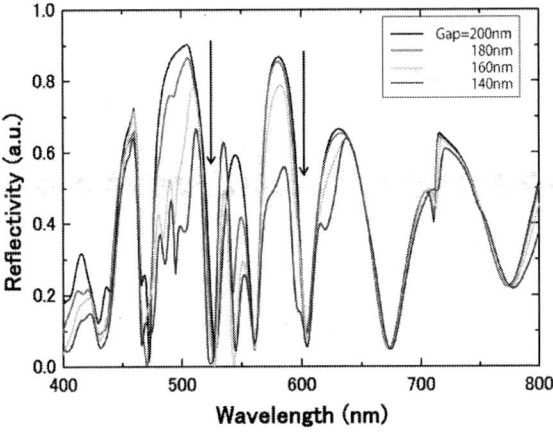

Figure 3. Reflectivity of the tunable color filter calculated by RCWA method..

In our design, the sub-micron grating width and height are fixed, and the air gap is controlled for changing the wavelength of the transmitted light by NEMS electrostatic actuators. Optical simulation using RCWA (Rigorous Coupled Wave Analysis) method indicates that the wavelength peak of the transmitted light is decreased by 40 % and 30 % at 580 nm and 500 nm wavelength with 60 nm displacement. The simulation results show not only three primary colors but intermediate color such as yellow and orange are obtained by several tens nm mechanical displacement.

Process overview is shown in Figure 4; an SOI thickness used in this work was 150 nm, and the BOX thickness was 1 μm. (a) EB resist (ZEP520A-7 ZEON) was spin-coated on the SOI wafer. The subwavelength gratings were patterned by EB direct writing. (b) Then, the SOI layer was etched by the RIE of SF6. (c) The electrostatic actuator was designed in an array of a bridge style with tiny anchors which were same width of sub-micron grating. Therefore, we should cover the anchor area in sacrificial etching. We used Parylene N as the passivation film because it was known to provide uniform coverage in nano gap. The parylene N was deposited and patterned by Oxygen plasma with alminum mask. (d) The sacrificial layer was etched by buffered HF, and then the supercritical CO_2 drying technique was used for reliable release the nano scale structures. Finally, the passivation film was removed by oxygen plasma.

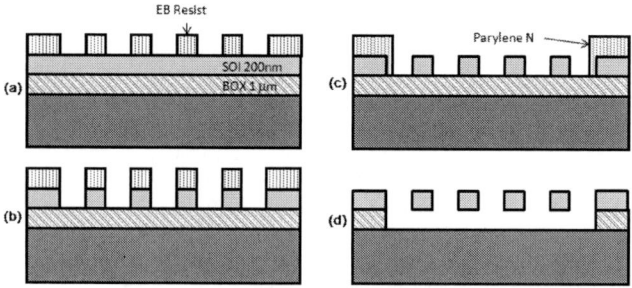

Figure 4. Fabrication process (a) Sub micron grating was patterned by EB direct writing. (b) The grating was etched by RIE. (c) Parylene N was deposited and patterned for passivation on anchor area. (d) Sacrifice layer was etched by BHF.

Figure 5 (a) shows an SEM image of the developed sub-micron grating integrated with the parallel actuator mechanism. The device size was 50 x 65 μm. The grating beam was 500 nm wide, 150 nm thick, and 65 μm long with the gap of 400 nm. Figure 5 (b) is close-up view of the NEMS grating. The gap of the suspension area was designed to be 600 nm wide gap and 400 nm narrow gap, which works the electrostatic attraction force for intended direction. The tiny anchors were 500 nm wide, as shown in Figure 5 (c). They were completely protected from the buffered HF by parylene N.

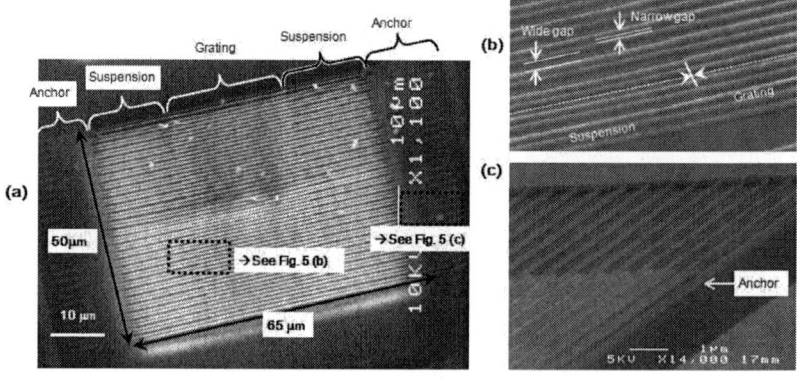

Figure 5. (a) SEM picture of the NEMS tunable color filter (b) close-up view of sub micron grating and (c) tiny anchors.

Experimental Result

Figure 6 is optical microscope images of the fabricated NEMS tunable color filter. The grating and suspension beam was found yellow and pink at the rest position, respectively. Drive voltage of 20 V and ground level were given to the grating slips every other place. The color of grating area was changed into green. On the other hand, the color of suspension area was found nonuniform color due to the nonuniform gap. In conclusion, the analog color tuning was demonstrated by electrostatic actuation of the sub-micron grating.

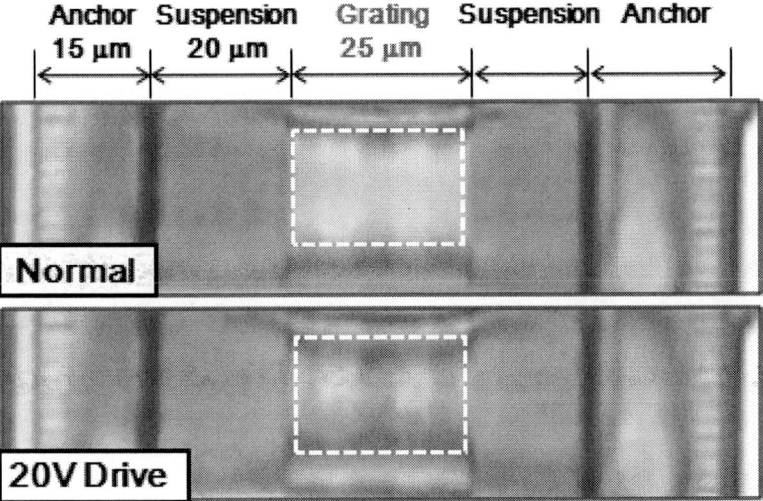

Figure 6. Optical microscope photograph of NEMS tunable color filter in operation. (a) OFF state (b) ON state

Conclusion

We have proposed the NEMS tunable color filter using sub-micron grating, which can modulate the reflected light by changing the period of the sub-micron grating pixel by means of electrostatic actuation. The NEMS electrostatic actuator could be designed a high mechanical resonance frequency with a relatively low drive voltage. The sub-micron grating was made in a top layer of an SOI (Silicon on Insulator) wafer. The tiny anchors were covered by parylene N during the sacrificial release process using buffered hydrofluoric acid. The color tuning from yellow to green was demonstrated at 20 V operation. The device performance such as resonant frequency and drive voltage can be improved by minimizing the grating period.

Acknowledgment

This work was supported in part by Global COE Program "Frontiers of Intelligent Sensing" from the Ministry of Education, Culture, Sports, Science and Technology, Japan.

References

1. Y. Kanamori, M. Shimono, and K. Hane, *IEEE Photonics Technology Letters*, **18**(20), 1136 (2006)
2. Y. Kanamori, H. Katsube, T. Furuta, S. Hasegawa, and K. Hane, *J. J. Appplied Physics*, **48**, 06FH04, (2009)
3. X. M. Zhang, Q. W. Zhao, T. Zhong, A. B. Yu, E. H. Khoo, C. Lu, A. Q. Liu, Proc. Tranceducers & Eurosensors '07, June. 10-14, 2007, Lyon, France, pp.2417-2420 (2007)
4. K. Hane, T. Kobayashi, F. R. Hu, and Y. Kanamori, *Applied Physics Letters*, 88, 141109 (2006)

CHAPTER 10

MEMS AND PHOTONICS-2

220

Nanomechanical testing of free-standing monocrystalline silicon beams

U. Bhaskar[a], S. Houri[a], V. Passi[a], T. Pardoen[b], J.-P. Raskin[a]

[a] Institute of Information and Communication Technologies, Electronics
and Applied Mathematics
[b] Institute of Mechanics, Materials and Civil engineering
Université catholique de Louvain, B-1348 Louvain-la-Neuve, Belgium

A fabrication process to characterize single crystalline silicon microbeams under uniaxial tension is presented. The microbeams subjected to an uniaxial tensile stress are released without stiction owing to a critical point drying step. Based on the deformation measured by scanning electron microscopy images, the corresponding strain and stress are extracted to provide the mechanical response of monocrystalline silicon. A complementary method based on the determination of the flexural resonance frequency of the released beams is discussed and compared to the results obtained with the first direct method.

Introduction

The study of the mechanical properties such as the Young's modulus, fracture strength and ductility of materials is a longstanding research topic. Polysilicon is extensively used as a structural material in current electromechanical systems (MEMS) that are manufactured by surface micromachining. Previous work presented by several researchers in the literature revealed considerable differences in the measured values of the Young's modulus and strength of polycrystalline silicon. Variations of the Young's modulus ranging from 123 GPa (1) to 190 GPa (2) and of the fracture strength ranging from 1 GPa (for measurements under tension) to 2.7 GPa (measurements under bending), suggested that the main reasons for such large variations involve: (i) the accuracy of the experimental setup and techniques and/or (ii) the difference in the nature of the materials being tested in these experiments. With the advent of Silicon-on-Insulator (SOI) substrates, monocrystalline silicon (Si) is finding an increased use not only in device fabrication but also in MEMS/NEMS systems (3-4) and would eventually replace polycrystalline silicon in many applications.

In this paper, we present a technique to fabricate microbeams from the top-silicon layer of a SOI wafer and to apply a mechanical load through the presence of a low pressure chemical vapor deposited silicon nitride (LPCVD-SiN), as well as two methods to extract the stress in the Si beams. The aforementioned technique (5-7) is designed to draw on the potential force reservoir of internal stresses present in silicon-nitride (referred to hereon as "actuator" material) to mechanically load/deform thin films of interest (single crystal silicon in our case). The first step to design an elementary tensile stage (micromachine) is to fabricate a cantilever beam in single crystal silicon followed by fabrication of another cantilever beam (relatively stiffer by construction) of silicon-nitride to partially overlap the silicon beam at the area of interest (Figure 1a). The essential idea here is to use the displacement induced upon release of the silicon-nitride beam, owing to its internal stress relaxation, to mechanically deform the attached silicon sample (Figure 1b). The intensity of mechanical loading or stress developed in the silicon beam is directly related to the ratio between the geometries (cross-sections and lengths) of the silicon-nitride and silicon beams and can be qualitatively modeled by simple equations (7). Hence, for a fixed geometry, every single tensile stage is designed to impose a pre-defined value of strain and by varying the lengths of these tensile stages, it is possible to determine the maximum load that can be sustained by the silicon beam.

Strain measurements are performed by measuring the displacement of cursors using scanning electron microscopy. Besides this static characterization technique, a complementary approach for determining stress based on the measurements of flexural resonance frequencies is proposed. Measuring the resonance frequencies of cantilevers and clamped-clamped beams is a widely used technique to determine the Young's modulus and the internal stresses in thin films (8). Dynamic characterization of MEMS devices based on resonance frequency measurements is adapted here to identify the stress inside each silicon beam structure.

Fabrication

SOI wafer with top-silicon thickness of 200 nm, buried-oxide of 1 μm and bulk-substrate of 780 μm-thick is considered as starting material. After cleaning the wafer in piranha mixture followed by de-ionized water (DIW) rinse, a short HF-2% dip is performed in order to remove the chemical oxide. Thermal oxide is grown on the wafers at 900°C to obtain a thickness of 20 nm. Positive resist is spin-coated on the wafer and baked at 120°C for 2 minutes. Using optical lithography patterns consisting of beams along with pads are exposed. After development of the resist, HF-5% dip is performed in order to etch the thermal oxide. The resist is removed by immersing the sample in acetone followed by rinsing in iso-propanol and DIW. Using thermal oxide as a mask the top-silicon is etched using chlorine chemistry. An over-etch time of 10 s is given to ensure complete etching of top-silicon. Ellipsometer measurement is made to measure the thickness of the buried-silicon-dioxide (SiO_2) after etching which was measured to be 1000 nm. Thermal oxide is removed by immersing the wafer in HF-2% for 2 minutes followed by DIW rinse. After cleaning the wafers using piranha mixture, SiO_2 of 10 nm is grown at 900°C followed by positive resist spin coating. Optical lithography is performed to define windows in the resist and etch openings into the thermal SiO_2 by dipping the wafer in BHF for 15 s. This SiO_2 acts as an etch stop layer during the SiN etching which is deposited in the coming steps. After resist stripping and standard cleaning, 360 nm of high stress silicon nitride (SiN) is deposited using low pressure chemical vapor deposition (LPCVD) process at 770°C. The value of the residual tensile stress measured by wafer curvature was around 1 GPa. Beams along with anchor pads are then exposed in order to overlap the region where the silicon beams are. Along with this, there are reference beams which are exposed in order to measure the displacement after releasing the wires. SiN is patterned using CF_4 chemistry. Release of silicon and SiN (by etching the buried-oxide) is carried out using HF-73% of concentration in order to reduce the etching of stoichiometric silicon nitride. Followed by wet release the samples are dried using critical point dryer in order to avoid any stiction, this fabrication process is schematically shown in Figure 1. 70 nm of aluminium was deposited on the released beams so as to have a good laser reflection for measurement of stress using resonance method described hereafter.

Figure 2a shows an optical image of a successfully released silicon structure with displacement of the cursors. Figure 2b shows fracture occurrence in silicon beams after release, indicating that the imposed deformation can be larger than the fracture strain of Si. Figure 3a-3b show SEM images of the micromachines without or with fracture. The buried silicon-dioxide is etched using HF-73% and released using critical point drying.

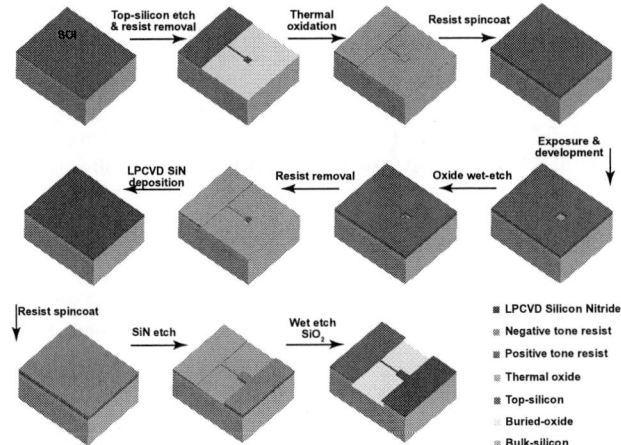

Figure 1. Complete process steps for the fabrication of microstructures.

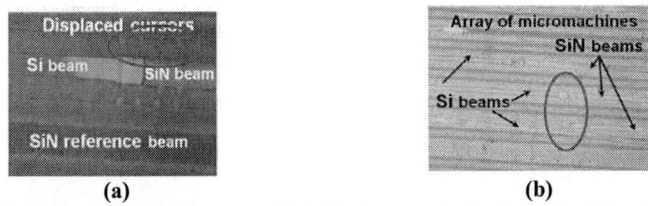

(a)   (b)

Figure 2. Optical microscope images of (a) displacement of a single test structure, (b) an array of test structures which have encountered fracture.

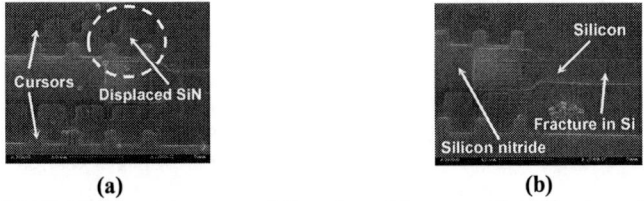

(a)   (b)

Figure 3. (a) Displacement in successfully released beams without fracture, (b) fracture occurrence in released silicon microbeams.

Measurement of stress using the resonance method

Resonance measurements are performed optically using an LDV (Laser Doppler Vibrometer), while the vibration excitation is induced via a piezoelectric disk on top of which the sample die is glued. Measurements are carried out in a vacuum chamber (<1 mTorr) involving an optical window. The choice of Al as a reflective layer stems from the facts that it has a low density, low Young's modulus, zero internal stress upon deposition, and a low deposition temperature which make the modeling of the reflective layer easier. Finally, the displacement of the cursors (u), measured by scanning electron microscopy, provides the value of the imposed strain $\varepsilon = u/L$, where L is the length of the silicon microbeam. Since the microstructures presented here are composite dual beam structures for which it is not possible to find a simple analytical formula relating the stress to the resonance frequencies as it is the case with clamped-clamped beams, modal analysis based on the Rayleigh-Ritz technique is done in order to find the axial stress in the structures. Rayleigh-Ritz technique is a widely used and well known method for the calculation of the natural frequencies of vibration for beams (9), plates and tapered plates (10). A modified Rayleigh-Ritz approach was developed in order to calculate the natural frequencies of structures that have both geometrical and material discontinuities (11). This approach was later on applied for MEMS modeling and analysis. The method first considers that the deflection of a beam, in our case a beam segment, can be described as the sum of a series of orthogonal basis functions that satisfies the boundary conditions of the structures (9), these basis functions are orthogonal polynomials generated using the Gram-Schmidt orthonormalization rule (12). Therefore, the deflection W^n of the segment n can be expressed as follows:

$$W^n = \sum_i A_i^n \Phi_i^n \qquad [1]$$

where Φ_i^n is the ith basis function for the n beam segments and A_i^n is the weighing factor corresponding to the specified segment and basis function. In the case treated here n takes only the values of 1 and 2 since there are only two beam segments for each structure. The natural frequencies of the system are those frequencies where the Rayleigh quotient goes to zero, expressed as follows:

$$\sum_{n=1}^{2} \sum_i \partial [U_{Total} - T_{Total}] / \partial A_i^n = 0 \qquad [2]$$

where U and T are the potential and the kinetic energies of the structure, formulated as described in ref (9). It is possible to separate the total potential and kinetic energies of the structures into the following terms.

$$U_{Total} = U_{Silicon} + U_{Silicon\ Nitride} + U_{Aluminium} + U_{Axial\ Force} \qquad [3]$$

where the first three terms on the right hand side represent the silicon, silicon nitride, and aluminum elastic potential energy, respectively, and the last term represents the potential energy created by the stress in the microstructure. And for the kinetic energy:

$$T_{Total} = T_{Silicon} + T_{Silicon\ Nitride} + T_{Aluminium} \qquad [4]$$

where the three right hand side terms represent, respectively, the kinetic energy of silicon, silicon nitride, and aluminum. Details regarding the theory, the systems of equation, and the modeling of these structures can be found in ref. (13). Once the resonance frequencies are measured, estimating the stress becomes an inverse problem regarding the system of equations described in [2] and having stress as a variable. At this point any numerical root finding algorithm will work with the resulting parametric matrix.

Results and discussion

The magnitude of strain ($\varepsilon = u/L$) imposed on each silicon beam is directly inferred from measuring the displacement u of the released beams using a scanning electron microscope (SEM). The relative difference in displacement, between a free SiN cantilever beam and a SiN-Si clamped-clamped beam yields an estimate of the mechanical constraint imposed by the presence of the silicon beam, see equations in (7). The static characterization technique exploits the direct relationship between the constraint imposed by silicon and the stiffness ratios of the SiN and Si beams, for evaluating the stress developed in the silicon beam. This is facilitated by the knowledge of the SiN and Si beam geometries (by design and by SEM measurements), along with the prior characterization of the Young's modulus of SiN. Figure 4a is a plot of the stress and strain measured from ~200 samples of 200 nm-thick silicon. A linear fit of the data yields a Young's modulus of 160 GPa in close correspondence with the literature (nix). Additionally, a maximum fracture strain of ~3.4% is measured, which shows a significant increase in comparison to the fracture strength of bulk silicon wafers (~0.2%).

Dynamic measurements are additionally performed to independently evaluate the stress developed in the released silicon beams. Figure 4b compares the results for both the static and dynamic measurement methods. The filled circles refer to points from the measurement of stress and strain performed using the static method, whereas the empty circles refer to points whose stress values are measured using the dynamic method. The dynamic measurements provide a slope (Young's modulus) of 162 GPa. When the dynamically extracted data plot is extrapolated to coincide with zero strain we obtain the residual stress in the SOI wafer which is calculated to be 24 MPa. This value is close to the values reported in the literature (14). For the specific test structures shown in Fig. 4b, there is a significant limitation on the calculation accuracy of the stress through static measurement largely due to the displacement measurement precision for low imposed strains, and due to the fact that for the specific stiffness ratio of the SiN and Si and the strain values imposed, the constraint imposed on the actuator is very small, see ref (7) for a comprehensive discussion on error analysis of the technique. Furthermore, the observed dispersion in the Young's modulus evaluated from static measurements is suspected to be due to the presence of an additional aluminum layer which has not been incorporated into the constitutive equations used to evaluate the stress in silicon (7).

The results obtained from the dynamic technique offer a much lower dispersion of values and thus a much lower experimental error in the evaluation of stress independent of the range of imposed strain. This can be explained by the fact that the static stress estimation technique depends on the deformation measurements to find the stress, while dynamic stress estimation offers an independent means where the stress estimation errors and the strain measurement errors are basically uncorrelated. Hence, in summary dynamic measurements serve as a complementary technique to static measurements towards more accurate estimation of the Young's modulus.

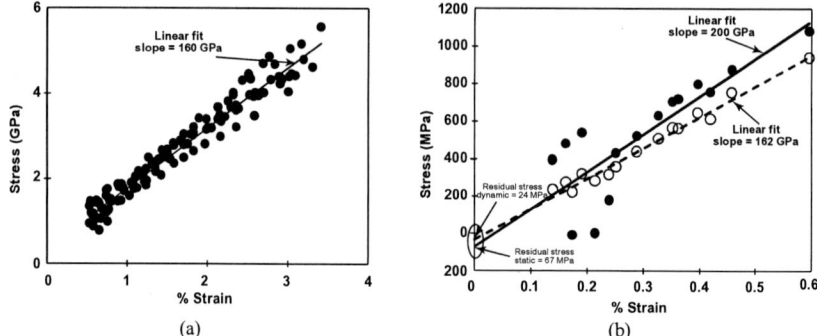

(a)   (b)

Figure 4. Comparison of stress measured using the static and dynamic methods. The filled circles represent the results obtained using the static method (SEM images), and the empty circles represent the stress measurements for the dynamic method (beam vibration).

Conclusions

Single crystalline silicon microbeams were fabricated using top-down approach and were subjected to uniaxial tensile stress with the use of silicon nitride. Critical point drying tool was used to release the microstructures without any stiction. Displacement measurements were performed by scanning electron microscopy and the stress was calculated. A statistically averaged plot of the imposed stress and strain (calculated by static measurements) yielded a Young's modulus value of 160 GPa. Additionally, dynamic measurements of the stress were performed using a resonance based method and the Young's Modulus of silicon was estimated to be 162 GPa. Dynamic extraction method is particularly interesting to overcome the intrinsic limitations of the static measurement for evaluating low values of applied stress.

References

1. Y.-C. Tai, R. S. Muller, *IEEE Journal of Microelectromechanical Systems*, pp. 147-152, 1990.
2. J. A. Walker, *et al., Journal of Electronic Materials*, vol. 20, pp. 665-670, 1991.
3. X. Li, *et al., Applied Physics Letters*, vol. 83, pp. 3081-3084, 2003.
4. Y. Zhu, *et al., Applied Physics Letters*, vol. 86, 2005.
5. A. Boe, *et al., Thin Solid Films*, 518, pp. 260-264, 2009.
6. T. Pardoen, *et al., Material Science Forum*, vol. 633-643, pp. 615-635, 2010.
7. S. Gravier, *et al., IEEE Journal of Microelectromechanical Systems*, vol. 18, no. 3, pp. 555-569, 2009.
8. J. S. Burdess, *et al., IEEE Journal of Microelectromechanical Systems*, vol. 6, no. 4, pp. 322-328, 1997.
9. G. J. Kynch, *British Journal of Applied Physics*, 8, pp. 64-73, 1957
10. R. B. Bhat, *Journal of Sound and Vibration,* 114, pp. 65-71, 1987.
11. K. Y. Lam, *et al., Applied Acoustics*, 28, pp. 49-60, 1989.
12. R. B. Bhat, *Journal of Sound and Vibration*, 102, pp. 493-499, 1985.
13. S. Houri, *et al.,* "Modal Analysis of Multi-Beams MEMS Structures for the Extraction of Stress-Strain Curves of Thin Films", *to be submitted.*
14. M. A. Hopcroft, et al., *IEEE Journal of Microelectromechanical Systems*, vol. 19, no. 2, 2010.
15. C. Maj, *et al.,Proc. of Eurosensors XXIII conf,.* pp. 429-432, 2009.

Silicon photonics devices based on SOI structures

S. Itabashi[1], K. Yamada[1], H. Fukuda[1], T. Tsuchizawa[1], T.Watanabe[1], H.Shinojima[1], H.ishi[1], R.Takahashi[1], Y.Ishikawa[2] and K.Wada[2]

[1] Microsystem Integration Laboratories, NTT Corp.,
3-1, Morinosato-Wakamiya, Atsugi, 243-0198 Japan
[2] Department of Materials Engineering, The University of Tokyo, Hongo 7-3-1, Bunkyo, Tokyo 113-8656, Japan

> This paper presents our recent progress with the development of silicon (Si) photonic devices utilizing silicon-on-insulator (SOI) wafers. Si photonics has emerged as an attractive technology for developing innovative and inexpensive optoelectronic devices. Si photonics enables us to make optical components small due to the high index contrast of Si and SiO_2. Therefore, ultrasmall electrical and optical components can be fabricated and integrated on SOI wafers. Si photonic devices will have various applications in telecommunications, optical interconnections, and quantum information systems.

Introduction

Silicon (Si) photonics has drawn considerable interest as an emerging technology for optical telecommunications and microelectronics. Photonic devices made of Si have many advantages, such as ultrasmall size, low cost, and convergence with electronic devices. This is because Si technology is an industrial standard that can provide low-cost, high-performance devices through high-density integration and mass-production techniques. And Si shows superior characteristics as a material, such as, ease of handling due to chemical stability and optical transparency in the C-band. Furthermore, fine pattern fabrication technology for Si is well developed. Si photonic devices are based on Si wire waveguides, which are composed of a submicrometer-scale Si core and SiO_2 cladding. The silicon-on-insulator (SOI) structure is very suitable for making these waveguides. A Si layer is fabricated as the core and A SiO_2-BOX layer works as the cladding. Because the Si layer of the SOI wafer is made of single crystal, the light scattering due to defects in the Si core or on the Si surface is suppressed, which reduces propagation loss. Various ultracompact passive optical and dynamic functional devices have already been developed using Si wire waveguides (1-7). In addition, optical branches, wavelength filters, modulators (8-12), and photodetectors (PDs) (13-15) can be fabricated on SOI wafers by using CMOS compatible processes. We can now make various individual Si photonics devices on a SOI wafer, and some of these devices are already at a level suitable for practical application. Integration is very important technique for Si photonics (16-20). Therefore, the next stage is to integrate them on a chip. In this paper, we describe our progress with the fabrication of Si photonic devices on SOI wafer and their integration. We also show a few telecommunications applications of monolithically integrated devices on SOI wafers.

Siliconphotonic Devices and Their Integration

Si Wire Waveguide based on SOI Structure

Si wire waveguides are the basic components of the Si photonics platform. Figure 1(a) shows the cross-sectional structure of our Si wire waveguide, which is a single-mode channel type. The waveguide consists of a Si core and silica-based cladding and is fabricated using SOI. We used SOI wafers composed of 200-nm-thick silicon layer and 3-μm-thick SiO_2 buried layer. The Si wire waveguide has a very high refractive index contrast, with delta of about 40%. This is very advantageous for high-density integration because it allows core width of less than a micrometer and a bending radius of just a few micrometers. The thickness and width of our waveguide core are roughly in the ranges of 200-300 and 300-600 nm. The overcladding SiOx, which is formed by chemical vapor deposition (CVD), is 3-6-μm thick. Serious issues for such waveguides had been large propagation loss and large coupling loss between the waveguides and external optical fibers.

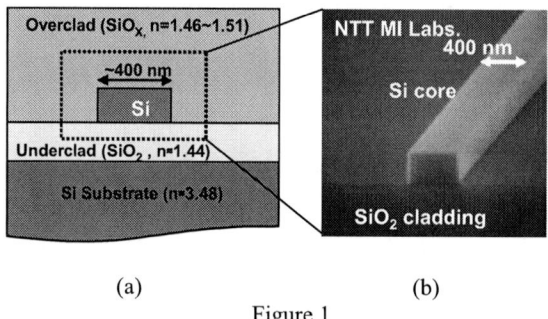

(a)   (b)

Figure 1
(a) Si wire waveguide structure based on SOI.
(b) Fabricated waveguide, showing smooth sidewalls.

The light propagation loss of the waveguides largely depends on the dimensions and sidewall roughness of the Si core. Therefore, we have made improvements to planar fabrication processes (21), such as lithography and etching, to reduce those losses. Electron beam (EB) and optical lithography equipment are used to form the waveguide patterns on resist. Although the LSI patterns consist largely of straight lines, waveguide patterns require curved ones. Therefore, we optimized the EB shots suitable for smooth curved lines. In the etching process, the smooth resist patterns must be transferred to Si faithfully without adding roughness. We have utilized electron cyclotron resonance (ECR) plasma etching with a fluoride gas mixture. Low-energy ions irradiated in the ECR plasma make smooth sidewalls because there is little damage and hardly any redeposition of reaction products during etching. As a result, our recent propagation loss is 1.2 dB/cm at Si waveguides, which enables various kinds of optical element fabrication and integration. Figure 1(b) shows a SEM micrograph of a fabricated waveguide.

The next problem was the large coupling loss between Si waveguides and optical fibers, which results from the large difference in core size. Then we have solved this problem by introducing a spot-size converter (SSC) consisting of Si and SiOx waveguides on SOI wafers. Figure 2 shows the structure of the SSC. The SSC we developed has a gradual adiabatic taper which is covered with a SiOx waveguide, which has a 3-μm-square core and an index contrast of about 3%. Finally, the coupling loss becomes around 0.5 dB per connection (1). These results enable us to evaluate ultrasmall optical elements and integrated ones.

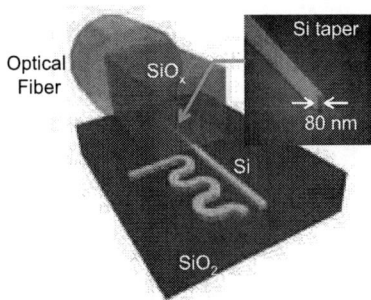

Figure 2 Spot-size converter, composed of tapered Si and a SiOx waveguide.

Passive and Active Functional Optical Components

In this section, we show the passive and active functional components we have developed. Branches are indispensable photonic components for constructing optical circuits. Several branch structures, such as the y-branch, multimode interferometer (MMI) and the directional coupler, are well known. The fabrication tolerance is an important factor for Si photonics because submicrometer waveguides are used. We adopted the MMI-type branch, which is easy to fabricate and has low sensitivity to geometrical error. The MMI branch consists of a simple rectangle and waveguide ports (1). The size of the rectangle is 2.6 × 1.8 μm². As this branch has no steep structures, accurate fabrication can be performed. The MMI branch shows flat transmission characteristics in a wide wavelength range of over 100 nm. Furthermore, the excess losses are as small as about 0.4 dB (1). Figure 3 denotes cascaded multimode interference (MMI) branches, which are one of the basic components. Each branch works as a 3-dB coupler well. The Si wire waveguide MMI branch already exhibits excellent characteristics.

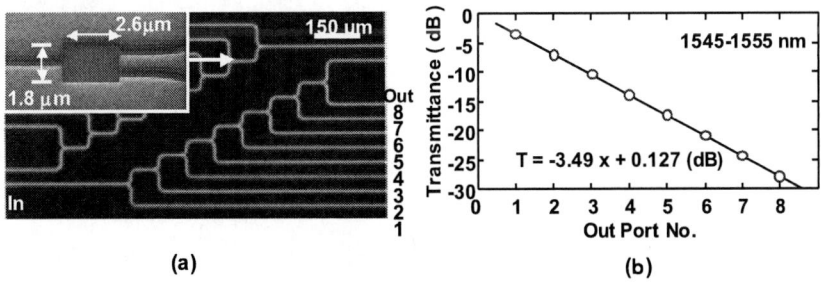

Figure 3
(a) Optical images of MMI structure and cascaded MMI branches.
(b) Transmittance of cascaded MMI branches which work as 3dB couplers.

Figure 4(a) shows a schematic of the channel-dropping lattice filters and their integration in parallel. We connected five filters. The basic filter structure consists of a

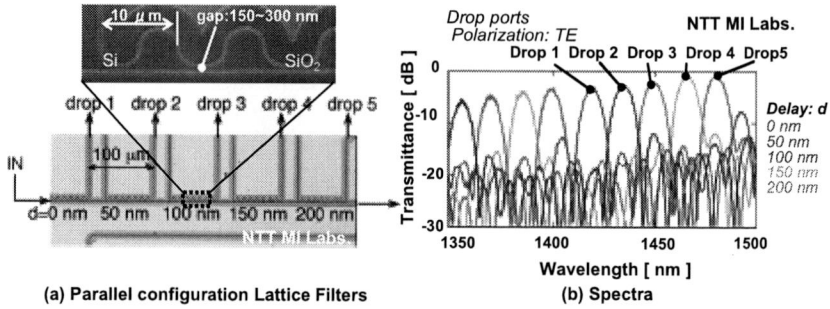

(a) Parallel configuration Lattice Filters   **(b) Spectra**

Figure 4
(a) Lattice filters in parallel (b) spectra at each drop port.

straight line and a snake-like delay line d. The cores of both are 400 x 200 nm. The delay line is made of semicircles with a radius of 2.5 μm and short straight segments d to adjust the delay. It encounters the straight line of the filter periodically and constructs directional couplers. Figure 4(b) shows the transmission spectra for the parallel configuration. The transmission spectra show that the filter can drop different wavelengths at the same time. The wavelength of the dropping channel is tuned by changing d. This is an example of integrations of passive components and proves the possibility of dense integration.

Dynamic functions in Si wire waveguides, such as optical attenuation, modulation, and switching, can be achieved by exploiting carrier-induced effects. We have developed a Si VOA based on a PIN carrier injection structure (22). Figure 5(a) shows a cross-sectional view of the Si VOA, which consists of a rib-type Si wire waveguide with the PIN structure. The rib waveguide has a 600 x 200-nm core and 100-nm-thick slab. The p^+ and n^+ regions are 1-mm long and defined in the slab section, and they are about 3 μm apart. Figure 5(b) shows transmission spectra for various injected currents. As the injected current increases, the carriers injected to the VOA absorb the guided light. We can see that the attenuation is over 30 dB and the wavelength dependence of the attenuation is very flat in the measured 40–nm bandwidth. Figure 5(c) shows the temporal response of a VOA. The 10%-90% rise and fall times are both about 2 ns. The fast response is due to its compactness.

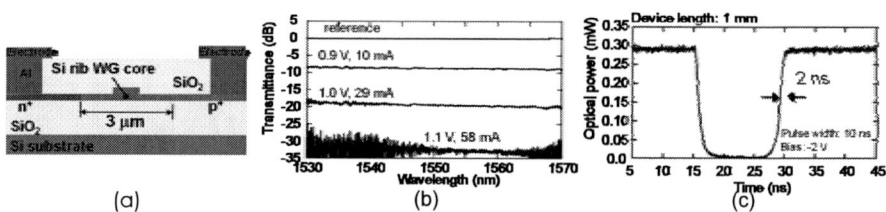

(a)   (b)   (c)

Figure 5
(a) Structure of VOA. (b) Transmission spectra. (c) Temporal response.

In Si photonics, a PD made with Ge is preferred because Ge has a high affinity with Si and can be formed monolithically on the SOI wafer. With the aim of achieving a Ge PD coupled with a Si wire waveguide, we developed a process for selectively growing a Ge film at the Si wire waveguide core (20). We use a two-step ultrahigh vacuum CVD (UHVCVD) method to grow the Ge selectively (23). After that, the wafer is annealed to reduce the dislocation density in the Ge film. The Ge for the PD is a few micrometers by a few ten micrometers in size. Our Ge PD is based on a vertical PIN diode. To obtain a Ge PD with a vertical PIN structure, the Ge film is grown on p-type Si and an n-type region is formed by implanting phosphorous ions into the upper section of the Ge.

Figure 6 show the current-voltage (I-V) characteristics of a fabricated Ge PD. The dark current is as low as around 60 nA at a reverse bias of 1 V, which corresponds to a dark current density of 15 mA/cm^2, which is a sufficiently small value. The low dark current in our PD probably results from the reduction of the dislocation in Ge by annealing and from the use of a high-quality Ge mesa with a flat top and faceted sidewalls. From the measured dark current, we estimated the minimum detectable light power of the photodetector to be just -41 dBm. The responsivity for the PD is 0.8 A/W, which is a favorable value. The PD operates at speeds of 1 GHz and above.

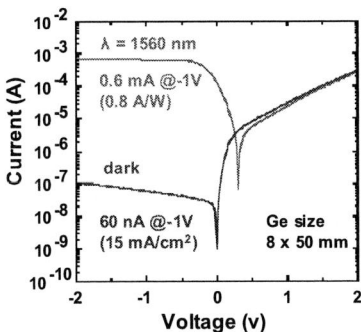

Figure 6 I-V characteristics of Ge PD

Integration

For Si photonics, an integration process is very important. The requirement in developing an integration process is that the fabrication process for one device should not interfere with that for another. In order to avoid such interference, an especially important factor for the design of the fabrication process is the thermal tolerability of each device. For example, a silica waveguide is usually made at high temperatures of more than 1000 ℃. At those temperatures, a Si wire waveguide is oxidized and impurities diffuse into Si, which degrades the optical properties. To obtain thermal tolerability, we use ECR-CVD with a mixture gas of SiH$_4$ and O$_2$. ECR plasma easily dissociates gas molecules and provides moderate energy to the substrate. This method enables fast deposition of high-quality film at temperatures lower than 200 ℃ (24).

In another integration case, we chose a process sequence that achieves monolithic integration of the Si VOA and Ge PD without adversely affecting either device. Ge epitaxial grouth is performed at 600 ℃, and the film is annealed at 900 ℃ to reduce the number of dislocations. The inplanted ions in Si and Ge have to be activated by annealing. We performed those processes in sequence to achieve monolithic integration. As a result, we can fabricate integrated devices in an array configuration. An optical micrograph of a chip integrating Si VOAs and Ge PDs in an array of eight device pairs is shown in Fig. 7(a). An MMI branch between the Si VOA and Ge PD routes light to the outside for device evaluation. Figure 7(b) shows the responsivity and dark current for each of the eight PDs in an array. The responsivity was calculated by estimating the optical power fed into a PD from the transmission losses. All the PDs have nearly the same responsivity and dark current

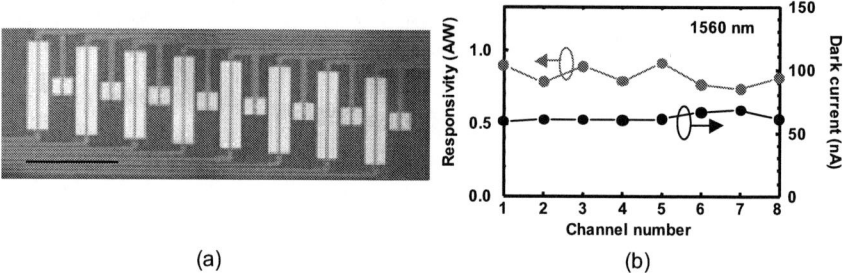

(a)   (b)

Figure 7
(a) Optical micrograph of a chip integrating Si VOAs and Ge PDs.
(b) Responsivity and the dark current of PDs.

Using the low-temperature silica deposition method, we managed to monolithically integrate Si VOAs and a SiO_x AWG (25). Figure 8 shows an optical microscope image of a fabricated VOA-AWG integrated device. The AWG for wavelength demultiplexing in 16 channels with 200-GHz spacing is made with $\triangle$ 3% SiO_x waveguides. Each output port of the AWG is connected to a Si VOA through SSCs. The integrated VOA-AWG was made on a SOI wafer with 3-μm-thick buried oxide layer. In this device, the same SiO_x layer is used for the cores of the waveguides of the AWG, the cores of the low-index waveguide over the Si taper of the SSCs, and the overcladding of the Si wire waveguide. The whole device is only about 15 by 8 mm^2 in size. The minimum AWG bending radius is 500 μ m.

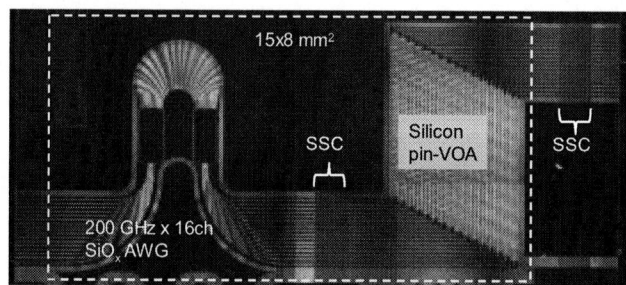

Figure 8 VOA-AWG integrated device

Applications

An integrated variable attenuator multiplexer/demultiplexer (VMUX/DEMUX) is an important component for photonic networks, such as a reconfigurable optical add/drop multiplexing (ROADM) system. The VMUX/DEMUX consists of VOAs for optical power-level control, PDs for power-level monitoring, and an AWG for WDM. As described in the previous section, Si photonics technology has the potential to integrate those components on a single chip with low cost. The Si VOA can operate with a high speed of a few nanoseconds, and the Ge PD can monitor the power attenuated by the VOA. Si wire AWGs with beautiful WDM filtering performance have also been developed by several other groups (2, 19). However, the loss and polarization dependence of the Si AWG are not sufficient yet for telecommunications applications. As a silica-based waveguide is suitable for constructing a high-performance AWG, we have developed the SiOx-AWG described above.

As the first step toward the development of a one-chip VMUX/DEMUX, we demonstrate the integration of high-speed Si VOA and Ge-PD for fast intensity equalization in a WDM transmission system with burst-mode packets.

A compact component integrating Si VOAs and Ge PDs is beneficial for fast optical stabilization. To demonstrate it, we attempted to equalize the optical intensity using the integrated VOA-PD device and an electronic feedback circuit. Figure 9(a) shows a schematic of the feedback system for optical power stabilization. Input light is guided to the VOA and then a half of the output of the VOA is guided to the Ge PD for power monitoring. The other half is guided to the output. The feedback circuit is a simple linear system. The excess output results in an increase of VOA injection current, by which the output optical level is stabilized. The feedback circuit is outside of the chip of the photonic devices. In the feedback circuit, photocurrent from the Ge PD is amplified by a transimpedance amplifier (TIA). The reverse bias to the Ge PD is 3.3 V. The output of the TIA is amplified by a differential amplifier with respect to a voltage reference, which represents the target power level for output stabilization. Then, delay and integral parameters are adjusted, and the signal is guided to the VOA driver. The feedback circuit covers a frequency bandwidth from DC to 100 MHz. Figure 9(b) shows the output optical power versus input power in the feedback operation. For a precise measurement with a large dynamic range, we performed the experiment in a DC operation. As shown in this figure, the output power was stabilized within a deviation of 2.7 dB for a 22-dB variation of input power. In spite of a large dynamic range in optical input, the feedback circuit works very stably. In a pulse operation, the 3-dB recovery time was about 50 ns, which would satisfy the timing criterion for burst optical packets in a 10-Gbps PON system. Further optimization would reduce the recovery time to a few ten nanoseconds.

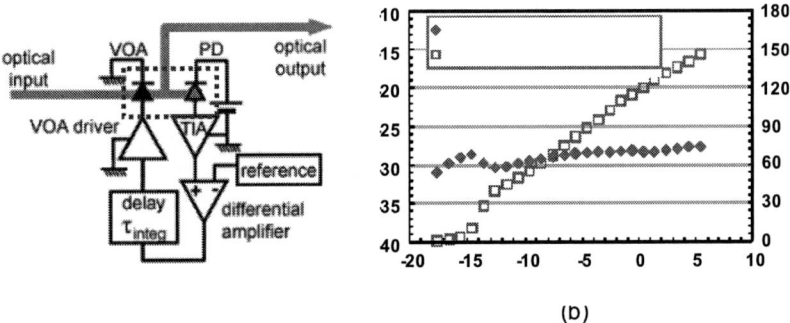

(b)

Figure 9 Optical power stabilization with VOA-PD device
(a) PI feedback circuit. (b) Stabilization of optical power output.

Next, we demonstrate the intensity adjustment for every channel by using the integrated VOA-AWG device. Figure 10 shows the intensity adjustment. For the measurement, we operated the VOAs individually and measured the transmission spectrum of each channel for the TE mode. We separated the channels into groups of four, and drove the VOAs so that all the channels in a group would have the same intensity. We can see that the VOAs independently set the optical level in a group. The rise and fall times are less than 15 ns. This speed is sufficient for the level adjustment of burst optical packets. This device exhibits wavelength demultiplexing owing to the AWG and high-speed power-level adjustment in individual channels owing to the Si-VOA. We believe that this integrated AWG-VOA will be useful in future telecommunications systems that combine WDM and burst systems.

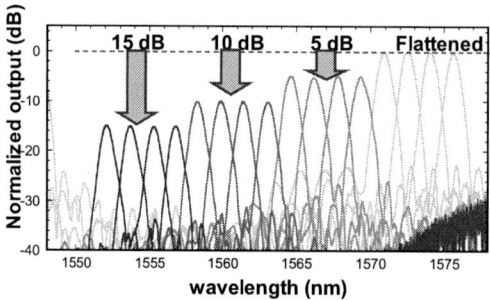

Figure 10 Optical intensity adjustment utilizing AWG-VOA device

Si photonic devices will also be available for quantum key distribution (QKD). If we are to construct 500-km QKD systems over fiber networks, we need multinode quantum communications systems such as a quantum repeater or a quantum relay. The generation of quantum entanglement in the 1.5-μm band is a key function for realizing such systems. Conventional QKD systems are composed of fiber-based entanglement light sources. However, there is a drawback with fiber-based sources, namely, noise photons generated by spontaneous Raman scattering (SpRS). Although cooling the dispersion-shifted fiber (DSF) reduces the number of noise photons, the need for cooling equipment complicates the system and is thus undesirable.

On the other hand, Si wire waveguides based on the SOI structure are now attracting attention as a nonlinear medium for quantum entanglement. Thanks to the large Kerr nonlinearity of silicon and its very small effective area. When signal and idler lights are injected in such waveguides, the four-wave mixing (FWM) process occurs very efficiently. We have proved that strong light confinement in the Si wire wavuguide induces FWM (26). In addition, noise photons resulting from SpRS can be significantly suppressed by setting signal and idler frequencies appropriately, since the Raman peak of silicon single crystal is 15.6 THz from the pump frequency and has a narrow bandwidth compared with that of DSF. As a result, we have succeeded in generating 1.5-μm band time-bin entangled photon pairs with 83% visibility without temperature control (27, 28).

The Si crystal structure on the SOI wafer enables us to fabricate devices with very low insertion loss and to avoid SpRS. We can therefore realize low-noise entangled photon generation without temperature control. Creation of the entangled photon pairs at room temperature will accelerate the development of QKD systems.

Conclusions

We demonstrated the monolithic integration of Si-wire-based devices on SOI wafers. To integrate photonic devices made of different materials, we developed new processes, such as the low-temperature deposition of silica waveguide film. These processes are compatible with the process used to make Si electronic devices and should be applicable to photonic-electronic integration. We also showed some applications of the monolithically integrated devices. The SOI structure enables us to achieve those applications.

Although the integration of many more devices is necessary for practical use, we believe that the integration of Si photonics devices will enable us to make low-cost, high-performance devices for telecommunications.

Acknowledgments

The QKE experiment was carried out at the NTT Basic Research Laboratories. We thank Dr. Takesue and Dr. Harada for their useful suggestions and discussions.

References

1. T. Tsuchizawa et al., *IEEE J. Sel. Top. Quant. Electron.*, 11, pp. 232–240 (2005)
2. W. Bogaerts etal., *IEEE J. Sel. Top. Quant. Electron.* 12, pp.1394-1401 (2006)
3. F. Xia et al., *Nature Photon.,* vol. 1, no. 1, pp. 65–71 (2007)
4. S. Xiao et al., *Opt. Express*, vol. 15, no. 12, pp. 7489–7498 (2007)
5. F. Xu and A. W. Poon, *Opt. Express*, vol. 16, no. 12, pp. 8649–8657 (2008)
6. J. Song et al., *Opt. Express* 16, pp. 8359-8365 (2008)
7. H. Fukuda et al., *Opt. Express*, Vol. 16, no. 7, pp.4872-4880 (2008)
8. L. Liao et al., *Opt. Express*, 13, pp. 3129-3135 (2005)
9. Q. Xu, et al., *Nature*, 435, pp. 325-327 (2005)
10. A. Liu, et al., *Opt. Express*, 15, pp.660-668 (2007)
11. W. M. J. Greeen, et al., *Opt. Express*, 15(25), pp. 17106-17113 (2007)
12. D. M. Morini et al., *Proceedings of the IEEE* 97, pp.1199-1215 (2009)
13. M. W. Geis et al., *IEEE Photon. Technol. Lett.* 19, pp.152-154 (2007)
14. D. Ahn et al., *Opt. Express,* 15, pp.3916-3924 (2007)
15. L.Vivien et al., *Opt. Express,* 15, pp.9843-9848 (2007)
16. L.Chen et al., *Opt. Express,* 17, pp. 15248-15256 (2009)
17. T. Pinguet et al., *in 5th IEEE International Conference on Group IV Photonics,* pp. 362–364 (2008)
18. Q. Fang et al., *Opt. Express,* 18(13), pp.13510-13515 (2010)
19. Q. Fang et al., *Opt. Express,* 18(5), pp. 5106-5113 (2010)
20. S. Park et al., *Optics Express,* 18 (8), pp. 8412-8421 (2010)
21. T. Tsuchizawa etal., *The 5th International Symposium on Advanced Science and Technology of Silicon Materials*, pp. 366-370, Kona (2008)
22. S. Park et al., *Opt. Express*, 18(11), pp.11282-11291 (2010).
23. S. Park et al., *IEICE Trans. Elect.*, E91-C, pp. 181–186 (2007)
24. S. Matsuo and M. Kiuchi, *Jpn. J. Appl. Phys.* 22, pp. L210-212 (1983)

25. H. Nishi et al., *Appl. Phys. Express,* 3, pp. 102203-1-3 (2010)
26. H. Fukuda et al., *Opt. Express,* 13(12), 4629 (2005)
27. H.Takesue et al., *Applied Physics Review Letter,* Vol. 91, 201108 (2007)
28. H.Takesue et al., *Opt. Express,* 16 (8), 5721 (2008)

CHAPTER 11

MATERIALS AND CHARACTERIZATION-2

238

ECS Transactions, 35 (5) 239-245 (2011)
10.1149/1.3570801 ©The Electrochemical Society

Ultra-thin film SOI/BOX substrate development, its application and readiness

W. Schwarzenbach, X. Cauchy, O. Bonnin, N. Daval, C. Aulnette, C. Girard,
B.-Y. Nguyen & C. Maleville

SOITEC, Parc Technologique des Fontaines, Bernin, F38926 Crolles, France

The Ultra-Thin SOI and BOX substrates are the foundation of Fully Depleted planar technology, a CMOS scaling solution for 20 nm node and beyond. Using the Smart Cut^{TM} technology, UTSOI substrates development, with SOI & BOX thickness reduced down to 12 & 25 nm respectively, is on the way to High Volume Manufacturing by the end of 2011. To improve device Vt variation control, SOI total layer thickness variation of less than +/- 1 nm for all the measured points and all the preproduction wafers is already achieved and +/- 0.5 nm variation is targeted. Tight SOI thickness variation at device scale and BOX thickness variation are also demonstrated.

Substrate Requirement for Next Technology Nodes

At small geometries, standard bulk CMOS technology scaling is facing severe issues (1). Shrinking of the gate length for the 20 nm node leads to deleterious short channel effects, random dopant fluctuations (RDF), and causes unacceptably threshold voltage (Vt) variability. There is consensus that Fully Depleted (FD) devices with undoped channel, also known as Ultra Thin Body Devices (2), are effective solutions to drastically reduce the variability and achieve performance improvement with low dynamic power, low leakage current & high transistor density.

A prerequisite for the Planar FD technology is the Ultra Thin Silicon-On-Insulator (UTSOI) substrate. The starting ultra-thin silicon layer sitting over the buried oxide has to be adapted to the subsequent FD CMOS processing. Cleaning, oxidation and etching steps remove a few monolayers of silicon and this silicon consumption has to be taken into account when specifying the initial UTSOI thickness. The targeted channel Si thickness is typically between 6nm – 7nm for 25nm gate length transistor (3). According to these processing requirements, starting wafers will need extremely thin SOI layer, down to 12 nm, to be compared with today's mass-produced Partially Depleted SOI wafers which have a top silicon layer 70 to 90 nm thick.

Thin to ultra-thin Buried Oxide (BOX) layer will be available, with thickness from 145 nm down to 25 nm then 10 nm in a second step. Reducing BOX layer thickness further improves electrostatic control and enables Vt modulation through back bias management of the channel (4).). The following nomenclature is used in this paper: wafers for FDSOI CMOS with 145nm BOX: "ETSOI", wafers for FDSOI CMOS with 25nm BOX: "UTSOI", wafers for FDSOI CMOS with 10nm BOX: "UTSOI – BOX10" (Figure 1).

239

The predominant source of Vt variability in today's CMOS technologies (Bulk and PD-SOI) is the Random Dopant Fluctuation (RDF). In contrast, the Fully Depleted technology, as it relies on undoped channel, is not subject to this problem. However it introduces a new dependency on silicon thickness (1) which becomes a key parameter to avoid re-introducing V_T variation in planar FDSOI devices. Typical uniformity requirements include on-wafer (WiW) uniformity and wafer-to-wafer (WtW) uniformity. Both of them combined are classified as layer total thickness variation (LTTV) and define the overall manufacturing process window for thickness uniformity. LTTV has to be achieved at the sub-nanometer range for the UTSOI layer for all wafers and all sites in order to meet the FD specifications.

Khakifirooz *et al* have shown an empirical correspondence between V_T variation on FD-SOI devices and SOI layer thickness variations, close to 25 mV/nm. (5, 6). SOI Uniformity requirements include on-wafer uniformity, from the sub-nanometer to the across- wafer range, and the wafer-to-wafer uniformity. In order to limit the Vt variation coming from the substrate non-uniformity to no more than 25mV, thereby offering a drastic reduction of variability compared to Bulk and PD-SOI alternatives, we specified for our FDSOI substrates a 1 nm maximum SOI thickness variation between any point on a wafer and any point on the same or a different wafer.

Ultra-Thin SOI/BOX Substrate Development

As shown in Figure 1, SOI product evolution offers a continuum of substrate solutions targeting high volume production capability end of 2011 for 20 nm CMOS platform. It includes ETSOI product (thin SOI layer and thick BOX layer, typically 145 nm) and UTSOI (thin SOI & thin BOX layers).

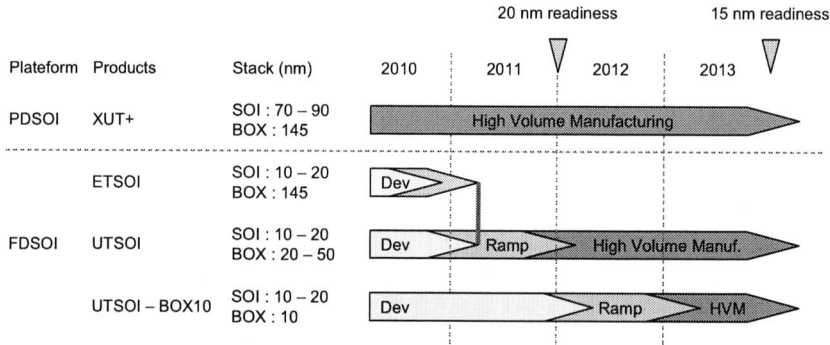

Figure 1. SOI Product evolution roadmap

Beyond 20 nm technology node, UTSOI buried oxide thickness will be extended from the initial 20 – 25 nm down to 10 nm, as shown in figure 2.

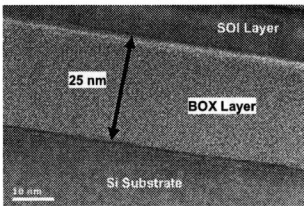 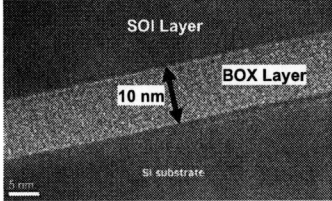

Figure 2. UTSOI (BOX25nm) & UTSOI – BOX 10nm TEM cross sections

Ultra-Thin SOI/BOX Substrate Preparation

Continuing SOI Substrate portfolio available to device manufacturers, ETSOI & UTSOI products are prepared using same process structure as current Partially Depleted products. Thanks to this synergy, FDSOI products will be able to use same toolset as PDSOI, thus taking benefits over its 10 years of learning and high-volume manufacturing expertise. Figure 3 shows schematically Smart Cut[TM] process as it is implemented. The Buried Oxide generation at the "oxidation" step is adjusted versus targeted BOX layer thickness. The Implantation, Splitting & Finishing steps are adjusted to ensure the FDSOI products meeting their tight thickness control requirements.

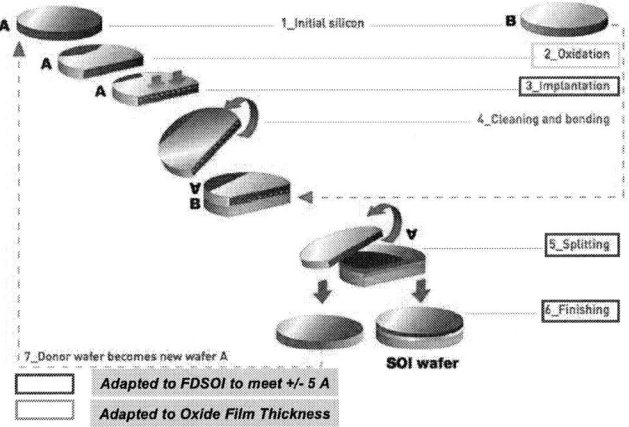

Figure 3. Schematics of Smart Cut[TM] process for FDSOI products

WtW thickness control is ensured by using standard tool to tool matching over the production line and by introducing dynamic processing in the finishing line. Final SOI WiW thickness uniformity includes several contributions from process steps and specific action plans focused on various contributors are on the way to improve WiW performance.

Ultra-Thin SOI/BOX Substrate Readiness

<u>SOI Thickness Performance @ Wafer Scale</u>

To match T_{Si} thickness variation requirement, tight Wafer to Wafer thickness control has to be implemented. As shown in Figure 4, maximum variation of less than ± 5 Å is already achieved on UTSOI pre-production volume. Path to ± 2 Å is identified and will be implemented before end of 2011 and start of the HVM phase.

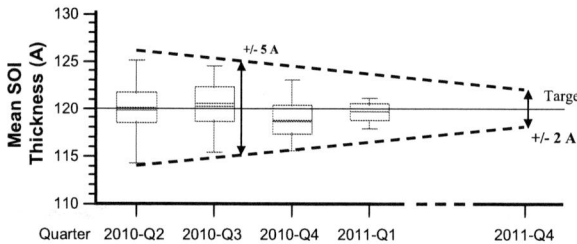

Figure 4. UTSOI Wafer to Wafer performance and improvement path

In addition, using 41 pts measurement F5X ellipsometer, On-Wafer uniformity less than 10 Å and down to 7 Å is currently achieved on 12 nm SOI / 25 nm BOX pre-production volume. Typical thickness mapping corresponding to these WiW uniformity values are shown in Figure 5. Improvement opportunities are identified to demonstrate 5 Å on-wafer uniformity at high volume phase.

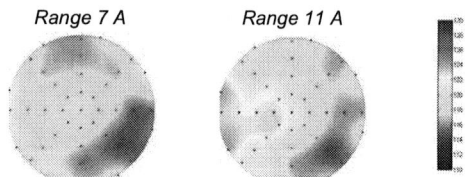

Figure 5. Typical UTSOI On-Wafer uniformity mapping

The total SOI thickness fluctuation for all measured points, all wafers, is obtained by combining wafer-to-wafer and on-wafer contributions. Figure 6 shows performance achieved on pre-production UTSOI volumes. Recent evolution allows total SOI thickness fluctuation to reach ± 10 Å. The combination of all mentioned improvements will guarantee a ± 5 Å control, corresponding to the 1 nm range required to meet device Vt variability target.

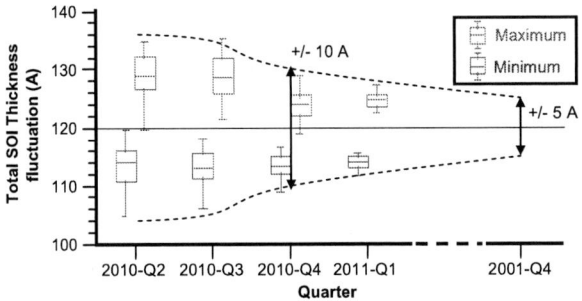

Figure 6. UTSOI Total SOI Thickness Variation performance and improvement path

SOI Thickness Performance @ Device Scale

Device scale thickness variation can be monitored using Atomic Force Microscopy (AFM). Using 30x30 μm^2 scan dimension, AFM gives height dimensions that translate into local silicon thickness variations for neighbor transistors. Figures 7A & 7B show typical AFM scan view on ETSOI & UTSOI products respectively. On both products, RMS of 2 Angstroms is demonstrated. Looking at a smaller inspection scale at the transistor level, 2x2 μm^2 AFM measurement demonstrate than UTSOI & polished bulk silicon surface, shown in Figures 7C & 7D respectively, are both close to 1.5 Angstrom RMS.

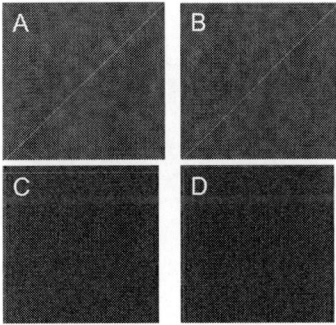

Figure 7. AFM Scan view, (A) 30x30 μm^2 ETSOI, (B) 30x30 μm^2 UTSOI, (C) 2x2 μm^2 UTSOI, (D) 2x2 μm^2 Polished Bulk. 30x30 μm^2 scan courtesy of IBM

BOX Thickness Performance @ Wafer scale

SOI final BOX Thickness variation is strongly correlated to buried oxide formation step. On UTSOI product, typical 25 nm BOX On-Wafer uniformity of less than 6 A is routinely achieved, as shown in Figure 8.

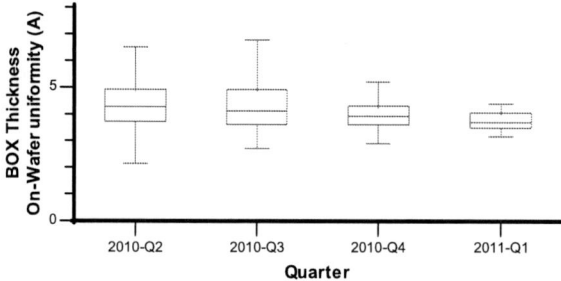

Figure 8. BOX Thickness On-Wafer uniformity performance

Combining On-Wafer uniformity & Wafer to Wafer thickness variation, as shown in Figure 9, total BOX Thickness fluctuation on all F5X measured points, all wafers, of less than +/- 10 A from the target is now obtained on pre-production UTSOI volumes.

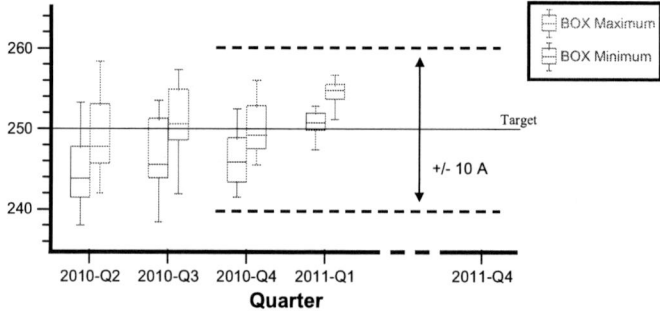

Figure 9. UTSOI Total BOX Thickness Variation performance and improvement path

Conclusion

Fully-depleted planar design is identified as an attractive option for small geometries CMOS technologies, 20nm node and beyond. It requires the UTSOI substrate with thin and precisely controlled thickness of the SOI and BOX layers, down to 12 and 25 nm respectively. Using the Smart Cut™ technology, these UTSOI substrates are already in pre-production mode and will be ready for High Volume Manufacturing by the end of 2011. SOI Thickness Wafer to Wafer variation of less than +/- 5 A and Total Thickness

Variation of less than +/- 10 A are already achieved. Additional improvements will allow to reach a total maximum SOI thickness variation of +/- 5 A, in adequation with device variation requirements.

Acknowledgments

Special thanks to IBM team and Richard Murphy for AFM 30x30 μm^2 measurement scans.

References

1. T. Skotnicki, C. Fenouillet-Béranger, C. Gallon, F. Boeuf, S. Monfray, F. Payet, A. Pouydebasque, M. Szczap, A. Farcy, F; Arnaud, S. Clerc, M. Sellier, A. Cathignol, J.-P. Schoellkopf, E. Perea, R. Ferrant, H. Mingam, *IEEE Trans. On Electron Devices*, **55**, pp. 96-130 (2008)

2. Changhwan Shin, Min Hee Cho, Y. Tsukamoto, B.-Y. Nguyen, C. Mazure, B. Nikolic, Tsu-Jae King Liu, *IEEE Trans. On Electron Devices*, **57**, pp. 1301-1309 (2010)

3. T. Skotnicki, IEDM Short Course "Low Power Logic and Mixed Signal Technology", IEDM Conference, Baltimore, 2009.

4. F. Andrieu, O. Weber, J. Mazurier, O. Thomas, J.-P. Noel, C. Fenouillet-Béranger, J.-P. Mazellier, P. Perreau, T. Poiroux, Y. Morand, T. Morel, S. Allegret, V. Loup, S. Barnola, F. Martin, J.-F. Damlencourt, I. Servin, M. Cassé, X. Garros, O. Rozeau, M.-A. Jaud, G. Cibrario, J. Cluzel, A. Toffoli, F. Allain, R. Kies, D. Lafond, V. Delaye, C. Tabone, L. Tosti, L. Brévard, P. Gaud, V. Paruchri; K. K. Bourdelle, W. Schwarzenbach, O. Bonnin, B.-Y. Nguyen, B. Doris, F. Boeuf, T. Skotnicki, O. Faynot, *VLSI Conference*, 2010

5. K. Cheng, A. Khakifirooz, P. Kulkarni, S. Ponoth, J. Kuss, D. Shahrjerdi, L. F. Edge, A. Kimball, S. Kanakasabapathy, K. Xiu, S. Schmitz, A. Reznicek, T. Adam, H. He, N. Loubet, S. Holmes, S. Mehta, D. Yand, A. Upham, S.-C. Seo, J. L. Herman, R. Johnson, Y; Zhu, P. Jamison, B. S. Haran, Z. Zhu, L. H. Vanamurth, S. Fan, D. Horack, H. Bu, P. J. Oldiges, D. K. Sadana, P. Kozlowski, D. McHerron, J. O'Neill, B. Doris, *IEDM Conference*, Baltimore, 2009

6. A. Khakifirooz, K. Cheng; P. Kulkarni, J. Cai, S. Ponoth, J. Kuss, B. S. Haran, A. Kimball, L. F. Edge, A. Reznicek, T. Adam, H. He, N. Loubet, S. Mehta, S. Kanakasabapathy, S. Schmitz, S. Holmes, B. Jagannathan, A. Majumdar, D. Yang, D. Horak, H. Bu, D. K. Sadana, P. Kozlowski, D. McHerron, J. O'Neill, B. Doris, W. Haensch, E. Leobondung, G. Shahidi, *VLSI-TSA Conference*, Hsinchu, 2010

Performance of SOI MOSFETs with Ultra-Thin Body and Buried-Oxide

A. Ohata[a], Y. Bae[b], S. Cristoloveanu[c], C. Fenouillet-Beranger[d,e],
P. Perreau[d,e], and O.Faynot[d]

[a]Osaka City University, Osaka, Japan
[b]Uiduk University, Gyeongju, Korea
[c]IMEP-LAHC (UMR 5130), Grenoble INP Minatec, France
[d]CEA-LETI-MINATEC campus, GRENOBLE, France
[e]STMicroelectronics, Crolles, FRANCE

The electrical characteristics in ultra-thin (8 nm) SOI-MOSFETs with 10 nm buried oxide thickness were studied. Threshold voltage was effectively controlled for both of N-channel and P-channel SOI-MOSFETs even in the short devices (45 nm). No degradation of the mobility due to the use of thin buried oxide could be detected. Through appropriate back-gate biasing, the performance of ultra-thin SOI-MOSFETs can be improved dramatically. In P-channel SOI-MOSFETs, the hole mobility measured in volume conduction regime is higher than when only one interface (Si/high-K or Si/SiO$_2$) is activated. This gain makes the hole mobility comparable with the universal mobility law and is promising for performance enhancement in CMOS circuits.

Introduction

Ultra-thin (UT) silicon-on-insulator (SOI)-MOSFETs can provide better control of short channel effects. The introduction of a high-K dielectric and metal gate enables to use a non-doped channel in UT SOI-MOSFETs. A single mid-gap metal gate in non-doped channels provides suitable threshold voltage for a low supply bias. Therefore, non-doped UT SOI-MOSFETs are now considered as one of the best candidates for low-power applications. For the further CMOS generations, the use of thin buried oxide (BOX) is needed to achieve better control of short-channel effects, relax self-heating issues, and prevent variability in the device performance (1). A thin BOX is also attractive because it allows dynamic threshold voltage operation with low voltage on the back gate. A dynamic threshold voltage operation can suppress leakage in off-state and enhance the channel current in on-state (2). This paper presents a study on the impact of back gate voltage (V_{g2}) on the channel current in 8 nm thick SOI MOSFETs with 10 nm BOX.

Device Fabrication

The test devices were fabricated at STMicroelectronics (Crolles) using (100) UNIBOND SOI wafers with initial doping of around 10^{15}/cm^3 and BOX thickness of 10 nm. The SOI films were thinned down using thermal oxidation and wet etching to a final thickness of around 8 nm. The transistor bodies were kept undoped during the entire process. A high-K dielectric (HfO$_2$) of approximately 2.5 nm, a metal gate (ALD TiN 10 nm), and a poly-Si (80 nm) were deposited. Effective oxide thickness (EOT) was determined to be 1.4 nm. A selective epitaxy of 10 nm film was carried out in the extension regions to reduce the external resistance. Raised extensions were then ion implanted. A typical device structure is shown in Fig.1.

P-channel and N-channel SOI-MOSFETs were fabricated with the channel lengths from 10 μm down to 40 nm as a target device size. The effective channel length (L_{eff}) was longer than the target size (L_g) by 5 nm. A voltage was applied to the substrate which operated as a back gate.

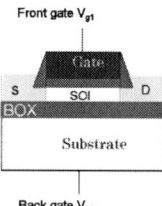

Figure 1. Schematic of the device structure used for this study. The thicknesses of the buried oxide and transistor body are 10 nm and 8 nm, respectively. The back-gate voltage (V_{g2}) is applied to the substrate.

For the comparison of the electrical characteristics in P-channel SOI-MOSFETs, additional P-channel devices with thick BOX (145 nm) and thin SOI (20 nm) using (110) UNIBOND SOI wafers were fabricated and measured. The channel direction was along <110>. A high-K dielectric (HfO$_2$ with a thickness of 3 nm) was deposited by atomic layer deposition on the chemically grown SiO$_2$ (approximately 1 nm thick). A metal gate (10 nm thick TiN) was deposited by physical vapour deposition. EOT was approximately 2 nm (3).

Electrostatic Behavior

Threshold Voltage Control

The advantage of thin BOX, is that the front-channel threshold voltage can be tuned effectively using V_{g2} bias. Figures 2 and 3 show V_{g2} voltage dependence of the threshold voltage (V_{TH}) that was determined by the use of a tangent line in drain current (I_d) versus front-gate voltage (V_{g1}) curve at the transconductance ($G_m = dI_d/dV_{g1}$) peak. Even for short channel such as $L_{eff} = 45$ nm, V_{TH} is controlled well for both P-channel and N-channel SOI-MOSFETs. For depleted back interface, the slopes of the curves in these figures were approximately 0.1 for both of P- and N-channels. This coupling slope is normally explained by Lim and Fossum theory (4)

$$\frac{dV_{TH}}{dV_{g2}} = \frac{C_{Si}C_{ox2}}{C_{ox1}(C_{ox2} + C_{Si} + C_{it2})} \qquad [1]$$

where C_{Si}, C_{ox1}, and C_{ox2} are the capacitances of SOI film, gate oxide, and BOX, respectively. $C_{it2} = qD_{it2}$ (q is the elementary charge.) is governed by the density of traps D_{it2} at the back interface. For thick BOX and usual low D_{it2} value, the slope reduces to C_{ox2}/C_{ox1}. For thin gate oxide, the inversion capacitance leads to a decrease in C_{ox1}, resulting in the increase in this slope. On the other hand, for thinner BOX, C_{ox2} becomes comparable with C_{Si} and affects differently the coupling slope. In addition, as $C_{Si} = 1.3$ x 10^{-6} F/cm^2, $C_{ox1} = 2.5$ x 10^{-6} F/cm^2, and $C_{ox2} = 3.5$ x 10^{-7} F/cm^2, the detection limit of D_{it2} becomes rather large (1.0 x 10^{12}/cm^2·eV), which means that the slope is insensitive to D_{it2}.

The subthreshold swing for front channel (S_1) and back channel (S_2) is also affected by the interface quality that is related to D_{it1} and D_{it2}. S_1 and S_2 are given by (5):

$$S_{1,2} = 2.3 \frac{k_B T}{q} \left(1 + \frac{C_{it1,2}}{C_{ox1,2}} + \frac{C_{Si}}{C_{ox1,2}} \frac{C_{ox2,1} + C_{it2,1}}{C_{Si} + C_{ox2,1} + C_{it2,1}} \right)$$ [2]

where k_B is the Boltzmann constant. S_2 for thin BOX is also less sensitive to D_{it2} than that for thick BOX. For N-channel SOI-MOSFETs, the experiment shows 230 mV/dec for L_g = 10 μm, while S_2 estimated by [2] is 200 mV/dec. This small difference is presumably due to the influence of the third interface (bonded Si-SiO$_2$, located underneath the thin BOX) rather than to an increase in C_{it2}

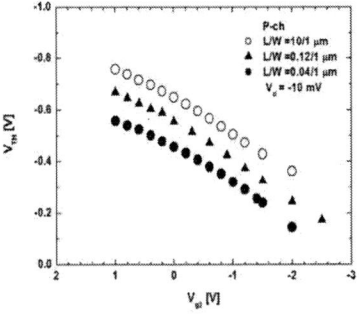
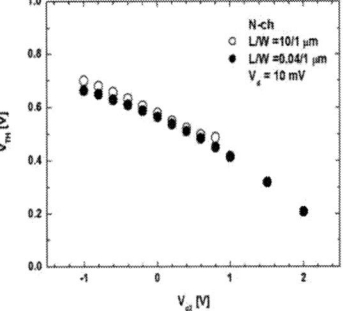

Figure 2. Threshold voltage V_{TH} versus back-gate bias V_{g2} for P-channel SOI-MOSFETs with different geometry. Channel length (L) and width (W) are the target size.

Figure 3. Threshold voltage V_{TH} versus back-gate bias V_{g2} for N-channel SOI-MOSFETs with different geometry. Channel length (L) and width (W) are the target size.

Carrier Mobility

Mobility was measured by the split-CV (capacitance-voltage) method and I_d-V_{g1} characteristics. Figure 4 shows the hole mobility versus V_{g1} voltage for various V_{g2} in P-channel UT SOI-MOSFETs with 10 nm BOX. The mobility is rather low at V_{g2} = 0 V, because of the high-K/metal-gate stack structure. The mobility peak (approximately 90 cm^2/Vs) shows a comparable value with several reported data in SOI-MOSFETs with high-K/metal gate and thick BOX (6,7). It means that no degradation due to the adoption of thin BOX could be detected.

The mobility increases under all negative V_{g2} conditions. In this case, the back channel tends to be activated and the centroid of the charge moves from the mediocre-quality front interface (Si/HfO$_2$) to the higher quality back interface (Si/SiO$_2$). As a result, the carrier transport is improved. Even for $V_{g1} \approx$ -0.5 V, the hole mobility remains reasonable, being limited by the scattering at the back interface.

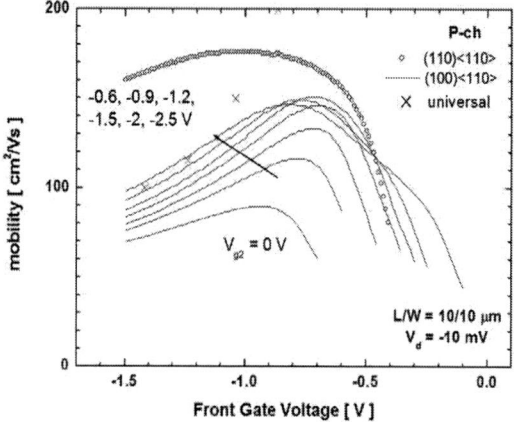

Figure 4. Mobility versus front-gate voltage for P-channel SOI-MOSFETs. L/W = 10/10 μm. V_d = -0.01 V. Lines show the mobility for several back-gate biases V_{g2} in (100) SOI-MOSFETs with thin (10 nm) BOX. Open circles show the mobility in (110) SOI-MOSFETs with thick BOX (145 nm). Cross marks show the 'universal mobility' assuming that (100) SOI-MOSFET with 30 nm SOI and 145 nm BOX has the universal mobility.

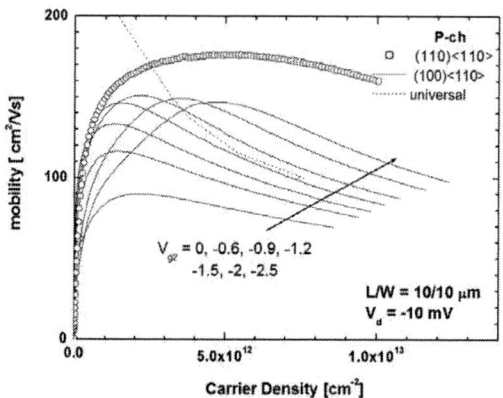

Figure 5. Mobility versus carrier density for P-channel SOI-MOSFETs. L/W = 10/10 μm. V_d = -0.01 V. Lines show the mobility for several back-gate biases (V_{g2}) in (100) SOI-MOSFETs. Open circles show the mobility in (110) SOI-MOSFETs with thick BOX (145 nm). Dashed lines show the 'universal' mobility.

From the calculation of $\delta C/\delta V_{g1}$, the front-channel threshold voltage under the condition of activated back channel, is estimated to be V_{TH} = -0.65 V. It means that for large negative V_{g2} and V_{g1} < -0.65 V, both channels are activated. Interestingly, this condition leads to maximum mobility values (see Fig. 4 for V_{g2} = -2 V and -2.5V).

The hole mobility becomes comparable to the universal mobility at more negative V_{g1}. Figure 5 (mobility versus carrier density) clearly confirms these effects. Cross marks correspond to the universal mobility assuming that (100) SOI-MOSFET with 30 nm SOI and 145 nm BOX has the universal mobility. This large enhancement of hole mobility is due not only to the gradually increased contribution of the high mobility channel at the SOI/SiO$_2$ interface, but also to the volume conduction in the thin SOI film (5,8). In the volume conduction for P-channel UT SOI-MOSFETs, the hole mobility is improved by the redistribution in the population from heavy holes to light holes. By these combined effects, the mobility at V_{g1} = -1 V corresponds to approximately 80 % of that in (110)/<110> SOI-MOSFETs (open circles) with high-K/metal gate stack. At the maximum point of the mobility (V_{g1} = -0.7 V for V_{g2} = -1.5 V), the hole mobility for (100)/<110> amounts to 90 % of the mobility in (110)/<110> channels.

Figures 6 and 7 show the mobility in N-channel UT SOI-MOSFETs with 10 nm BOX. A large enhancement in electron mobility is observed at positive V_{g2}. Although mobility is rather low at V_{g2} = 0 V, again due to the high-K/metal gate stack structure, the maximum value (280 cm^2/Vs at V_{g2} = 0 V) is comparable to the reported data for UT SOI-MOSFETs with high-K/metal gate and thick BOX (7,9). This confirms that thin BOX induces no damage to the front channel even in UT SOI. The mobility degradation

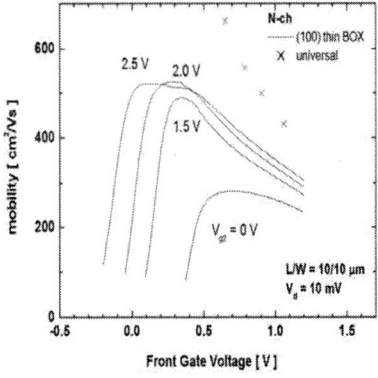 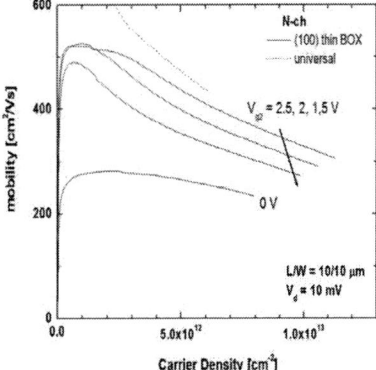

Figure 6. Mobility versus front-gate voltage for N-channel SOI-MOSFETs. L/W = 10/10 μm. V_d = 0.01 V. The front-channel threshold voltage determined from $\delta C/\delta V_{g1}$ while the back channel is inverted (large positive V_{g2}) is 0.39 V. Cross marks show the 'universal mobility'.

Figure 7. Mobility versus carrier density for N-channel SOI-MOSFETs. L/W = 10/10 μm. V_d = 0.01 V. Dashed lines show the 'universal mobility'. The mobility peaks appear only in the low carrier density regions.

resulting from the high-K/metal-gate stack can be improved by 35 % at $V_{g1} = 1$ V when the back and front channels are simultaneously activated. The threshold voltage for activating the front channel, while the back channel is inverted (large positive V_{g2}), is estimated from $\delta C/\delta V_{g1}$ to be $V_{TH} \approx 0.39$ V. The maximum mobility is observed only when the back channel is activated. Cross marks correspond to the universal mobility when this ultrathin SOI is assumed to feature the same mobility on the universality curve. By contrast with P-channels, the electron mobility cannot exceed the universality even for volume inversion conditions.

Conclusion

The electrical characteristics in SOI-MOSFETs with 8 nm SOI and 10 nm BOX were studied. The threshold voltage was effectively controlled for both of N-channel and P-channel transistors even in very short devices (45 nm). There is no apparent degradation of the mobility induced by the use of thin buried oxide. Through appropriate back-gate biasing, the performance of UT SOI-MOSFETs can be improved dramatically. In particular, the hole mobility in volume inversion regime is higher than the values obtained for front-channel or back-channel conduction. The hole mobility can even become comparable to the universal mobility. These results are promising for enhancing the performance of CMOS circuits.

Acknowledgments

This work was performed in the frame of the CEA–STMicroelectronics–SOITEC-IMEP collaboration. This work was also supported by Eurosoi, Nanosil and WCU projects, and by a Grand-in-aid for Scientific Research from the Ministry of Education, Science, Sports and Culture, Japan.

References

1. C. Fenouillet-Beranger, P. Perreau, S. Denorme, L. Tosti, F. Andrieu, O. Weber, S. Monfray, S. Barnola, C. Arvet, Y. Campidelli, S. Haendler, et al., *Solid-State Electronics*, 54, 849 (2010).
2. C. Fenouillet-Beranger, O. Thomas, P. Perreau, et al., *VLSI Symp. Tech. Dig.*, p. 65 (2010).
3. T. Signamarcheix, F. Andrieu, B. Biasse, M. Cassé, A-M Papon, E. Nolot, B Ghyselen, O. Faynot, L. Clavelier, submitted.
4. H.K. Lim and J.G. Fossum, *IEEE Trans. Electron Devices*, 30, 1244 (1983) .
5. S. Cristoloveanu, S. Li, in *Electrical Characterization of Silicon-On-Insulator Materials and Devices*, Kluwer (1995).
6. B. Doris, et al, *VLSI Symp. Tech. Dig.*, p. 214 (2005).
7. C. Fenouillet-Beranger, S. Denorme, P. Perreau, C. Buj, O. Faynot, et al., *Solid-State Electronics*, 53, 730 (2009).
8. S. Kobayashi, M. Saitoh, and K. Uchida, *Tech. Dig. Int., Electron Dev. Meet*, p. 707 (2007).
9. A. Vandooren et al., *IEEE International SOI Conference Tech. Dig.*, p. 221 (2005).

TiN/HfSiON for Analog Applications of nMuGFETs

M. Rodrigues[a], M. Galeti[a], J. A. Martino[a], N. Collaert[b], E. Simoen[b], C. Claeys[b,c]

[a]LSI/PSI/USP, University of Sao Paulo, Sao Paulo 05508-010, Brazil
[b]imec, B-3001 Leuven, Belgium
[c]E.E. Dept., KU Leuven, B-3001 Leuven, Belgium

This work presents an analysis of the impact of the implementation of a hafnium gate dielectric on triple-gate structures for analog applications. A reduced Early voltage V_{EA} is observed for the hafnium dielectric. This was attributed to a slightly higher physical gate dielectric thickness observed on HfSiON dielectric that could be leading to a increasing on the horizontal electric field contribution on the drain current. As a result, lower intrinsic voltage gain was observed for high-k dielectrics when compared with a SiON gate insulator.

Introduction

In order to improve the MOSFET transistor performance, aggressive scaling of devices has continued including the gate oxide thickness. However, at sub-1nm, SiO_2 has reached its scaling limit due to the high tunneling current (1). The use of high dielectric constant (k) gate insulators may circumvent this limitation (2). Thicker dielectrics can be used to reduce the gate leakage while maintaining the same level of inversion charge. The incorporation of nitrogen into these high-k materials can improve their thermal stability, reduce the dopant penetration and allow further equivalent oxide thickness (EOT) scaling (3). However, in order to maintain a good interface with the silicon substrate for high carrier mobility, SiO_2 has been implemented as interfacial layer (IL) between the high-k dielectric and the silicon film (4). Additionally, the use of titanium nitride (TiN) as metal gate material has also been studied to tune the threshold voltage (V_T) for CMOS applications (5-7).

In combination with high-k/metal gate materials SOI Multiple-gate devices (MuGFETs) offer an excellent gate control over the channel charges and reduce short-channel effects (8). They also present an attractive behavior for analog operation, showing a reduced drain output conductance and a large Early voltage. Moreover, a quasi-ideal subthreshold slope and a better ratio between on-off current were also observed (9-12).

The aim of this work is to analyze the impact of the implementation of a hafnium dielectric on triple-gate structures for analog applications.

MuGFET Fabrication

The devices under investigation are 5-fin SOI triple-gate nFETs with 65nm Si film on a 150nm buried oxide. The different gate oxide options studied are: Fig. 1a) 2.0 nm HfSiON (with an interfacial layer of 1 nm chemical SiO_2); Fig. 1b) 1 nm ISSG (in-situ stream generated) SiO_2 with 2.3 nm HfSiON on top and Fig. 1c) 1.9 nm SiON. The gate stack consists of 5 nm MOCVD TiN covered by 100 nm polysilicon. More details on the device fabrication and characteristics can be found in (13).

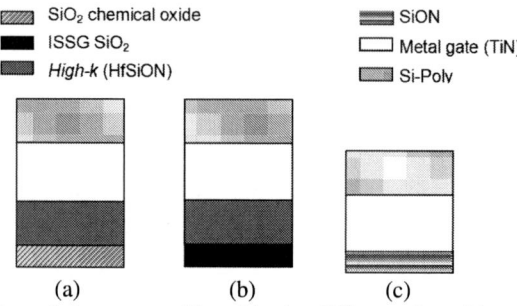

(a)   (b)   (c)

Figure 1. Schematics of the gate oxide with the different deposition sequences: (a) SiO_2/HfSiON/TiN; (b) ISSG SiO_2/HfSiON/TiN and (c) SiON/TiN.

Results and discussions

A summary of the V_T and the maximum transconductance ($g_{m,max}$) behavior is presented in Figure 2 as a function of the effective channel width. A higher V_T is observed for the HfSiON dielectric and this is related to the increased V_{FB} (due to an increase on the interface trap density) even for a reduced equivalent oxide thickness (EOT). The low EOT for HfSiON can be related to the larger dielectric constant observed for these gate stacks (14). The increase in interface trap density for HfSiON dielectric can also be derived from the reduced maximum transconductance that is related with a lower mobility.

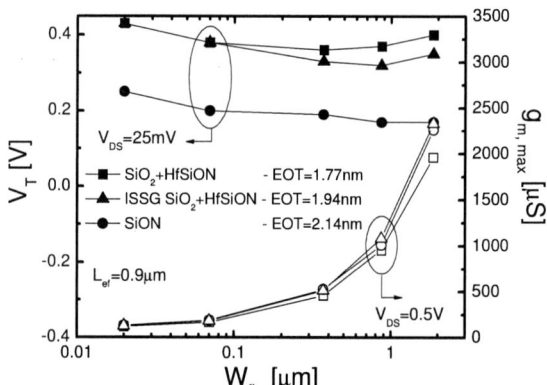

Figure 2. Threshold voltage and maximum transconductance as a function of W_{fin} for the different gate dielectrics.

The Early voltage ($V_{EA} \cong I_D/g_D$) was extracted from the output conductance (g_D) at a gate voltage overdrive ($V_{GT} = V_{GF} - V_T$) of 200mV for the different gate dielectrics, as shown in Figure 3. As can be seen, a reduced V_{EA} is observed for the hafnium dielectric, while the SiON gate oxide gives the highest values. Comparing both high-k dielectrics, a lower V_{EA} is found for the devices with an ISSG SiO_2 compared to a chemical oxide as interfacial layer.

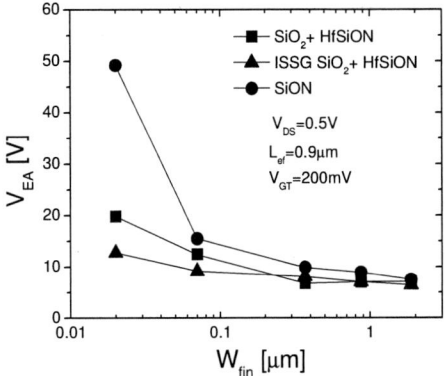

Figure 3. Extracted V_{EA} (V_{GT}=200mV) as a function of W_{fin} for the different gate dielectrics at V_{DS}=0.5V.

This behavior could be related with a possible variation in the electric field components with gate oxide thickness (t_{ox}), as illustrated in Figure 4, giving a schematic view of a transistor and the different electric field components. The slightly higher physical gate dielectric thickness for a HfSiON dielectric (3 nm) compared to SiON (1.9 nm) could lead to a reduction of the vertical electric field (E_S) influence on the drain current, increasing in that way the horizontal electric field contribution. As a result a smaller V_{EA} is observed for the devices with HfSiON as gate dielectric.

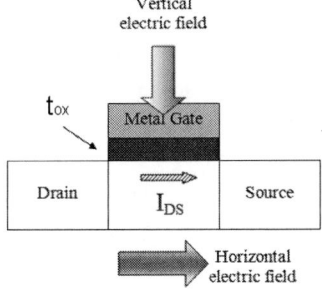

Figure 4. Schematic of the electric field components in a transistor.

This behavior can also be seen in Figure 5 showing the gate-induced drain leakage (GIDL) current extracted in saturation (15). The current under negative gate voltage is dominated by a component of the off-state leakage current that can be modeled by eq. 1, showing its dependence on the vertical electric field (eq. 2).

$$J_{GIDL} = A \cdot E_s \cdot \exp(-B/E_s) \qquad [1]$$

$$E_S = \frac{(V_{DG} - V_{FB} - 1.2)}{3 \cdot t_{ox}} \qquad [2]$$

where A is a pre-exponential parameter, B (typically 23–70MV/cm) is a physics-based exponential parameter (15), V_{FB} is the flatband voltage and V_{DG} is the drain gate voltage. As can be seen in Figure 5, a lower GIDL is observed for a hafnium dielectric than for SiON and this result is in agreement with an expected reduced E_S. According to eq. 2, it is also possible to correlate this electric field reduction with a possible V_{FB} increase for devices with a HfSiON gate dielectric. A lower V_{EA} and GIDL are observed for ISSG SiO_2+HfSiON when compared with the chemical SiO_2+HfSiON dielectric, which can be related with a higher EOT for ISSG SiO_2+HfSiON.

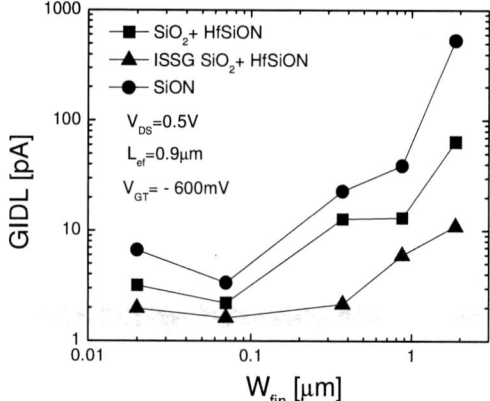

Figure 5. Extracted GIDL (V_{GT}= - 600mV) as a function of W_{fin} for the different gate dielectrics at V_{DS}=0.5V.

Finally, the device intrinsic voltage gain (A_v=V_{EA}*g_m/I_{DS}) has been extracted and is presented as a function of the fin width at V_{DS}=0.5V and V_{GT}=200mV in strong inversion (Fig. 6). A SiON dielectric offers a higher A_V due to the increased V_{EA} and mobility (higher transconductance). For both hafnium dielectrics, even with the reduced transconductance observed for HfSiON, a higher A_V is seen thanks to the improved V_{EA}.

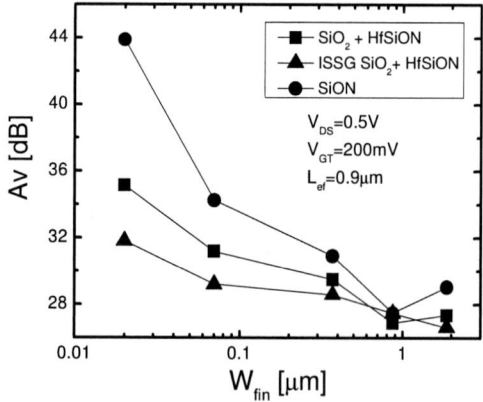

Figure 6. Calculated intrinsic gain as a function of W_{fin} for the different gate dielectrics at V_{DS}=0.5V and V_{GT}=200mV.

The noise spectra versus frequency (Figure 7) are typically of the $1/f^\gamma$ type for all the cases studied. It is clear that from a noise perspective the gate dielectric with ISSG SiO$_2$ as interfacial layer yields a lower noise due to a better quality interface with the silicon channel.

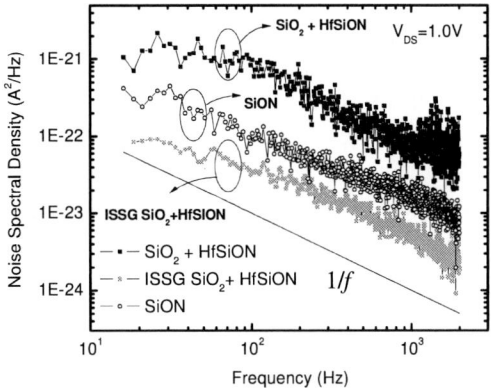

Figure 7. Noise spectral density versus frequency.

Conclusions

N-channel MuGFETs with high-k dielectric show a lower intrinsic voltage gain when compared with counterparts with a SiON insulator. This was assigned to a lower V_{EA} obtained due to the higher V_{FB} and gate oxide thickness. The dielectric with an ISSG silicon oxide as interfacial layer before the hafnium deposition presents the poorest gain but has a lower 1/f noise.

Acknowledgments

The authors would like to thank Prof. Marcelo Pavanello and the Centro Universitário da FEI for support with the noise measurements. Michele Rodrigues, Milene Galeti and Joao Antonio Martino would like to thank the Brazilian research-funding agencies FAPESP and CNPq for the support for developing this work. Part of the work has been performed within the frame of the CNPq-FWO Brazil-Flanders cooperation agreement.

References

1. S.-H. Lo, D. A. Buchanan, Y. Taur, W. Wang, *IEEE Electron Device Letters*, **18**, 209 (1997).
2. G. Vellianitis et al., in *IEEE International Electron Devices Meeting/IEDM 2007*, p. 681 (2007).
3. M. R. Visokay, J. J. Chambers, A. L. P. Rotondaro, A. Shanware, L. Colombo, *Applied Physics Letters*, **80**, 3183 (2002).

4. Min Yang, Evgeni P. Gusev, Meikei Ieong, Oleg Gluschenkov, Diane C. Boyd, Kevin K. Chan, Paul M. Kozlowski, Christopher P. D'Emic, Raymond M. Sicina, Paul C. Jamison, and Anthony I. Chou, *IEEE Electron Device Letters*, **24**, 339, (2003).
5. S. Yongxun Liu Kijima, E. Sugimata, M. Masahara, K. Endo, T. Matsukawa, K. Ishii, K. Sakamoto, T. Sekigawa, H. Yamauchi, Y. Takanashi, E. Suzuki, *IEEE Transactions on Nanotechnology*, **5**, 729 (2006).
6. I. Ferain, N. Collaert, B. O'Sullivan, T. Conard, M. Popovici, S. Van Elshocht, J. Swerts, M. Jurczak, K. De Meyer, in *38th European Solid-State Device Research conference/ESSDERC 2008*, Stephen Hall and Anthony Walton, Editors, p. 202 (2008).
7. K. Choi, H.-C Wen, H. Alshareef, R. Harris, P. Lysaght, H. Luan, P. Majhi, B. H. Lee, in *35th European Solid-State Device Research conference/ESSDERC 2005*, Gérard Ghibaudo, Thomas Skotnicki, Sorin Cristoloveanu and Michel Brillouët, Editors, p. 101 (2005).
8. J. P. Colinge, *Solid-State Electronics*, **48**, 897 (2004).
9. V. Kilchytska, N. Collaert, R. Rooyackers, D. Lederer, J-P Raskin, D. Flandre, in *34th European Solid-State Device Research conference/*ESSDERC 2004, R.P. Mertens and C.L. Claeys, Editors, 65 (2004).
10. D. Lederer, V. Kilchytska, T. Rudenko, N. Collaert, D. Flandre, A. Dixit, K. De Meyer, J-P Raskin, *Solid-State Electron*, **49**, 1488 (2005).
11. V. Subramanian, B. Parvais, J. Borremans, A. Mercha, D. Linten, P. Wambacq, J. Loo, M. Dehan, N. Collaert, S. Kubicek, R.J.P. Lander, J. C. Hooker, F. N. Cubaynes, S. Donnay, M. Jurczak, G. Groeseneken, W. Sansen, S. Decoutere, in *IEEE International Electron Devices Meeting/IEDM 2005*, 898 (2005).
12. M. A. Pavanello, J. A. Martino, E. Simoen, R. Rooyackers, N. Collaert, C. Claeys, *Solid-State Electron*, **51**, 285 (2008).
13. I. Ferain, N.J. Son, L. Witters, N. Collaert, B. Onsia, B. Kaczer, T. Kauerauf, C. Adelmann, P. Favia, O. Richard, H. Bender, S. Van Elshocht, P. Lehnen, H.T. San, K. De Meyer, S. Biesemans, M. Jurczak, *in Proceeding of the IEEE SOI Conference 2007*, 141 (2007).
14. M. Saitoh, M. Terai, N. Ikarashi, H. Watanabe, S. Fujieda, T. Iwamoto, T. Ogura, A. Morioka, K. Watanabe, T. Tatsumi and H. Watanabe, *Japanese Journal of Applied Physics,* **44**, 2330 (2005).
15. Y.-K Choi, D. Ha, T.-J King, J. Bokor, *Japanese Journal of Applied Physics*, **42**, 2073 (2003).

X-Ray Radiation Effects in the Circular-Gate Transistors

K. H. Cirne[a,b], M. A. G. Silveira[c], J. A. De Lima[b], L. E. Seixas Jr.[b] and S. P. Gimenez[a]

[a] Electrical Engineering Department, FEI University Center, Sao Bernardo do Campo, Sao Paulo 09850-901, Brazil
[b] Center for Technology for Information (CTI) - IC Design House, Campinas, Sao Paulo 13069-901, Brazil
[c] Physics Department, FEI University Center, Sao Bernardo do Campo, Sao Paulo 09850-901, Brazil

This work performs two experimental comparative analyses of the x-ray radiation effects in the Conventional, Wave and Overlapping-Circular-Gate nMOSFETs. In the first experiment, the x-ray radiation influence is studied without biasing the devices during the irradiation process, considering two channel lengths and after they have been exposed up to a x-ray irradiation of 1.5 Grad and with a dose ratio of 22 Mrad/min. The second one performs an experimental comparative study of the x-ray radiation influence between the Conventional and Overlapping-Circular Gate nMOSFET for a channel length equal to 12 μm, when they are submitted to the x-ray irradiation of 60 Mrad and maintaining the same bias conditions (overdrive gate and drain voltages) during the irradiation process. In both studies, we observe that the Overlapping-Circular Gate layout style presents higher x-ray irradiation robustness than those found in the other transistors studied, due to the absence of the bird's beak in Overlapping-Circular Gate MOSFET.

I. Introduction

The electronic devices used for space applications are strongly influenced by the irradiation effects that exist in the natural space environment (charged particles like electrons (up to 30 MeV) and protons (up to 500 MeV), heavy ions (up to 10 MeV/nucleon), cosmic rays and x-ray irradiation, etc) (1). These high-energy ionizing particles in the space environment are responsible by degrading of the performance of the electronics devices. The Total Ionization Dose (TID) effect is generated by accumulation of ionizing dose deposition over a long time in insulators (1) and can cause displacements in the threshold voltage (V_{TH}), increase of the subthreshold slope and off-state drain current (I_{OFF}) (1-2). These effects are mainly caused by the accumulation of charges in the bird beak structure that exist in rectangular-geometry MOSFET (2).

Low cost layout techniques, such as Enclosed Layout Transistors (ELT) (Hexagonal gate shape), can be used to reduce the radiation effects (3). The ELTs can also be implemented with Circular Gate Transistor (CGT) (4), Wave Transistors (5) and Overlapping-Circular Gate Transistors (O-CGT) (6).

Knowing that the Partially-Depleted (PD) SOI MOSFET presents similar behavior than the one found in the conventional (bulk) MOSFET (7), this work performs and experimental study of the TID effects in different layout structures (Conventional, Wave and O-CGT) in order to verify which geometry features higher radiation hardness. Experimental analyses performed are twofold: i) with devices unbiased during the x-ray radiation process and ii) with the transistors under same bias conditions (overdrive gate and drain voltages).

II. The MOSFETs

Figure 1 shows an example of the Wave MOSFET (WM) (5) (Fig. 1.a) and an example of the O-CGT (6) (Fig. 1.a), respectively. Both structures present higher integration capability than the one found in the conventional (rectangular) transistor (6, 7).

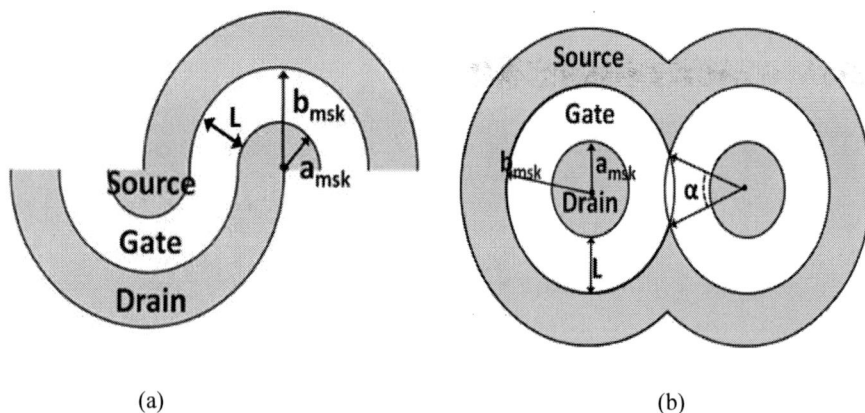

(a)   (b)

Figure 1. Examples of the Wave MOSFET (a) and O-CGT's layouts (b), respectively.

In Figure 1, L is the channel length, W is the channel width, a_{msk} and b_{mak} are respectively the internal and external radius that define the transistor channel length, and α is the overlapping angle corresponding to the region where the gate interconnection of the O-CGT occurs.

The relations between the geometric factors (W/L) of the CM and WM (5) and CM and O-CGT (6) are given by the equations [1] and [2], respectively.

$$\left(\frac{W}{L}\right)_{CM} = \left[\frac{2\pi}{\ln\left(b_{msk}/a_{msk}\right)}\right]_{WM}$$ [1]

$$\left(\frac{W}{L}\right)_{CM} = \left(\frac{360^o - 2\alpha}{360^o}\right)\left[\frac{2\pi}{\ln\left(b_{msk}/a_{msk}\right)}\right]_{O-CGT} \qquad [2]$$

The nMOSFETs were manufactured by using the standard (bulk) CMOS technology from 0.35 µm AMI (On-Semiconductor) manufacturing process, via MOSIS (MEP).

Table I presents the dimensions of the CMs, WMs and O-CGTs counterparts. The α angles of the O-CGTs are equivalent to 46.8° and 56.2° for channel length of 2.3 µm and 12 µm, respectively.

TABLE I. Dimensions of the CMs, WMs and O-CGTs for L equal to 2.3 µm and 12 µm, respectively.

	CM		O-CGT		WM	
L (µm)	2.3	12	2.3	12	2.3	12
W/L	10.6	3.5	11.4	4.7	10.2	3.5

III. Radiation Source

The X-Ray Diffractometer XRD-7000 of Shimadzu was used to irradiate the nMOSFETs. (FEI University Center in São Bernardo do Campo, São Paulo, Brazil). The Diffractometer was configured with beam current and voltage of 30 mA and 40 keV, respectively, providing a constant radiation rate of 22 Mrad/min to irradiate the unbiased devices. Alternately, the x-ray source was configured with beam current and voltage of 5 mA and 20 keV to irradiate biased transistors, at a rate of 1.15 Mrad/min. Unbiased and biased devices were irradiated up to 1.5 Grad and 60 Mrad, respectively.

IV. Experimental Results

For the case of unbiased devices during the x-ray exposure, Figures 2 displays the $I_{DS}/(W/L)$ as a function of V_{GT} curves of the CM, WM and O-CGT with L=2.3 µm (Fig. 2.a) and L=12 µm (Fig. 2.b) respectively, for V_{DS}=10 mV.

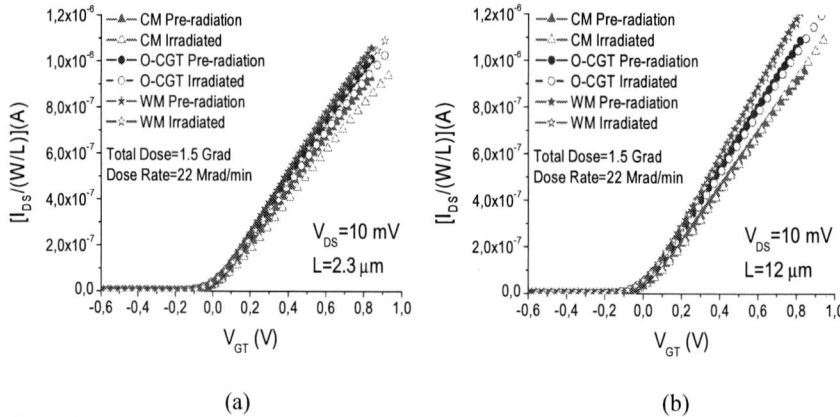

(a)   (b)

Figure 2. $I_{DS}/(W/L)$ as a function of V_{GT} curves of the devices with L=2.3 μm (a) and L=12 μm (b) respectively, for V_{DS}=10 mV (the transistors were not biased during the x-ray irradiation process).

When the transistors are not biased, we observe that V_{TH} variations were practically the same for all devices after the x-ray irradiation process and presented an average V_{TH} variation around 15 %. Therefore, we can assume that the presence of the bird's beak structure of the CMs and WMs mitigates the TID effect caused by the x-ray irradiation.

Considering now the biased transistors during irradiation, Figure 3 presents the $I_{DS}/(W/L)$ as a function of V_{GS} curves of the CM (Fig. 3.a) and O-CGT (Fig. 3.b), respectively, for V_{DS} equal to 100 mV, for L equal to 12 μm and different doses (10, 30 and 60 Mrad, respectively). These devices were biased with V_{GT} and V_{DS} equals to 0 V and 50 mV, respectively, during the x-ray radiation process. The CM and O-CGT V_{TH} were equals to 0.7 V.

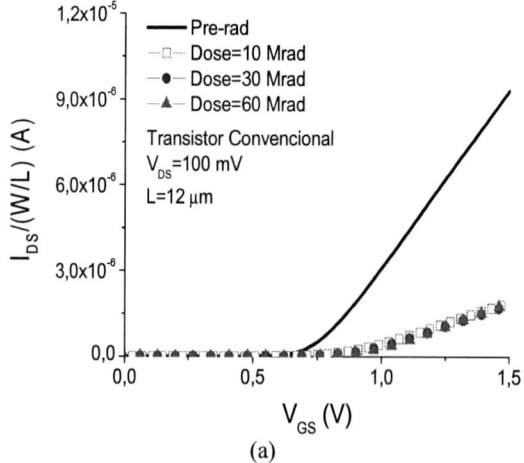

(a)

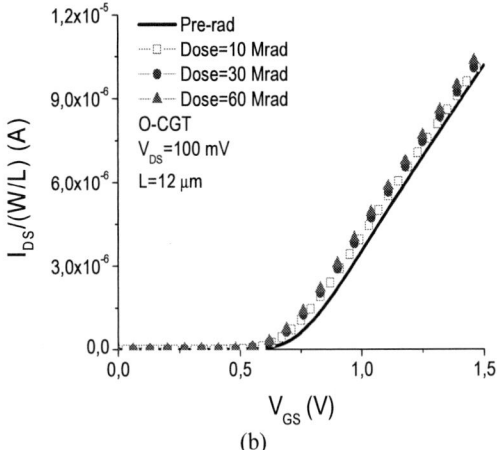

(b)

Figure 3. $I_{DS}/(W/L)$ as a function of V_{GS} curves of the CM (a) and O-CGT (b) where the devices were biased with V_{GT} and V_{DS} equal to 0 V and 50 mV respectively, during the x-ray irradiation process.

Table II presents the V_{TH} variations of the devices biased during the x-ray irradiation process as a function of the irradiation dose.

TABLE II. The V_{TH} variations of the MOSFETs for the different doses and the devices were biased during the x-ray irradiation process.

CM				O-CGT		
Dose (Mrad)	V_{TH} (V)	Var (%)		Dose (Mrad)	V_{TH} (V)	Var (%)
0	0.74			0	0.74	
10	0.88	18.92%		10	0.68	-8.11%
30	0.92	24.23%		30	0.65	-12.16%
60	0.96	28.38%		60	0.65	-12.16%

We can remark that the O-CGT presents smaller variations on parameters V_{TH} and I_{DS} with respect to those found in the CM counterpart. Such an improved behavior upon irradiation may be attributed to the presence of the bird's beak structure in the CM.

V. Conclusions

This paper analyzed the x-ray irradiation effects (TID) taking into account two different approaches regarding the bias of the devices.

In the first study the devices (Conventional, Wave and Overlapping-Circular Gate MOSFETs) were not biased during the x-ray radiation exposition in contrast to the second one, where we kept the same bias conditions (overdrive gate and drain voltages) in the MOSFETs during the irradiation process. In this second study, we only considered two devices (Conventional and Overlapping-Circular Gate MOSFETs) to perform the experiment.

From the collected data, it was verified that, when the devices were unbiased during the x-ray radiation process, TID effects in the Conventional, Wave and Overlapping-Circular Gate MOSFETs are around 15% for all devices studied. Besides that, the presence of the bird's beak structure in the Conventional and Wave MOSFETs does not have relevant influence to affect significantly these devices when they are exposed to the radiation when we compare to the influence found in the Overlapping-Circular Gate MOSFETs.

When we compare the conventional and Overlapping-Circular Gate MOSFETs, considering that they are biased during the x-ray irradiation process, we observe that the O-CGT presents higher x-ray irradiation robustness than the one of the Conventional MOSFET, keeping the same bias conditions (overdrive gate and drain voltages).

Therefore, we conclude that O-CGT layout style is an alternative device to increase the x-ray radiation robustness of the analog and digital integrated circuits for the applications in space electronics.

VI. Acknowledgments

The authors would like to thank CNPQ (grant 311149/2009-0) and CI Brazil program by the financial support and MOSIS by manufacturing the ICs. The authors also acknowledge the CTI for providing the measurements equipments and for packaging the MOSIS ICs.

References

1. Duzzelier, S., Aerospace Science and Technology, **9**, p. 93 (2005).
2. H. J. Barnaby, *IEEE Transactions on Nuclear Science*, **53** (6), p. 151 (2006).
3. Jie Liu, et al., *Microelectronics Reliability*, **50**, p. 45 (2010).
4. S. P. Gimenez et. al., in *SBMicro 2006*, J. Diniz et. al. Editors, ECS Trans. 4 (1), p. 309, Ouro Preto, Brazil (2007).
5. S. P. Gimenez, in *215th ECS Meeting*, Y. Omura et. al. Editors, ECS Trans. 19 (4), p. 153, San Francisco, CA (2009).
6. J. A. De Lima and S. P. Gimenez, in SBMicro 2009, ECS Trans. 23 (1), p. 361, Natal, Brazil (2009).
7. J. P. Colinge, *Silicon-On-Insulator Technology: Materials to VLSI*, p. 156, Kluwer Academic Publishers, Boston (2004).

CHAPTER 12

NANOSCALE SIMULATIONS

ECS Transactions, 35 (5) 267-276 (2011)
10.1149/1.3570805 ©The Electrochemical Society

Thermoelectric Properties of Silion-On-Insulator Nanostructures

Z. Aksamija and I. Knezevic

Department of Electrical and Computer Engineering, University of Wisconsin-Madison,
Madison, Wisconsin 53706, USA

Hotspots in integrated circuits degrade the performance and reliability of devices and pose a challenge to further device scaling (1). Thermoelectric (TE) refrigeration using Si-based nanowires and nanoribbons is an attractive approach for targeted cooling of local hotspots (2) due to the ease of on-chip integration and the nanowires' enhanced TE figure of merit (3), given by $ZT=S^2T\sigma/\kappa$ (4). Silicon-on-insulator (SOI) membranes (5) and membrane-based nanowires and ribbons show promise for application as efficient thermoelectrics, which requires both high power factor $S^2\sigma$ and low thermal conductivity κ. The design of efficient semiconductor thermocouples requires a thorough understanding of both charge and heat transport; therefore, thermoelectricity in silicon-based nanostructures requires that both electronic and thermal transport are treated on equal footing.

Introduction

Thermoelectric (TE) materials can achieve conversion of thermal-to-electrical and electrical-to-thermal energy with no moving parts. A desirable property of solid-state thermoelectric conversion is its scalability to very small devices, even down to the nanoscale, where it can be incorporated into integrated circuits, lab-on-a-chip solutions, and other small electronics. Semiconductor nanostructures are promising candidates for efficient thermoelectric energy conversion, with applications in solid-state refrigeration and power generation. Thermoelectric refrigeration using Si-based nanowires and nanoribbons is especially attractive for targeted cooling of local hotspots in densely packed integrated circuits (1,2) due to the ease of on-chip integration and the nanowires' enhanced TE figure of merit (3,4). Silicon-on-insulator (SOI) membranes (5) and membrane-based nanowires and ribbons (4) show promise for application as efficient thermoelectrics, which requires both high power factor and low thermal conductivity.

Semiconductors are the class of thermoelectric materials with the highest dimensionless figure of merit **ZT** (1). ZT is a composite figure of merit, equal to $S^2T/\rho\kappa$, where S is the Seebeck coefficient (thermopower), σ is the electrical conductivity, κ is the thermal conductivity, and T is the average operating temperature. In semiconductors, the power factor S^2/ρ and thermal conductivity κ are largely decoupled: the power factor is governed by charge carrier transport and can be enhanced by doping, while thermal transport is governed by phonon scattering (6).

Bulk silicon has been studied as a thermoelectric material for many years, and it is well known that its ZT of 0.01 is insufficient for practical applications. Enhancement of ZT can be achieved through a selective reduction in thermal conductivity κ, while at the same time maintaining electronic properties, such as the power factor $S^2\sigma$. This approach is usually referred to as the 'electron-crystal, phonon-glass' approach (1). Reduced thermal conductivity κ is achievable through nanostructuring (such as making nanoscale wires, ribbons, and membranes), alloying, surface roughening, or incorporating structural heterogeneities. Recently, silicon nanowires and nanoribbons have emerged as promising thermoelectric materials (3,4). They experience an enhancement of nearly two orders of magnitude in the thermoelectric figure of merit ZT over bulk, because of reduced thermal conductivity caused by phonon scattering at rough boundaries (7). Therefore, it is expected that nanoengineering of semiconductor structures can lead to improved thermoelectric performance, especially by manipulating surfaces and interfaces in order to achieve the desired reduction of thermal conductivity.

In this paper, we discuss both charge and thermal transport in silicon nanomembranes and nanoribbons, and treat thermoelectric power factor and lattice thermal conductivity, both key ingredients of the TE figure of merit. We show that confinement in gated silicon nanoribbons allows us to tune the Seebeck coefficient of the structure and brings about an increase in the total TE power factor over what is commonly achievable in doped nanostructures (8). We also calculate the lattice thermal conductivity in such thin silicon nanomembranes and show that the interaction between lattice vibrations and surfaces dominates thermal transport and results in a strong anisotropy of thermal conductivity, with thermal conductivity differing by up to a factor of 2 for different surface orientations (9). This anisotropy allows us to tailor the thermal conductivity of nanostructures to suit particular applications, such as choosing a low thermal conductivity orientation in order to improve the figure of merit for thermoelectric applications.

The paper is organized as follows: first we present the phonon thermal conductivity model, discuss the calculation of scattering rates, and present results for the in-plane thermal conductivity of 20 nm silicon nanomembranes with different surface orientation and transport directions. Next we present our calculation of the Seebeck coefficient, including both diffusion and phonon drag components, and power factor of gated silicon nanoribbons. We demonstrate that thermoelectric properties can be tuned using the gate voltage and a very high power factor can be achieved. Finally we conclude with a brief summary and a few final remarks.

Modeling thermal conductivity of ultrathin SOI nanomembranes

Thermal conductivity in thin SOI layers, as well as in thin wires and ribbons, is dominated by boundary scattering even at room temperature. Therefore, surface orientation and the direction of heat flow are expected to play a significant role in thermal transport and they offer additional degrees of freedom to control the thermal conductivity in confined systems.

We demonstrate the sensitivity of the lattice thermal conductivity in thin SOI to the surface crystalline orientation and the direction of heat flow, based on solving the phonon Boltzmann transport equation under the relaxation time approximation (9). In this work, in order to capture the full anisotropy of lattice heat conduction, we calculate the thermal conductivity tensor

$$\mathbf{K}^{\alpha\beta}(T) = k \sum_{j} \sum_{\vec{q}} \left[\frac{\hbar\omega_j(\vec{q})}{kT} \right]^2 \tau_j(q) v_j^{\alpha}(\vec{q}) v_j^{\beta}(\vec{q}) \frac{e^{\hbar\omega_j(\vec{q})/kT}}{\left[e^{\hbar\omega_j(\vec{q})/kT} - 1 \right]^2}, \qquad [1]$$

where $v_j^{\alpha}(\vec{q})$ and $\omega_j(\vec{q})$ are the α-th component of the phonon velocity vector and the frequency in branch j, respectively, while k is the Boltzmann constant and T is the temperature. We calculate the full phonon dispersion from the adiabatic bond charge model (10) that accurately captures the phonon dispersions in all directions. Bulk dispersions give an accurate description of phonon states in SOI as thin as 5 nm (11).

The calculation is made more tractable by expressing the thermal conductivity in terms of the generalized transport matrices (12)

$$\mathbf{K}^{\alpha\beta}(T) = \frac{1}{T} \left[L^{(2)} \right]^{\alpha\beta}$$

$$\left[L^{(2)} \right]^{\alpha\beta} = \sum_j \int_0^{\omega_j} d\omega \left(\frac{dN_0}{d\omega} \right) \sigma_b^{\alpha\beta}(\omega) (\hbar\omega)^2, \qquad [2]$$

where N_0 is the equilibrium Bose-Einstein phonon distribution function, given by

$$N_0(\omega) = \left[\exp\left(\frac{\hbar\omega}{k_B T} \right) - 1 \right]^{-1}, \qquad [3]$$

and the limit of integration is ω_j $$, the highest phonon frequency of each phonon branch j. The energy dependent thermal conductance is given by

$$\sigma_j^{\alpha\beta}(\omega) = \int \frac{d\vec{q}}{(2\pi)^3} \tau_j(\vec{q}) \vec{v}_j^{\alpha}(\vec{q}) \vec{v}_j^{\beta}(\vec{q}) \delta\left[\omega - \omega_j(\vec{q}) \right], \qquad [4]$$

where the integral of the energy-conserving delta function over the whole first Brillouin zone is calculated using the numerical integration method of Gilat and Raubenheimer (12).

In the calculation of the relaxation time $\tau_j(\vec{q})$ for a phonon in mode j and with wave vector $\vec{q}$, we consider normal (N) and umklapp (U) three-phonon scattering, isotope scattering (I), and boundary surface roughness (B) surface roughness scattering, with branch- and momentum-dependent relaxation times $\tau_{j,N}(\vec{q})$, $\tau_{j,U}(\vec{q})$, $\tau_{j,I}(\vec{q})$, and $\tau_{j,B}(\vec{q})$, respectively. The total relaxation time $\tau_j(\vec{q})$ is given by

$$\frac{1}{\tau_j(\vec{q})} = \frac{1}{\tau_{j,N}(\vec{q})} + \frac{1}{\tau_{j,U}(\vec{q})} + \frac{1}{\tau_{j,I}(\vec{q})} + \frac{1}{\tau_{j,B}(\vec{q})}. \qquad [5]$$

We employ a momentum-dependent specularity parameter (6)

$$p(\vec{q}) = \exp\left(-4\Delta^2 q^2 \cos^2 \Theta_B \right), \qquad [6]$$

where Θ_B is the angle between the incident phonon and the surface normal. This formulation allows us to connect the specularity parameter directly to the magnitude of surface roughness. The momentum-dependent specularity parameter can be averaged over the energy isosurface in order to plot the energy dependence, shown in Figure 1. The plot shows that surface roughness scattering is nearly specular for phonons with small energy (and correspondingly large wavelenght/small momentum) and nearly diffuse (p=0) for larger energy phonons (due to their small wavelength). Diffuse boundary scattering gives a phonon a lifetime that is proportional to the smallest distance between the boundaries; in the case of thin SOI, this is the thickness L of the layer, producing

$$\tau_{boundary}(\omega) = \left[\frac{1+p(\vec{q})}{1-p(\vec{q})}\right]\frac{L}{v(\vec{q})},$$ [7]

where $v(\vec{q})$ is the phonon group velocity for a given branch. In the calculation of the mean free path Λ, surface roughness, Umklapp phonon-phonon, and isotope scattering have been considered.

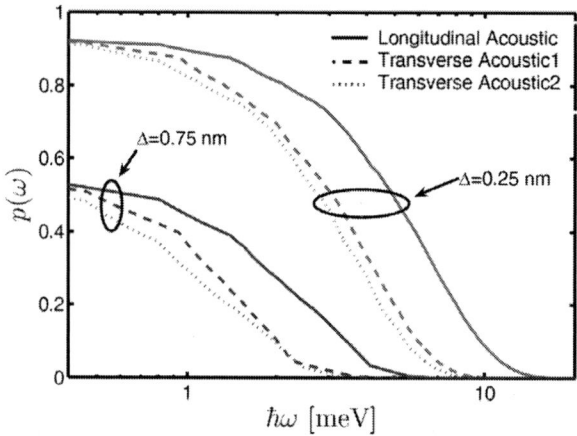

Figure 1. Dependence of the specularity parameter p on the phonon energy for two values of surface roughness rms amplitude Δ =0.25 nm and Δ =0.75 nm. The momentum-dependent specularity parameter was averaged over the energy isosurface in order to plot the energy dependence. The plot shows that surface roughness scattering is nearly specular for phonons with small energy (and correspondingly large wavelenght/small momentum) and nearly diffuse (p=0) for larger energy phonons (due to their small wavelength).

With the exception of $1/\tau_{j,B}$, the scattering rates do not depend on the phonon momentum direction explicitly but rather on its energy and are thus constant on energy isosurfaces. The dependence of N and U relaxation times on the phonon energy and lattice temperature can be expressed through

$$\frac{1}{\tau_{j,N/U}^{LA/TA}(\omega)} = A_{N/U}^{TA/LA}\omega_j^n g(T),$$ [8]

where n=2 for normal scattering and n=4 for umklapp scattering (13). This model, based on *ab initio* calculations, gives a natural cross-over between normal scattering, dominant among low-energy phonons, and umklapp scattering, dominant among higher-energy phonons (13). The temperature function is given by

$$g(T) = T[1 - \exp(-3T/\Theta_D)],$$ [9]

with $\Theta_D = 645$ K being the Debye temperature of silicon. The constants $A_{N/U}^{TA/LA}$ are dependent on the polarization of the phonon branch. Based on *ab initio* calculations, their values for the normal scattering process are $A_N^{TA} = 253322\,(meV^2 Ks)^{-1}$ for the transverse acoustic modes, and $A_N^{LA} = 163921$ (meV^{2}Ks)$^{-1}$ for the longitudinal acoustic mode. For the case of umklapp scattering, the constants are $A_U^{TA} = 2012$ (meV^{4}Ks)$^{-1}$ and $A_U^{LA} = 507$ (meV^{4}Ks)$^{-1}$ (13).

Isotope scattering has no temperature dependence, and the energy dependence of the isotope scattering rate is calculated from (14)

$$\frac{1}{\tau_I(\omega)} = \frac{\pi V_0}{6}\Gamma_{Si}\omega^2 \, VDOS(\omega), \qquad [10]$$

where V_0 is the volume per atom and VDOS(ω) is the vibrational density of states calculated from the full phonon dispersion using the method of Gilat and Raubenheimer (12). $\Gamma_{Si} = \sum_i f_i(1 - m_i / \bar{m})^2$, where f_i is the natural abundance of isotope i with mass m_i, and the average mass is $\bar{m} = \sum_i f_i m_i$. The value of Γ_{Si} is 2.0×10^{-4} (15).

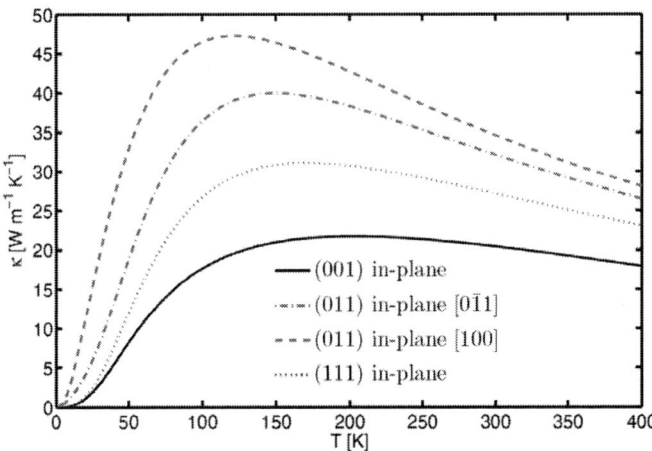

Figure 2. Plot of the lattice thermal conductivity for a 20 nm thick SOI with rms surface roughness Δ =0.5 nm. Strong dependence on both the surface orientation and the direction of heat propagation is apparent: surface orientation dictates the roughness-limited phonon scattering rate, while the component of the phonon group velocity in the direction of the thermal gradient controls how fast each phonon mode carries heat.

Thermal conductivity along a particular direction expressed by the unit direction vector $\hat{t}$ is then computed by pre- and post-multiplying the thermal conductivity tensor by the direction vector $\kappa_t(T) = \sum_{\alpha\beta} \hat{t}^\alpha \mathbf{K}^{\alpha\beta}(T)\hat{t}^\beta$. Doing so reduces thermal conductivity to a scalar by using the projection of the phonon velocity vector on the direction of transport $v_t(\vec{q}) = \vec{v}(\vec{q})\cdot\hat{t} = \| \vec{v}(\vec{q}) \| \cos\Theta_t$, where Θ_t is the angle between the phonon velocity vector and the transport direction. This reduces $\sigma_j(\omega)$ to

$$\sigma_j(\omega) = \int \frac{d\vec{q}}{(2\pi)^3}\tau_j(\vec{q})v_t(\vec{q})^2 \delta\big(\omega - \omega_j(\vec{q})\big), \qquad [11]$$

and allows us to compute the value of thermal conductivity along a particular transport direction given by $\hat{i}$. Consequently, thermal conductivity is also controlled by the direction of the heat gradient through the angle between the velocity vector and transport direction, in addition to being controlled by the surface orientation.

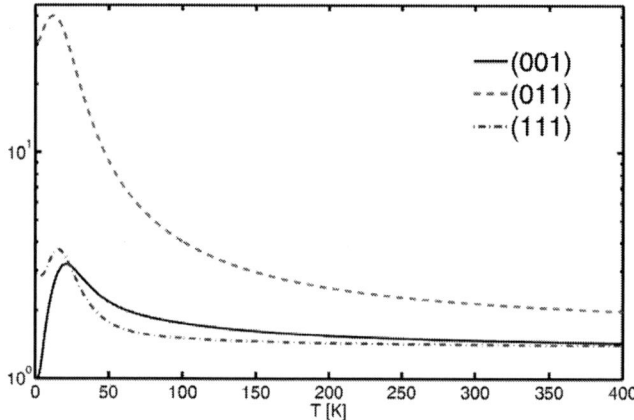

Figure 3. Plot of the ratio of the in-plane and the out-of-plane thermal conductivity as a function of temperature. At each temperature, the 3x3 tensor has distinct diagonal and off-diagonal components, leading to one thermal conductivity eignevalue for the out-of-plane transport direction, and a pair of values for the in-plane thermal conductivity eigenvalue. The in-plane and cross-plane values differ by a factor of 2 or more at all temperatures.

The computed values of κ show excellent agreement with the measurements (16) on thin SOI at both high and low temperatures. Results for 20 nm SOI with different surface and transport orientations (Figure 2) show a strong anisotropy of thermal conductivity due to the directional dependence of the phonon velocity and boundary scattering. Lowest thermal conductivity is achieved on (001) and the highest is on the (011) surface in the [100] direction (9). At each temperature, the 3x3 thermal conductivity tensor $\mathbf{K}^{\alpha,\beta}(T)$ has distinct diagonal and off-diagonal components, leading to one thermal conductivity eigenvalue for the cross-plane transport direction, and a pair of values for the in-plane thermal conductivity eigenvalue. Figure 3. shows the ratio of the in-plane and out-of-plane thermal conductivity eigenvalues. The in/out-of-plane ratio is around 2 at room temperature on the (011) surface, and increases at lower temperature. The surface orientation and the in-plane/out-of-plane anisotropy allows additional control over the magnitude of thermal conductivity in SOI membranes and presumably quasi-1D structures such as nanowires and ribbons.

Seebeck coefficient and power factor in gated silicon nanoribbons

We present the calculation of the thermopower in thin silicon-on-insulator nanoribbons, in which a back gate is used to tune the charge density (8). As the gate voltage increases, the sheet density increases and the thermopower decreases. Total thermopower can be expressed as a sum of diffusion and phonon-drag components (S_d

and S_{ph}, respectively). Charge carriers in wide and thin (~20 nm) gated nanoribbons experience 1D confinement, shown in Figure 4. The expression for the diffusion thermopower S_d is derived from the 2D Boltzmann transport equation (BTE) under the effective-mass and relaxation time approximations (RTA). At a temperature T, the conductivity σ and the diffusion thermopower S_d are given by (6)

$$\sigma = L^{(0)}$$

$$S_d = \frac{1}{qT}\frac{L^{(1)}}{L^{(0)}},$$ [12]

where the generalized transport matrix element $L^{(j)}$ is given by

$$L^{(j)} = \sum_{v,i}\frac{2q^2}{m_i^v}\int dE \frac{\partial f_0(E)}{\partial E}(E - E_F + E_i^v)^j E \tau_i^v(E) g_i^v(E).$$ [13]

Here, $f_0(E)$ is the Fermi-Dirac distribution function, and E_F is the Fermi level. E_i^v is the energy of the bottom of subband i in valley v, $g_i^v(E)$ is the density of states, and $\tau_i^v(E)$ is the relaxation time of a hole with effective mass m_i^v and kinetic energy E in that subband. The subband energies and wavefunctions were calculated by solving the 1D Schroedinger equation in the effective mass approximation self-consistently with the Poisson equation (17). Both equations were discretized using standard finite-difference methods. The discretized Poisson problem is tri-diagonal for this 1-dimensional problem, and can be solved efficiently by using the Thomas algorithm. Transport is dominated by scattering from acoustic phonons, optical phonons, and boundary roughness. $1/\tau_i^v(E)$, the scattering rate, is a sum of the rates due to all the important scattering mechanisms: acoustic phonons, optical phonons, surface roughness, and charged interface traps (18).

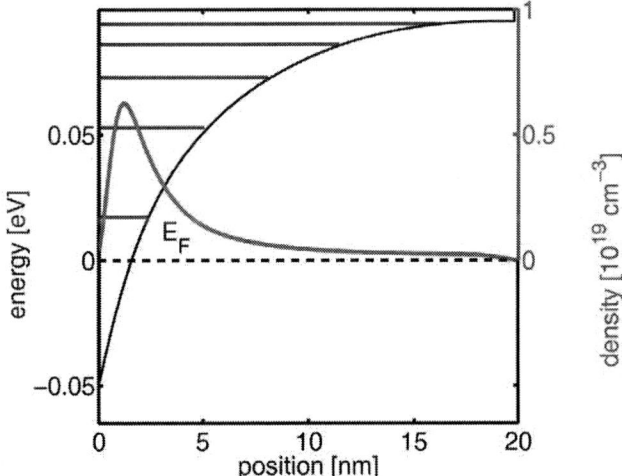

Figure 4. Schematic representation of the vertical cross-section of the gated nanoribbon, showing the potential well produced by the gating, and the subband levels inside the well (only the heavy hole subbands are depicted for clarity). The total charge density profile (thick line, right axis) shows that the inversion layer is strongly confined close to the Si/SiO₂ interface.

The phonon-drag contribution to the thermopower, S_{ph}, depends on the fraction of the total electron or hole scattering rate that corresponds to electron/hole-acoustic phonon scattering, the velocity and mean-free-path of acoustic phonons (v_{ph} and Λ_{ph}, respectively), and the hole mobility μ (19):

$$S_{ph} = \frac{\gamma v_{ph}\Lambda_{ph}}{\mu T}.$$ [14]

Because the hole mobility in inversion layers is largely phonon-limited up to very high densities (20), we assume $\gamma = 1$. The phonon mean-free-path is calculated from the phonon velocity and relaxation time by averaging over all phonon wavevectors using the same method as described in the previous section for the thermal conductivity tensor. The resulting temperature and rms roughness height dependence of the phonon mean-free-path Λ_{ph} is shown in Figure 5, where we note that the mean-free-path is larger at low temperatures and for small roughness rms heights because at low temperatures, most of the occupied phonon states have large wavelength and therefore undergo mostly specular scattering with the rough boundaries, leading to larger mean-free-paths. Conversely, at high temperatures, small wavelenght (large $\bar{q}$) phonons dominate, and they scatter diffusely with the boundaries (as shown in Figure 1), leading to a smaller relaxation time, and, consequently, smaller Λ_{ph}. This translates into a temperature dependence of the phonon drag component of the thermopower. Phonon drag accounts for ~40% of the thermopower at 100 K and 200 K, and the contribution drops with increasing temperature (phonon mean free path decreases) and increasing carrier density (roughness scattering overshadows phonon scattering).

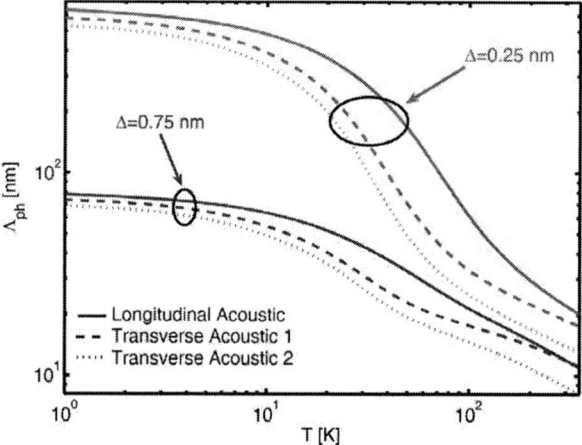

Figure 5. Dependence of the phonon mean-free-path Λ_{ph} on temperature and surface roughness rms amplitude $\Delta = 0.25$ nm and $\Delta = 0.75$ nm. The phonon mean-free-path is larger at low temperatures and for small rms roughness because at low temperatures, most of the occupied phonon states have large wavelength and therefore undergo mostly specular scattering with the rough boundaries, leading to larger mean-free-paths. At high temperatures, small wavelength phonons dominate, and they scatter diffusely with the boundaries, leading to a smaller relaxation time, and, consequently, smaller Λ_{ph}.

The hole power factor (S^2/ρ, ρ being the resistivity) of the nanoribbon is shown in Figure 6. At room temperature, the hole power factor is ~50% higher than that of heavily doped, ungated silicon nanowires (3,4). It clearly demonstrates the tunability of the power factor in nanoribbons by the gate bias, and the importance of the phonon drag contribution to thermopower, especially at high inversion layer densities. Increased power factor can be attributed to confinement in the nanoribbon and absence of dopant scattering (8).

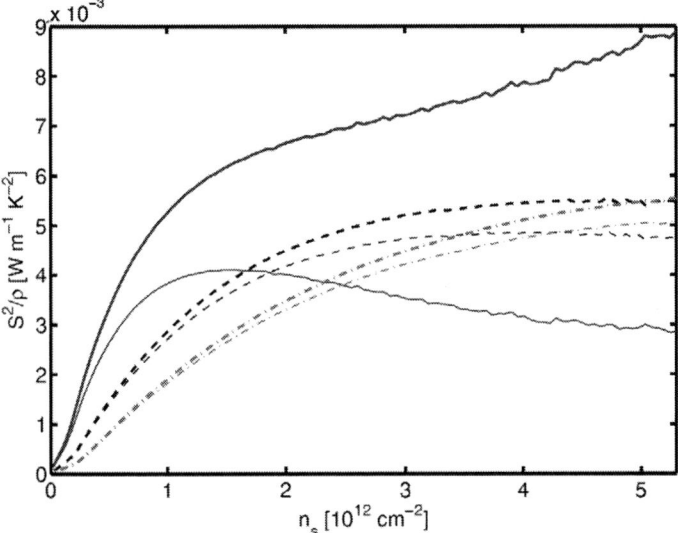

Figure 6. Power factor of the gated nanoribbon at 100 K (solid), 200 K (dashed), and 300 K (dash-dotted). The thin lines represent diffusion power factor only, and thick lines are diffusion and phonon drag together, demonstrating that the phonon drag plays a significant role in enhancing the power factor, especially at high inversion layer charge densities.

Conclusions

Silicon-based thermoelectric devices hold great promise for future applications such as on-chip cooling, requiring efficient thermoelectric conversion. Improved thermoelectric efficiency is achieved by boosting the power factor and reducing thermal conductivity. Gated silicon nanoribbons show improved power factors and offer tunability of thermoelectric properties by controlling the gate bias. Ultrathin silicon membranes have reduced thermal conductivity due to the increased scattering of phonons with rough boundaries. Thermal transport in ultrathin silicon is highly anisotropic. By changing surface orientation and direction of heat flow, we can achieve a further reduction of thermal conductivity. The improvements in the thermoelectric figure-of-merit coming from the anisotropy of the thermal conductivity can be coupled together with the effects of anisotropy in electronic transport. It has been shown that both electron and hole mobility is highly anisotropic in a wide variety of nanostructures, ranging from inversion layers to nanowires (21). Thermoelectric power factors have also been shown to depend on orientation (22). By utilizing anisotropy, the effects of increased mobility and power factor can be combined with reduced thermal conductivity in order to lead to more efficient silicon thermoelectric devices.

Acknowledgments

This work has been supported by the AFOSR (YIP program, award No. FA9550-09-1-0230) and by the NSF and the CRA (award No. 0937060) through ZA's Computing Innovation Fellowship, subaward No. CIF-146.

References

1. A. Majumdar, *Nature Nanotechnology*, vol. 4, no. 4, pp. 214–215 (2009).
2. I. Chowdhury, R. Prasher, K. Lofgreen, G. Chrysler, S. Narasimhan, R. Mahajan, D. Koester, R. Alley, and R. Venkatasubramanian, *Nature Nanotechnology*, vol. 4, no. 4, p. 235 (2009).
3. A. Hochbaum, R. Chen, R. Delgado, W. Liang, E. Garnett, M. Najarian, A. Majumdar, and P. Yang, *Nature*, vol. 451, p. 163, 2008.
4. A.I.Boukai, Y. Bunimovich, J. Tahir-Kheli, J. Yu, W. A. G. III, and J. R. Heath, *Nature*, vol. 451, p. 168, 2008.
5. M. Huang, C. S. Ritz, B. Novakovic, D. Yu, Y. Zhang, F. Flack, D. E. Savage, P. G. Evans, I. Knezevic, F. Liu, and M. G. Lagally, *ACS Nano*, vol. 3, no. 3, p. 721 (2009).
6. J. Ziman, *Electrons and Phonons: The Theory of Transport Phenomena in Solids*. Oxford University Press Inc., 1960.
7. P. Martin, Z. Aksamija, E. Pop, and U. Ravaioli, Phys. Rev. Lett., vol. 102, p. 125503 (2009).
8. H.-J. Ryu, Z. Aksamija, D. M. Paskiewicz, S. A. Scott, M. G. Lagally, I. Knezevic, and M. A. Eriksson, Phys. Rev. Lett., vol. 105, p. 256601 (2010).
9. Z. Aksamija and I. Knezevic, Phys. Rev. **B**, vol. 82, p. 045319 (2010).
10. W. Weber, Phys. Rev. **B**, vol. 15, p. 4789 (1977).
11. N. Mingo, Phys. Rev. **B**, vol. 68, 113308 (2003).
12. G. Gilat and L. Raubenheimer, Phys. Rev., vol. 144, p. 390 (1966).
13. A. Ward and D. A. Broido, Phys. Rev. **B**, vol. 81, p. 085208 (2010).
14. D. T. Morelli, J. P. Heremans, and G. A. Slack, Phys. Rev. **B**, vol. 66, p. 195304 (2002).
15. S.-I. Tamura, Phys. Rev. **B**, vol. 27, p. 858 (1983).
16. W. Liu and M. Asheghi, J. Heat Transf., vol. 128, p. 75 (2006).
17. T. Ando, A. B. Fowler, and F. Stern, Rev. Mod. Phys., vol. 54, p. 437 (1982).
18. I. Knezevic, E. B. Ramayya, D. Vasileska, and S. M. Goodnick, J. Comput. Theor. Nanosci., vol. 6, p. 1725 (2009).
19. M. Tsaousidiou, P. N. Butcher, and G. P. Triberis, Phys. Rev. **B**, vol. 64, p. 165304 (2001).
20. L. Donetti, F. Gamiz, N. Rodriguez, and A. Godoy, IEEE Electron Dev. Lett., vol. 30, p. 1338 (2009).
21. M. V. Fischetti, Z. Ren, P. M. Solomon, M. Yang, and R. Kim, J. Appl. Phys., vol. 94, p. 1079 (2003).
22. T. T. Vo, A. J. Williamson, V. Lordi, G. Galli, Nano Lett., vol. 8, p. 1111 (2008).

Properties of Silicon Ballistic Spin Fin-Based
Field-Effect Transistor

D. Osintsev, V. Sverdlov, Z. Stanojevic, A. Makarov, J. Weinbub, and S. Selberherr

Institute for Microelectronics, TU Wien, Gußhausstraße 27-29, A-1040 Wien, Austria

We investigate the properties of ballistic fin-structured silicon spin field-effect transistors. The spin transistor suggested first by Datta and Das employs spin-orbit coupling to introduce the current modulation. The major contribution to the spin-orbit interaction in silicon films is of the Dresselhaus type due to the interface-induced inversion symmetry breaking. The subband structure in silicon confined systems is obtained with help of a two-band $k \cdot p$ model and is in good agreement with recent density functional calculations. It is demonstrated that fins with [100] orientation display a stronger modulation of the conductance as function of spin-orbit interaction and magnetic field and are thus preferred for practical realizations of silicon SpinFETs.

Introduction

The spectacular increase of computational speed and power of modern integrated circuits is supported by the continuing miniaturization of semiconductor devices' feature size. With scaling approaching its fundamental limits, however, the semiconductor industry is facing the challenge to introduce new innovative elements and engineering solutions and to improve MOSFET performance. Employing spin as an additional degree of freedom is promising for boosting the efficiency of future low-power nanoelectronic devices, with high potential for both memory (1) and logic (2) applications.

Silicon, the main element of microelectronics, possesses several properties attractive for spin-driven applications: it is composed of nuclei with predominantly zero spin and is characterized by small spin-orbit interaction. Because of that the spin relaxation in silicon is relatively weak, which results in large spin life time a (3,4). In experiments, coherent spin propagation through an undoped silicon wafer of 350µm thickness was demonstrated (5). Coherent spin propagation over such long distances makes the fabrication of spin-based switching devices in the near future increasingly likely.

The original proposal for the spin transistor by Datta and Das (6) employs the spin-orbit coupling for current modulation. The current modulation appears due to spin precession in an effective magnetic field caused by the spin-orbit interaction. Due to the structural inversion asymmetry induced by the effective electric field in the conducting channel, the strength of the spin-orbit interaction becomes a function of the gate voltage.

The electrons with a non-zero spin polarization are injected from the ferromagnetic source contact into the channel. The total current through the device depends on the relative angle between the magnetization direction of the drain contact e_D and the electron spin polarization at the drain end of the conducting channel. Because the angle of the spin precession depends on the gate voltage, the total current through the device is modulated by the gate voltage.

The effective Hamiltonian of the spin-orbit interaction due to the structural-induced inversion asymmetry along the z-axis is usually considered to be of the Rashba type:

$$H_R = \alpha(p_x\sigma_y - p_y\sigma_x)/\hbar,$$ [1]

where α is the effective electric field dependent parameter of the spin-orbit interaction, $p_{x,y}$ are the electron momentum projections, and $\sigma_{x,y}$ are the Pauli matrices.

The weak strength of the spin-orbit interaction would be an obstacle to employ silicon for building a spin field-effect transistor similar to the one suggested by Datta and Das. As it is demonstrated in recent papers (7,8), the Rashba term [1] is indeed relatively small in silicon films inside SiGe/Si/SiGe heterostructures. Interestingly, in both perfect (001) silicon structures (7) and the silicon structures with interfacial disorder (8) there is another contribution to the spin-orbit interaction. Compared to the Rashba term, this contribution is approximately ten times larger, it depends strongly on the electric field, and it is described by the effective Hamiltonian

$$H_D = \beta(p_x\sigma_x - p_y\sigma_y)/\hbar,$$ [2]

which is of the Dresselhaus type. The value of the spin-orbit interaction β is estimated as $0.5 \ 10^{-12}$ eVcm at the built-in field $0.5 \ 10^5$ V/cm, in agreement with the experimental value (9). The spin-orbit interaction of such strength is sufficiently strong to investigate the possibility to build a silicon spin FET.

The stronger spin-orbit interaction, however, leads to an increased spin relaxation. The D'yakonov-Perel' mechanism is the main spin relaxation mechanism in systems with the degeneracy between the electron dispersion curves for the two spin projections lifted. In quasi-one-dimensional electron structures, however, the complete suppression of the spin relaxation was predicted (10).

Indeed, in case of the elastic scattering only back-scattering is allowed. Reversal of the electron velocity and momentum results in the inversion of the effective magnetic field direction in [2]. Therefore, the precession angle does not depend on scattering along the carrier trajectory in the channel, but is a function of the channel length only. Thus, spin-independent elastic scattering does not result in additional spin decoherence.

Model

We investigate the properties of ballistic fin-structured silicon spin field-effect transistors (SpinFETs). The SpinFET consists of the ferromagnetic source and drain electrodes connected by a silicon fin. The strength β of the spin-orbit interaction [2] depends on the electric field which is induced by applying a voltage to the gate. The Hamiltonian in the ferromagnetic regions which sandwich the silicon region are (11)

$$H_F^L = \frac{p_x^2}{2m_F} + h_0\sigma_z, \qquad x \le 0,$$ [3]

$$H_F^R = \frac{p_x^2}{2m_F} \pm h_0\sigma_z, \qquad x \ge L,$$ [4]

where L is the channel length, m_F the effective mass in the contacts, σ_z is the Pauli matrix, and $h_0 = 2PE_F/(1+P^2)$, with $P<1$ being the spin polarization and E_F the Fermi energy. The plus/minus sign in [4] stands for parallel/anti-parallel configuration of the contact magnetization.

In order to circumvent the impedance mismatch problem between the metal electrodes and the semiconductor channel and to facilitate spin current injection in the channel (5) the delta-function barriers of strength U are introduced at the interfaces between the contacts and the channel (11). Contrary to (11), the spin-orbit interaction is taken in the Dresselhaus form [2] relevant for silicon (7,8). The Hamiltonian in the silicon region $0<x<L$, for [100] and [110] fin orientations, is

$$H_S = \sum_n \frac{p_x^2}{2m_n} + \delta E_n - \frac{\beta}{\hbar} p_x \sigma_x + \frac{1}{2} g\mu_B B\sigma^*, \qquad \text{[100] fin,} \qquad [5]$$

$$H_S = \sum_n \frac{p_x^2}{2m_n} + \delta E_n - \frac{\beta}{\hbar} p_x \sigma_y + \frac{1}{2} g\mu_B B\sigma^*, \qquad \text{[110] fin,} \qquad [6]$$

where m_n is the nth subband effective mass, δE_n is the band mismatch between the nth subband in the channel and the source and drain contacts, B is the magnetic field, μ_B is the Bohr magneton, g is the Landé factor, and $\sigma^* = \sigma_x \cos\gamma + \sigma_y \sin\gamma$, with γ defined as the angle between the magnetic field and the fin direction.

Results

In our studies silicon fins have a square cross-section with (001) horizontal faces. The parabolic band approximation is not sufficient in thin and narrow silicon fins. In order to compute the subband structure in silicon fins we employ the two-band **k·p** model proposed in (12), which has been shown to be accurate up to 0.5eV above the conduction band edge (13). The resulting Schrödinger differential equation with the Hamiltonian (12), with the confinement potential appropriately added, is discretized using the box integration method and solved for each value of the conserved momentum p_x along the current directions using efficient numerical algorithms available through the Vienna Schrödinger-Poisson framework (VSP) .

Fig.1 demonstrates the dependence of the subband minima as function of the fin thickness t, for the lowest four subbands. The fin orientation is along [110] direction. The dependence of the splitting between the unprimed subbands with decreasing t, which are perfectly degenerate in the effective mass approximation, is clearly seen. Splitting between the valleys in a [100] fin can be ignored (14). In contrast, the dependence of the effective mass of the ground subband in [100] fins on t is more pronounced as compared to [110] fins. Results of density-functional calculations (14) confirm the mass dependences obtained from the **k·p** model (Fig.2).

With the values of the effective masses and subband offsets obtained we study the conductance G through the system, for parallel and anti-parallel configurations of the contacts. Fig.3 shows the dependence of tunneling magnetoresistance (TMR) defined as

$$\text{TMR} \equiv \frac{G_{\uparrow\uparrow} - G_{\uparrow\downarrow}}{G_{\uparrow\downarrow}}, \qquad [7]$$

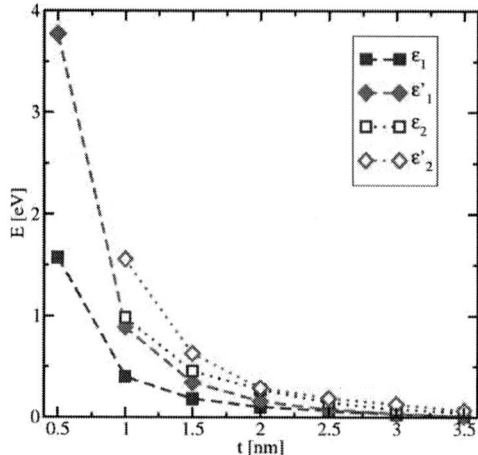

Figure 1. Subband minima as a function of [110] fin thickness t.

for [100] and [110] oriented fins with t=1.5nm on the value of spin-orbit interaction. Fins of [100] orientation display stronger dependence on β and are thus preferred for practical realizations of silicon SpinFETs. This is due to the fact that the scale of the TMR dependence on the spin-orbit interaction is determined by the characteristic wave vector $k_D = m_n \beta / \hbar^2$. Because the effective mass in [110] fins is substantially smaller than in [100] structures, one needs a larger variation of β in order to acquire the same variation of k_D.

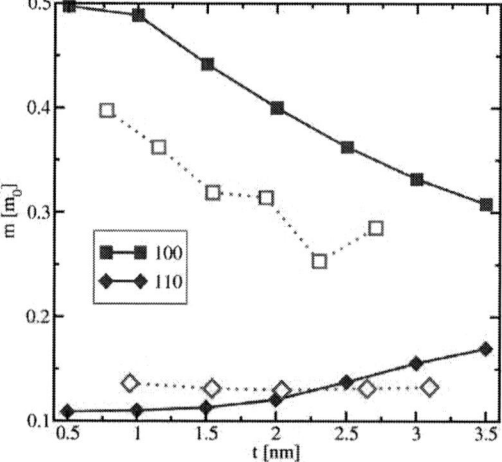

Figure 2. Ground subband effective mass dependence on t in [100] and [110] fins obtained with the $\mathbf{k \cdot p}$ method (filled symbols) and with the first-principle calculations (14) (open symbols). The discrepancy between the curves is due to surface passivation and structure relaxation present in the first principles calculations (14) but not in $\mathbf{k \cdot p}$.

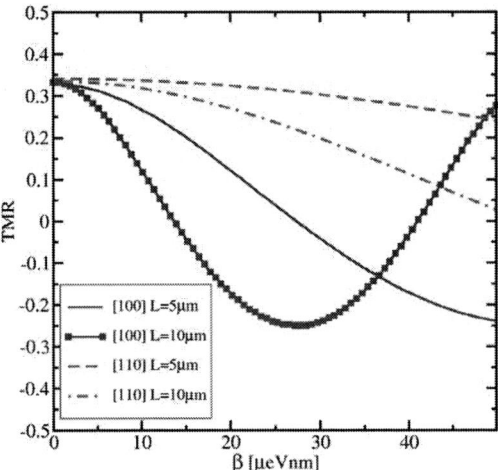

Figure 3. TMR dependence on the value of the Dresselhaus spin-orbit interaction parameter for t=1.5nm, B=0T, P=0.4, z=5 ($z = U\sqrt{2m_F/E_F}/\hbar$).

Thanks to the Dresselhaus form of the spin-orbit interaction, the TMR of [110] fins is most affected by the magnetic field along the transport direction (Fig.4), while the magnetic field orthogonal to the transport direction influences the TMR of [100] fins (Fig.5). Also, the TMR in [100] fins is most modified by the external magnetic field, which provides an additional option to tune the performance of the silicon SpinFET.

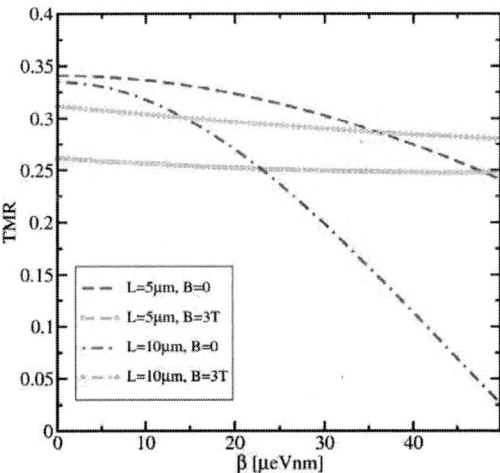

Figure 4. TMR dependence on the value of the Dresselhaus spin-orbit interaction parameter for a [110] fin with t=1.5nm, (P=0.4, z=5) in a magnetic field B=3T parallel to the transport direction.

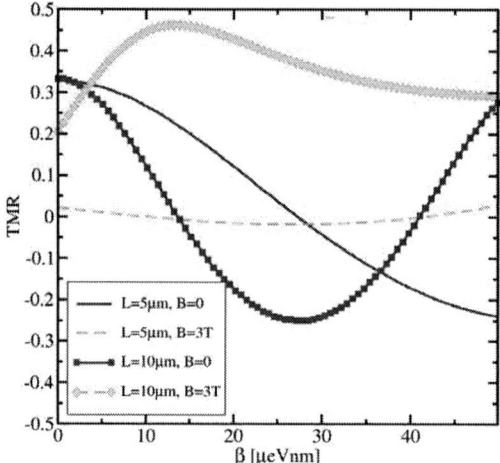

Figure 5. TMR dependence on β for a [100] fin with t=1.5nm, (P=0.4, z=5) in a magnetic field B=3T in [010] direction.

Conclusion

A possibility to build a SpinFET by using silicon fins is investigated. The spin-orbit interaction due to the interface-induced inversion symmetry breaking is taken in the Dresselhaus form. It is shown that [100] fins are more suitable for practical realizations of silicon SpinFETs.

Acknowledgments

This work is supported by the European Research Council through the grant #247056 MOSILSPIN.

References

1. S.Parkin, M.Hayashi, L.Thomas, *Science* **320**, 190 (2008).
2. T.Marukame, T.Inokuchi, M.Ishikawa, H.Sugiyama, Y.Saito, IEDM 2009, pp.1-4.
3. J.L.Cheng, M.Wu, J.Fabian, *Phys. Rev. Lett.* **104**, 016601 (2010).
4. S.P.Dash, S.Sharma, J.C.Le Breton, H.Jaffrès, J.Peiro, J.-M.George, A.Lemaître, R. Jansen., arXiv:1101.1691 (2011).
5. B.Huang, D.Monsma, I.Appelbaum, *Phys. Rev. Lett.* **99**, 177209 (2007).
6. S.Datta, B.Das, *Appl.Phys.Lett.* **56**, 665 (1990).
7. M.O.Nestoklon, E.L.Ivchenko, J.-M.Jancu, P.Voisin, *Phys.Rev.B* **77**, 155328 (2008).
8. M.Prada, G.Klimeck, R.Joynt, *Birk &NCN Publications*, paper 516 (2009).
9. Z.Wilamowski, W. Jantsch, *Phys. Rev. B.* **69**, 035328 (2004).
10. A.Bournel, P.Dollfus, P.Bruno, P.Hesto, *European Phys. J. Appl. Phys* **4**, 1 (1998).
11. K.M.Jiang, R.Zhang, J.Yang, C.-X.Yue, Z.-Y.Sun, *IEEE T-ED* **57**, 2005 (2010).
12. G.L.Bir, G.E.Pikus, Symmetry and Strain-Induced Effects in Semiconductors, J.Willey & Sons, NY, 1974.
13. V.Sverdlov, O.Baumgartner, T.Windbacher, S.Selberherr, *J.Computational Electron.*, **8**, 192 (2009).
14. H.Tsuchiya, H.Ando, S.Sawamoto, T.Maegawa, T.Hara, H.Yao, M.Ogawa, *IEEE T-ED* **57**, 406 (2010).

ECS Transactions, 35 (5) 283-288 (2011)
10.1149/1.3570807 ©The Electrochemical Society

The Roles of the Electric Field and the Density of Carriers in the Improved Output Conductance of Junctionless Nanowire Transistors

R. T. Doria[a,*], M. A. Pavanello[a,b], R. D. Trevisoli[a], M. Souza[b], C. W. Lee[c], I. Ferain[c], N. Dehdashti Akhavan[c], R. Yan[c], P. Razavi[c], R. Yu[c], A. Kranti[c], J. P. Colinge[c]

[a] LSI/PSI/USP, University of Sao Paulo, Sao Paulo, Brazil
*e-mail: rdoria@lsi.usp.br

[b] Department of Electrical Engineering, Centro Universitário da FEI, São Bernardo do Campo, Brazil

[c] Tyndall National Institute, University College Cork, Cork, Ireland

This paper evaluates the roles of the electric field (E) and the density of carries (n) in the drain conductance of Junctionless Nanowire Transistors (JNTs). The behavior of E and n presented by JNTs with the variation of the gate and the drain voltages has been compared to the one presented by Inversion Mode (IM) Trigate devices of similar dimensions. It has been shown that the lower drain output conductance exhibited by Junctionless transistors with respect to the IM ones is correlated not only to the differences in the mobility and its degradation but also to the electric field, the density of carries and the first order derivative of these variables with respect the drain voltage.

Introduction

For the scaling of MOSFET in the sub-22 nm era multi-gate transistors are of huge importance since they present reduced short-channel effects in comparison with planar transistors due to the better control of the channel electrostatics (1). However, in extremely short-channel devices the formation of source/drain to channel junctions become especially challenging as the doping concentration has to vary by several orders of magnitude within a few nanometers. Thus, ultrafast dopant activation processes are needed to avoid lateral diffusion of impurities into the channel region. To solve this problem, a novel structure called the Junctionless Nanowire Transistor (JNT) was recently developed (2).

The JNT presents no source/drain to channel junctions since it has constant heavy doping concentration through source, channel and drain (N-type in an nMOS device and P-type in a pMOS device) as shown in Figure 1 where a scheme of the JNT is shown, as well as its cross-section and the cross-section of an inversion-mode (IM) device. Due to its constant heavy doping concentration from source to drain, the JNT can be considered as a gated resistor (2). Although the JNT structure is similar to that of an accumulation-mode (AM) transistor (3), the conduction occurs only in the center of the nanowire whereas in AM transistors surface accumulation channels are formed.

283

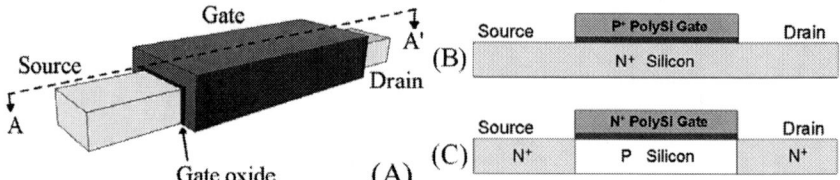

Figure 1: (A) Scheme of a Junctionless, (B) Cross-section of a Junctionless (JNT) and (C) Cross-section of an inversion mode transistor.

JNTs present better subthreshold slope and DIBL and than IM devices of similar dimensions as stated in (4). Also, a recent study on the analog behavior of JNT devices showed that they can provide even better analog properties than IM Trigate devices such as a larger intrinsic voltage gain (A_V) (5). In Figure 2(A), the A_V of JNT and IM Trigate devices is shown as a function of the transconductance over the drain current ratio (g_m/I_{DS}). The better gain of the JNT is attributed to its lower drain output conductance (g_D) (5) since g_D is associated to the A_V of the devices ($A_V = g_m/g_D$).

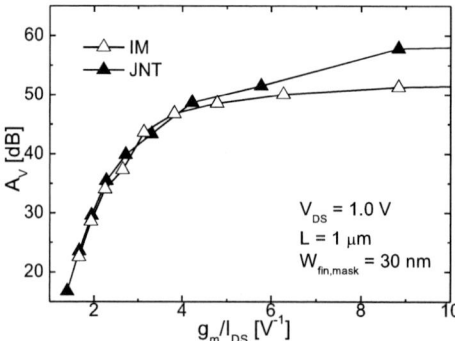

Figure 2: Measured A_V as a function of g_m/I_{DS} for both IM and JNT transistors.

This work presents a comparative analysis of the electric field (E) and the density of carriers (n) in the channel region of JNT and IM Trigate devices aiming at understanding their role in g_D. This analysis was performed through experimental results and 3D numerical simulations.

Devices Characteristics and Simulations

Initially, JNT and IM devices had their g_m and g_D extracted as shown in Fig. 3 as a function of g_m/I_{DS}. The IM and JNT devices measured present similar physical characteristics such as gate oxide and silicon thicknesses of 10 nm, channel length (L) of 1 μm and mask fin width ($W_{fin,mask}$) of 30 nm. The JNT presents constant doping concentration of $N_D = 3\times10^{19}$ cm^{-3} whereas the IM has channel doping concentration of $N_A = 10^{18}$ cm^{-3}. The devices were fabricated according to the process described in (2). Through the Sentaurus tools (6), 3D simulations of JL and IM transistors were performed. The simulated devices present similar physical characteristics to the measured ones except for the simulated fin width ($W_{fin,sim}$), which was set to 10 nm since in experimental devices the effective W_{fin} is expected to be reduced of 10~20 nm by the fabrication process. Also, IM devices were simulated for two channel doping concentrations: 10^{18} cm^{-3} and 10^{15} cm^{-3}. As the simulated results present similar behavior to the measured ones, the simulations could be validated.

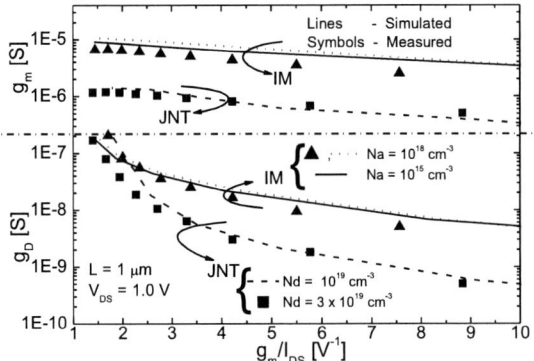

Figure 3: Measured and simulated curves of g_m and g_D as a function of g_m/I_{DS} for both IM and JNT transistors.

Output Conductance Analysis

According to Figure 3, JNT devices present both lower g_m and lower g_D than IM devices. This trend could be expected since JNT devices present lower drain current at room temperature due to the reduced carrier mobility (μ) associated to larger doping concentration. Besides the mobility, other parameters could affect g_m and g_D. The drift expression shown in [1] correlates the current (I) in a semiconductor not only to the mobility, but also to E and n integrated in the cross-section area (S) of the device ($N(x)$, being x the position along the semiconductor perpendicular to the current flow). In expression [1], q is the electron charge.

$$I = q\mu \int_S n(S) E(x)dS = qN(x)\mu E(x) \quad [1]$$

As g_D is defined as $\partial I_{DS}/\partial V_{DS}$, expression [1] can be differentiated with respect V_{DS} resulting in [2].

$$g_D = \mu q \left[N(x)\frac{\partial E(x)}{\partial V_{DS}} + E(x)\frac{\partial N(x)}{\partial V_{DS}} \right] \quad [2]$$

To verify the importance of $E(x)$, $N(x)$ and their derivatives in g_D of JNT and IM transistors, the drain conductance curves were simulated considering no mobility degradation model and a similar value for the low-field mobility between the devices (100 cm^2/V.s). Even though, the g_D curves of JNT and IM devices were different as exhibited in Figure 4 indicating that $E(x)$ and $N(x)$ play an important role in g_D.

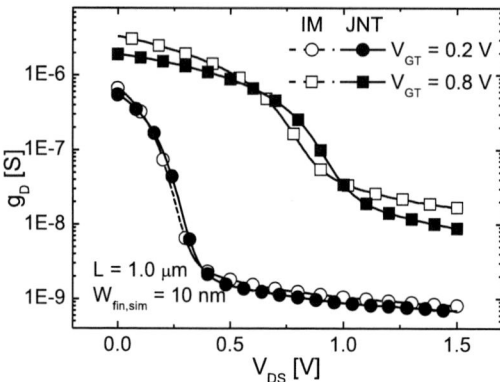

Figure 4: Simulated curves of g_D as a function of V_{DS} for both IM and JNT transistors considering a similar low field mobility (100 cm^2/V.s) and no mobility degradation between the devices.

Figure 5 shows the extracted curves of $N(x)$, $E(x)$, $\partial E(x)/\partial V_{DS}$ and $\partial N(x)/\partial V_{DS}$ for both devices in the active layer through a longitudinal line in the simulated structures biased at $V_{GT} = V_G - V_T$ of 0.2 V and 0.8 V.

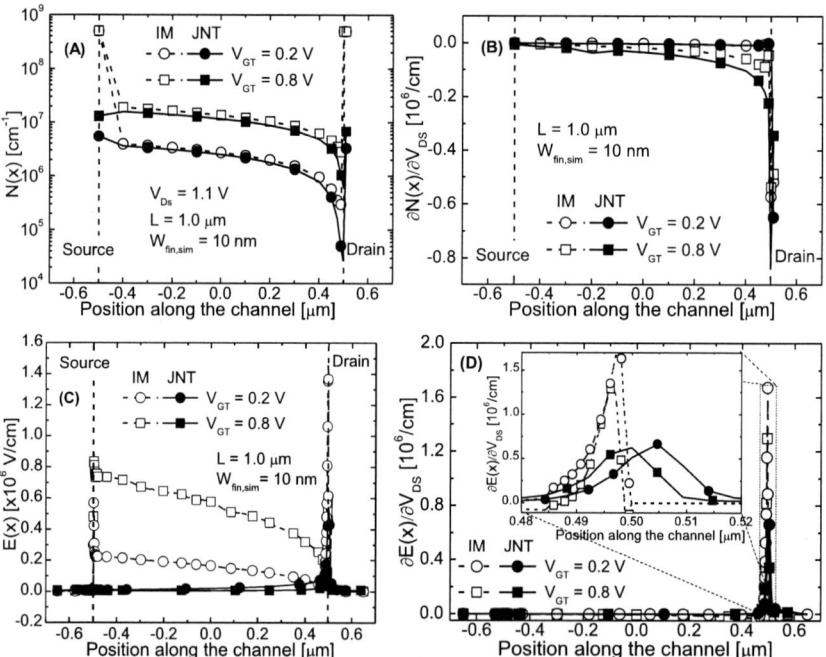

Figure 5: (A) $N(x)$, (B) $\partial N(x)/\partial V_{DS}$, (C) $E(x)$ and (D) $\partial E(x)/\partial V_{DS}$ in the channel length direction (x) obtained from the simulated JL and IM transistors.

The curves of $N(x)$ are shown in Figure 5(A) along the channel for IM and JNT devices. In this figure, the channel begins at $x = -0.5$ μm and ends at $x = 0.5$ μm. The source and the drain of the devices are located at x ≤ -0.5 μm x ≥ 0.5 μm, respectively. As it can be noted through the figure, Junctionless devices present lower amount of carriers in comparison to the Inversion-Mode ones along the entire channel. Despite the small difference in the amount of carriers between the devices, the reduced $N(x)$ of the Junctionless yields a lower drain current which affects g_D. The dependence of the carrier density on drain voltage is expressed by the term $\partial N(x)/\partial V_{DS}$ in Figure 5(B), where $\partial N(x)/\partial V_{DS}$ is plotted along the channel length for IM and JNT devices. From Figure 5(B) one can note the variation of carrier concentration near the drain is more sensitive to V_{DS} in Junctionless devices than in IM transistors since, at similar V_{GT}, JNT devices exhibit a larger negative derivative of the density of carriers integrated than IM transistors. Therefore, one can conclude that the term $\partial N(x)/\partial V_{DS}$ contributes to a reduction of g_D in Junctionless transistors when compared to IM devices. When observing the dependence of $N(x)$ and $\partial N(x)/\partial V_{DS}$ on V_{GT} one can notice a similar variation of $N(x)$ in both IM and JNT transistors. In addition, $\partial N(x)/\partial V_{DS}$ decreases more in Junctionless transistors than in IM devices when V_{GT} is increased.

Reference (7) reports that the electric field perpendicular to current flow is much lower in the channel of Junctionless transistors electric field than in that of Inversion-Mode Trigate devices. This observation is confirmed for the devices evaluated in Figure 5(C), in which $E(x)$ is presented along the channel length in Junctionless and in Inversion-Mode devices. The electric field was extracted in both devices in the region of their cross-section that presents the highest density of electrons. Thus, in the JNT transistors, $E(x)$ was extracted at the center of the silicon layer due to the bulk conduction whereas in IM devices it was taken 1 nm below the gate oxide/silicon interface because of surface channel conduction. According to Figure 5(C), IM devices always present higher electric field than JNT transistors and this field increases when increasing V_{GT}. In Junctionless transistors, however, the opposite behavior is observed: a reduction of the electrical field is observed with the raise of V_{GT}. In JNTs, increasing the gate voltage induces a reduction of the depletion region, decreasing $E(x)$, whereas in IM devices, any increase of the gate bias increases the inversion charge, thereby increasing the electric field. Figure 5(D) considers the dependence of the electrical field with V_{DS} through the term $\partial E(x)/\partial V_{DS}$ along the channel length of both devices. From the figure, one can perceive that for the two V_{GT} values under consideration, IM devices show a larger variation of $E(x)$ near the drain as a function of V_{DS} than the JNTs. In fact, along the rest of the channel, the derivative of electric field is very small in both transistors and does not significantly influence g_D, since the shift of the drain conductance of both devices can be associated to channel length modulation, which takes place only near the drain. In addition to the smaller derivative of $E(x)$ with V_{DS}, the dependence of $\partial E(x)/\partial V_{DS}$ on V_{GT} is lower in Junctionless devices than in Inversion-Mode transistors. Indeed, in JNT transistors a similar maximum value of $\partial E(x)/\partial V_{DS}$ is obtained at both analyzed V_{GT} and only a slight shift of this peak along the channel is observed when V_{GT} is increased.

Through the numerical analysis of the terms $N(x).(\partial E(x)/\partial V_{DS})$ and $E(x).(\partial N(x)/\partial V_{DS})$ it could be seen that the former is at least ten times larger than the latter. Thus, $N(x).(\partial E(x)/\partial V_{DS})$ is the dominant term, besides μ, contributing to the reduction of g_D in the JNT when compared to the IM transistor. For this reason, the lower values of

$N(x)$ and $\partial E(x)/\partial V_{DS}$ observed for the JL transistor in Figure 5 contribute to lowering the g_D. On the other hand, the higher g_D dependence on V_{GT} obtained in JNT devices can be related to the lower $\partial E(x)/\partial V_{DS}$ dependence on V_{GT}. In IM devices, the increase of $N(x)$ resulting from increasing gate bias is compensated by the decrease of $\partial E(x)/\partial V_{DS}$ observed at higher V_{GT}. Although the Junctionless devices also exhibit an increase of $N(x)$ with V_{GT}, the $\partial E(x)/\partial V_{DS}$ of these transistors is almost independent of V_{GT}. As a result, g_D of JNT devices is more sensitive to the gate bias variation than in IM devices.

Conclusions

This work aimed to verify the influence of the electric field and the density of carriers along the channel on the drain output conductance of Junctionless Nanowire Transistors once JNTs showed lower drain conductance than Inversion-Mode Trigate devices of similar dimensions. Along the work it was shown that g_D is not only affected by the electric field and the density of carriers integrated in the cross-section area of the devices but also their first order derivative with respect to V_{DS}. Indeed, the term $N(x).(\partial E(x)/\partial V_{DS})$ is the most influent on the drain conductance of the different devices. The lower g_D of the JNTs with respect to the Inversion-Mode ones is given mainly by the smaller values of $N(x)$ and $\partial E(x)/\partial V_{DS}$. Also, the higher g_D dependence on V_{GT} of JNTs is related to the smaller variation of $\partial E(x)/\partial V_{DS}$ with the gate bias as $N(x)$ increases with V_{GT}. In IM devices, the raise of $N(x)$ with the gate bias is compensated by the decrease of $\partial E(x)/\partial V_{DS}$ at larger V_{GT}.

Acknowledgments

The authors Rodrigo T. Doria, Marcelo A. Pavanello, Renan D. Trevisoli and Michelly de Souza thank the Brazilian research-funding agencies FAPESP and CNPq for the financial support.

References

1. J. P. Colinge, *FinFETs and Other Multi Gate Transistors*, Springer, 340p., 2008.
2. J. P. Colinge, C. W. Lee, A. Afzalian, N. Dehdashti Akhavan, R. Yan, I. Ferain, *et al.*, In: *Proceedings of International SOI Conference*, **1**, 1(2009).
3. C. W. Lee, I. Ferain, A. Afzalian, R. Yan, N. Dehdashti Akhavan, P. Razavi, J. P. Colinge, *Solid-State Electronics*, **54**, 97(2010).
4. J. P. Colinge, *IEEE Transactions on Electron Devices*, **37**, 718(1990).
5. R. T. Doria, M. A. Pavanello, R. D. Trevisoli, M. de Souza, C.W. Lee, I. Ferain, *et al.*, In: *Proceedings of International SOI Conference*, **1**, 72(2010).
6. *Sentaurus Device User Guide*, Version C-2009.06, 2009.
7. J. P. Colinge, C. W. Lee, I. Ferain, N. Dehdashti Akhavan, R. Yan, P. Razavi, R. Yu, A. Nazarov, R. T. Doria, *Applied Physics Letters*, **96**, 073 510(2010).

ECS Transactions, 35 (5) 289-294 (2011)
10.1149/1.3570808 ©The Electrochemical Society

Stress Relaxation Empirical Model for Biaxially Strained Triple-Gate Devices

R. D. Trevisoli[a], J. A. Martino[a], E. Simoen[b], C. Claeys[b] and M. A. Pavanello[a,c]

[a] LSI/PSI/USP, University of Sao Paulo, Av. Prof. Luciano Gualberto,
trav. 3, n. 158, 05508-900, São Paulo, Brazil
[b] imec, Kapeldreef 75, B-3001 Leuven, Belgium
[c] Department of Electrical Engineering, Centro Universitário da FEI,
Av. Humberto de Alencar Castelo Branco 3972, 09850-901,
São Bernardo do Campo, Brazil

Multiple gate devices provides short channel effects reduction,
been considered promising for sub 20 nm era. Strain engineering
has also been considered as an alternative to the miniaturization
due to the boost in the carrier mobility. The stress non-uniformity
in Multiple gate devices cannot be easily considered in a TCAD
device simulation without the coupled process simulation which is
a cumbersome task. This work analyses the use of an analytical
function to compute accurately the dependence of the strain on the
device dimensions. The maximum transconductance gain and the
threshold voltage shift are used as key parameters to compare
simulated and experimental data.

Introduction

Multiple-gate devices provide a reduction in the short channel effects due to the better
electrostatic control of the charge in the channel region in comparison with planar
devices (1). A schematic view of a triple-gate device, which is constituted of a stripe of
silicon surrounded by gate material, is presented in Fig. 1, in which the fin width (W_{Fin}),
the fin height (H_{Fin}), the channel length (L), the gate oxide thickness (t_{ox}) and the buried
oxide thickness are indicated. In triple-gate devices, multiple inverse layer channels are
formed in surface with different crystallographic orientations. As the carriers mobility
depends on this orientation, the electrons in the top gate (surface (100)) have a higher
mobility than the electrons in the sidewall gate (surface (110)) (2).

Strain engineering boosts the carrier mobility, increasing the drain current.
Mechanical stress can be applied either globally (3) or locally (4). The former technique
provides stress in both length and width directions (biaxial stress), while the later
provides stress only in the length direction (uniaxial stress). The devices studied in this
work are biaxially strained. During the fabrication, the biaxial stress suffers relaxation (or
reduction) in devices with reduced dimensions (5-6), resulting in a non-uniform stress
profile. Ref. (5) analyzes the influence of the fin width reduction in a stressed silicon
island, showing that stress in the fin width direction reaches a maximum in the center of a
wide fin and suffers a relaxation at the 100 nm next to the sidewall interfaces. In this
reference, the stress in the fin width direction varies from 1.5 GPa (centre) to 0 GPa
(sidewall interface). This reference also shows that when reducing the fin width, there is
also a slight decrease of the stress in the channel length direction suggesting that the
relaxation occurs in the strain, not in the stress.

This relaxation is not easily accounted in a TCAD simulation, so the purpose of this work is to study the use of an analytical function based on the stress profile available in the literature to compute the stress accurately. The stress analyses are performed through the maximum transconductance gain and the threshold voltage shift. These two parameters were adopted once they are related to two different effects of the stress: the mobility increase and the bandgap reduction, respectively (3).

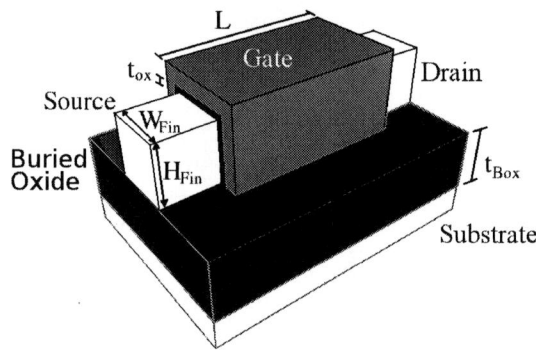

Figure 1. Schematic view of a triple-gate device.

Device Characteristics

The measured devices were fabricated according to ref. (7) and present buried oxide thickness of 145 nm, gate stack composed by 1 nm of thermal oxide, 2 nm of HfO_2 oxide and 5 nm of TiN. The channel length ranges from 100 nm up to 10 μm and fin width ranges from 20 nm up to 1 μm. The fin height is 60 nm for unstrained devices and 55 nm and strained devices. Three-dimensional simulations were performed using Sentaurus tool (8). The simulated devices present gate oxide thickness of 2 nm, L and W_{Fin} varying in a similar range as the fabricated devices. It was considered a mid-gap material at the gate with a workfunction of 4.7 eV and a p-type doping concentration in the channel region of 10^{15} cm^{-3}. As the carriers present different mobility at the top and the sidewalls, the methodology described in (9) was applied. The stress effects in the mobility, in the conduction and valence energies and in the density-of-states were taken into account.

Stress Relaxation Empirical Model

In order to consider the stress dependence on the device dimensions, a function was adopted for the strain (10). This function was chosen based on the stress profile described by ref. (5). The stress is constant in the central region of the device and reduces next to the sidewall interfaces (at the 100 nm closer to the sidewall) as shown in Fig. 2. For a device narrower than 200 nm, the stress at the centre of the device is lower than 1.5 GPa.

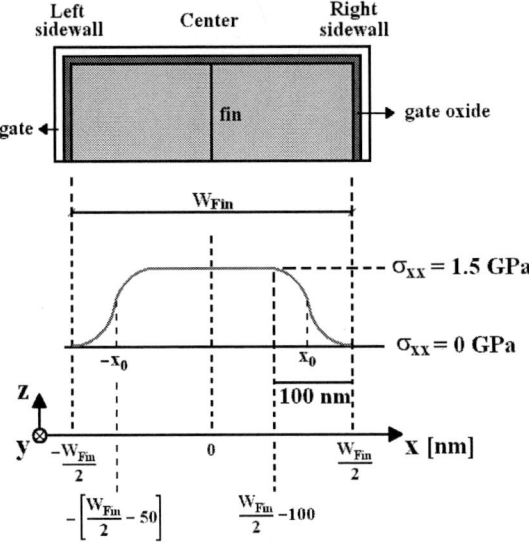

Figure 2. Schematic view of the stress component in the fin width direction for a wide device (10).

The equation for the strain in the fin width direction, in order to have a profile as described in Fig. 2, can be given by a Fermi-Dirac-like equation:

$$\varepsilon_{xx} = A_{Min,xx} + \frac{A_{Max,xx} - A_{Min,xx}}{1 + e^{(x-x_0)/dx}} - \frac{A_{Max,xx} - A_{Min,xx}}{1 + e^{(x+x_0)/dx}}$$ [1]

where $A_{Min,xx}$ is the strain at the sidewalls, dx controls the transition between the maximum and the minimum, x_0 is the distance from the sidewall interface in which the strain reaches half of the variation ($A_{Max,xx} - A_{Min,xx}$) and $A_{Max,xx}$ is the strain at the center of the channel described by eq. [2]:

$$A_{Max,xx} = \frac{A_{Max,xxW} - A_{Min,xx}}{1 + e^{(-x_{0,w})/dx_w}} + A_{Min,xx}$$ [2]

The constant $A_{Max,xxW}$ represents the strain at the center of a wide device (in this work $A_{Max,xx}$ is the strain required to generates 1.5 GPa biaxial stress). The variables x_0 and dx have a maximum value for wide devices ($x_{0,w}$ and dx_W, respectively) and vary with the dimensions of the device according to expressions [3] and [4].

$$x_0 = x_{0,w} + \left(1 - \frac{A_{Max,xxW} - A_{Min,xx}}{A_{Max,xx} - A_{Min,xx}}\right)\left(0.05 - \frac{W_{Fin}}{4}\right)$$ [3]

$$dx = dx_W - \left(1 - \frac{A_{Max,xx}W - A_{Min,xx}}{A_{Max,xx} - A_{Min,xx}}\right)dx_W\left(\frac{W_{Fin}}{2 \cdot 0.1}\right) \qquad [4]$$

The constants $x_{0,W}$, dx_W, $A_{Max,xx}W$ and $A_{Min,xx}$ were adjusted as $(W_{Fin}/2 - 0.05)$ [μm], 0.01 [μm], 0.009 and -0.002, respectively, in order to fit the experimental data. The term $\left(1 - \frac{A_{Max,xx}W - A_{Min,xx}}{A_{Max,xx} - A_{Min,xx}}\right)$ in equations [3] and [4] controls the transition between a wide and a narrow device. For a wide device, $A_{Max,xx}W = A_{Max,xx}$ and the term equals zero while for a narrow device $A_{Max,xx}W \gg A_{Max,xx}$ and the term tends to the unit.

The stress in the channel length direction can be obtained in a similar way to the one in the fin width direction, being given by equations [5-8]:

$$\varepsilon_{yy} = A_{Min,yy} + \frac{A_{Max,yy} - A_{Min,yy}}{1 + e^{(y-y_0)/dy}} - \frac{A_{Max,yy} - A_{Min,yy}}{1 + e^{(y+y_0)/dy}} \qquad [5]$$

$$A_{Max,yy} = \frac{A_{Max,yy}L - A_{Min,yy}}{1 + e^{(-y_{0,L})/dy_L}} + A_{Min,yy} \qquad [6]$$

$$y_0 = y_{0,L} + \left(1 - \frac{A_{Max,yy}L - A_{Min,yy}}{A_{Max,yy} - A_{Min,yy}}\right)\left(0.05 - \frac{L}{4}\right) \qquad [7]$$

$$dy = dy_L - \left(1 - \frac{A_{Max,yy}L - A_{Min,yy}}{A_{Max,yy} - A_{Min,yy}}\right)dy_L\left(\frac{L}{2 \cdot 0.1}\right) \qquad [8]$$

where the constants $y_{0,L}$, dy_L, $A_{Max,yy}L$ and $A_{Min,yy}$ were adjusted as $(L/2 - 0.23)$ [μm], 0.03 [μm], 0.009 and -0.002, respectively, to fit the experimental data.

Results

The comparison between experimental and simulated data was performed through the maximum transconductance ($g_{m,max}$) gain, i. e. the difference between the $g_{m,max}$ of strained and unstrained device, and the threshold voltage. Two of the main effects of the stress application are the rise in the carrier mobility and the reduced bandgap (3). As the g_m is proportional to the mobility and the threshold voltage is related to the bandgap, these two parameters are adequate to analyze if the stress effects have been accounted accurately.

In Fig. 3, the maximum transconductance gain extracted for a drain bias of 100 mV is exhibited as a function of the channel length. From this figure, it is possible to see that the simulation described adequately the relaxation of the stress for shorter devices, reducing the $g_{m,max}$ gain. For the experimental data with L = 0.1 μm, the strained devices present a lower $g_{m,max}$ in relation to unstrained ones. This is related to the slightly higher series resistance of the strained transistors due to the lower H_{Fin} (6). In Fig. 4, the maximum transconductance extracted at a drain bias of 600 mV is presented as a function of the fin width, which shows an agreement between the analytical function approach and the experimental results. The reduction of the stress for narrow devices is accurately described by the model.

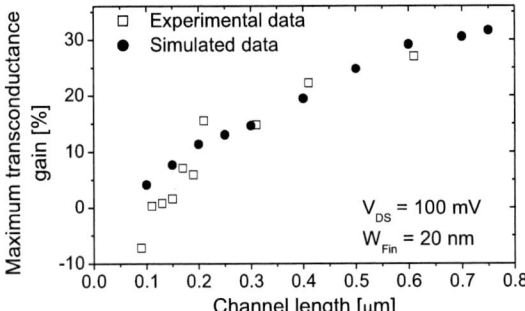

Figure 3. Maximum transconductance gain as a function of the channel length.

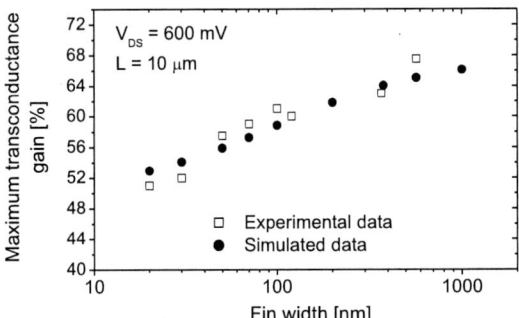

Figure 4. Maximum transconductance gain as a function of the fin width.

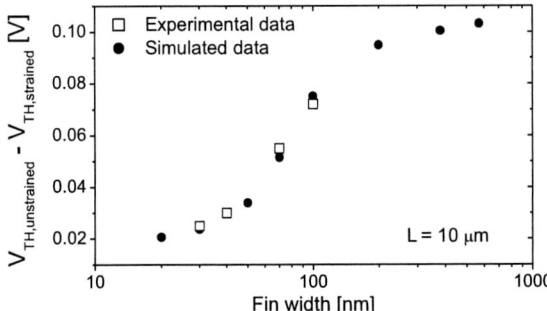

Figure 5. Difference of the threshold voltage between strained and unstrained devices.

In Fig. 5, a comparison of the threshold voltage shift due to the stress application is presented. The threshold voltage was extracted for V_{DS}=50 mV, using the g_m/I_D method (11). Results show that the effect of the stress in the bandgap has been adequately accounted according to the experimental data. In this figure, a long device has been considered (L = 10 μm). For a short device, the threshold voltage shift is negligible due to the stress relaxation.

Conclusions

This work analyzed the use of an empirical analytical model to account for the stress/device dimensions dependence in 3D TCAD simulations. The equations were defined based on the stress profile available in the literature, resulting in a Fermi-Dirac-like equation. The stress analyses were based on the maximum transconductance gain and the threshold voltage which are related to the boost in the mobility and the bandgap narrowing, respectively. The proposed equations have shown to describe accurately the stress effects for devices of any dimensions in the studied range.

Acknowledgments

The authors would like to thank to FAPESP, CNPq and CAPES for the financial support.

References

1. J. P. Colinge, *FinFET and Others Multi-Gate Transistors*, Springer (2008).
2. E. Landgraf, W. Rösner, M. Städele, L. Dreeskornfeld, J. Hartwich, F. Hofmann *et al., Solid State Electronics,* **50**, 38 (2006).
3. C. W. Liu, S. Maikap and C.-Y. Yu, *IEEE Circuit and Devices Magazine,* **21**, 21 (2005).
4. I. Lauer and D. A. Antoniadis, *IEEE Electron Device Letters,* **26**, 314 (2005)
5. F. Andrieu, C. Dupré, F. Rochette, O. Faynot, L. Tosti, C. Buj *et al., Symposium on VLSI Technology Digest of Technical Papers 2006,* 134 (2006).
6. M. A. Pavanello, J. A. Martino, E. Simoen, R. Rooyackers, N. Collaert and C. Claeys, *Solid State Electronics*, **52**, 1904 (2008).
7. N. Collaert, M. Demand, I. Ferain, J. Lisoni, R. Singanamalla, P. Zimmerman *et al., Symposium on VLSI Technology Digest of Technical Papers, 108 (2005).*
8. Sentaurus Device, SYNOPSYS (2010).
9. J. E. Conde, A. Cerdeira and M. A. Pavanello, *ECS Transactions,* **14** (1), 197 (2008).
10. R. D. Trevisoli and M. A. Pavanello, *ECS Transactions,* **31** (1), 377 (2010).
11. A. I. A. Cunha, M. A. Pavanello, R. D. Trevisoli, C. Galup-Montoro and M. C. Schneider, *Solid State Electronics,* **56** (1), 89 (2011).

ECS Transactions, 35 (5) 295-300 (2011)
10.1149/1.3570809 ©The Electrochemical Society

Transport-Confined Multi-Barrier FETs: A New Paradigm For Low-Leakage High On-Current Transistors

A. Afzalian and D. Flandre

Electrical Engineering department, ICTEAM Institute, Université catholique de Louvain, Louvain-La-Neuve, 1348, Belgium

Physics and performances of a new concept of nanoscale MOSFET, the Gate-Modulated Resonant-Tunneling (RT)-FET, are investigated through 3D Non-Equilibrium Green's Function simulations. Owing to the additional barriers and the related longitudinal confinement, the density of states in a RT-FET is reduced in its off state, while remaining comparable, in its on state, to that of a MOS transistor without barriers. The RT-FET thus features both a lower RT-limited off current and a faster increase of the current with gate voltage, i.e. an improved slope characteristic, and hence an improved I_{ON}/I_{OFF} ratio, along with high on current and therefore good speed performance. RT-FETs could therefore be promising devices for future generation low power, high speed applications owing to superior delay-power trade-off than a MOSFET. In addition, RT-FETs are intrinsically immune to source-drain tunneling and appear promising candidate for extending the roadmap below10nm.

INTRODUCTION

As transistors are scaled down in the nanoscale regime, scaling alone is not sufficient to achieve performance improvement and new boosters and device concepts are needed. For instance, the trade-off between power and performance in electronics is one of the most limiting factors to push further technology scaling and development. With scaling, the reduction of supply voltage to keep power density under control (1-3), the rise of source and drain resistance due to film thickness reduction in order to keep good electrostatic control (3), and finally source-drain (SD) tunneling that degrades subthreshold slope and increases leakage of transistors below 10nm (3-4), are major roadblocks that degrade on- and/or off-current, and I_{ON}/I_{OFF} ratios and therefore the power-delay trade-off of transistors. This is governed by the gate-controlled single-barrier paradigm on which present field-effect transistors (FET) are based. In a standard transistor, there is only one barrier from channel-to-source and the density of state close to the top of the channel barrier is about constant with V_G (5). Supposing ideal gate coupling and N-MOSFET operation, when increasing the gate voltage the current increases therefore at a rate dictated by Fermi-Dirac statistics only. In subthreshold, this current increase is exponential with an optimal minimal inverse subthreshold slope (SS) of kT/q.log10, i.e. about 60mV/decade at T=300K. Above threshold, when the channel barrier passes below the source Fermi level, E_{FS}, enabling the source highly-occupied states to drive a significant current density, and thus good delay performance, the current increase is much slower and the inverse slope reaches much higher values. We have recently shown the

295

possibility of achieving better slopes than that dictated by Fermi-Dirac, both in subthreshold and above threshold, together with high on-current, by using a Si "Multi-barrier boosted" CMOS transistor, the gate modulated resonant tunneling (RT)-FET (5-6). It is a MOSFET boosted with additional tunnel barrier(s) (TB) (i.e. barriers of a few nanometers width and less than 10nm) near the gate edge(s) and under electrostatic control of the gate that creates additional longitudinal confinement in the device. Such TBs can be created for instance in a planar technology from a local reduction, or constriction, of the device cross-section, resulting from a local oxidation that can be well controlled (7-8), or from Schottky Barriers and dopant segregation techniques (6) in this case allowing for steep slope and low source and drain resistance. In this paper, physics and performances of this new device are further investigated through 3D self-consistent Non-Equilibrium Green Function (NEGF) quantum simulations.

Physics of RTFET

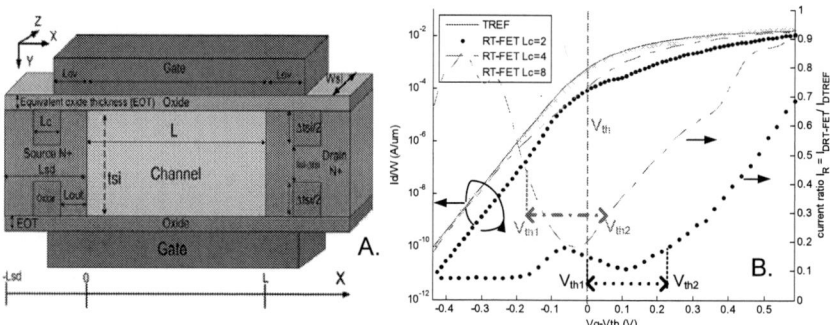

Figure 1. A) [100] Gate-all-around SOI N nanowire with constrictions (the lateral gates are not shown). $t_{si}=w_{si}=2nm$. Channel: L=10nm, doping $N^-=10^{15}cm^{-2}$. S/D extensions: $L_{sd}=L_{ov}+7nm$, doping $N^+=10^{20}cm^{-3}$ Oxide: EOT=0.5nm. T=300K. B) Drain current $I_D(V_G)$ vs. V_G-V_{th} of TREF (reference nanowire MOSFET transistor, i.e. without constrictions) (1) and RT-FETs with, $L_C=L_{ov}=$ 2nm (2), $L_C=L_{ov}=4nm$ (3) and $L_C=L_{ov}=8nm$ (4). $V_d=1V$. L=10nm. The current ratio, $I_R(V_G)$, is also shown for cases (2) and (3). The threshold voltage, V_{th}, of each device is extracted by considering the maximum of the second derivative of its $I_D(V_G)$ curve. V_{th} is in between V_{th1}, the channel barrier related threshold voltage and V_{th2} the tunnel barrier (TB) related threshold voltage. $TB_S=0.2eV$, $TB_D=0.41eV$. $L_{out}=1nm$ (5).

As a case of study, we will use planar Silicon-on insulator (SOI) nanowires with rectangular tunnel barriers arising from a local reduction, or constriction, of the wire cross-section (7-8). The SOI nanowire transistors were simulated using a Fast Coupled Mode Space (FCMS) self-consistent three-dimensional Non-Equilibrium Green Function (NEGF) quantum simulator. The Hamiltonian is written in the effective mass formalism but it includes a non-parabolic correction for transport effective mass and subband levels that has been shown to compare well with atomistic simulations down to $1.36\times1.36nm^2$ cross-section (8). Fig.1.A shows a schematic device representation and gives the parameters of the SOI Gate-all-around n-channel nanowire with constrictions of width L_C and section reduction Δt_{si}. Using constrictions, barriers between a few meV to several hundreds of meV can be obtained by tuning Δt_{si} (7-8). An overlap covering the constrictions is required in order to keep adequate electrostatic control of the gate over the tunnel barriers (TBs) (5).

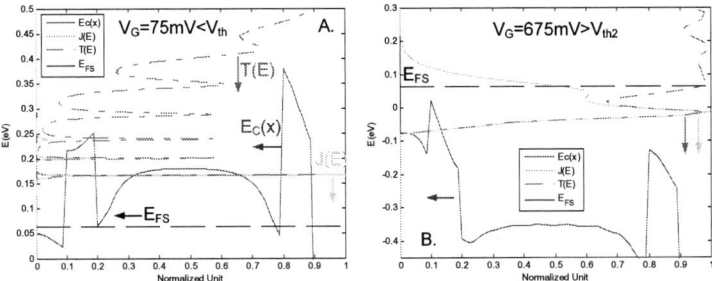

Figure.2: Conduction band profile E_C vs. normalized distance $((x-L_{sd})/(L+2*L_{sd}))$, normalized current density spectrum J(E) and transmission spectrum T (non normalized) vs. energy (i.e. T and J curves are rotated by 90°) for RT-FET with $L_{out}=1$ and $L_{ov}=2nm$. The position of the source Fermi level, E_{FS}, is also shown for comparison. A) Below threshold: the non-confined transmission states above the well are filtered by the Fermi-Dirac statistics. The current is therefore flowing through the few first quasi-bound or resonant tunneling states in the well and becomes very low compared to a standard MOSFET. B) Above V_{th2}, when the tunnel barriers pass below E_{FS}, the filtering action of the Fermi-Dirac statistics vanishes and non-confined transmission states above the well start to drive a significant amount of thermionic current (5).

As observed in Fig.2, owing to the additional barriers and the related longitudinal confinement, the density of states (DoS) in a RT-FET is reduced in its off state, while remaining comparable in its on state, to that of a MOS transistor without barriers. The RT-FET thus features both a lower RT-limited off-current and a faster increase of the current with V_G, i.e. an improved slope characteristic, and hence an improved I_{ON}/I_{OFF} ratio. The DoS being function of position, the equivalent and more rigourous concept of transmission T(E) will be used for the current, i.e. the source-injected current spectrum J(E) (we neglect the drain injected current here for simplicity assuming V_D greater than a few kT/q) is proportional to $T(E).f_{FD}(E-E_{FS})$. In a well optimized RT-FET, in subthreshold regime (Fig. 2.A), the channel and tunnel barriers being above E_{FS}, the high non-confined transmission states above the well are filtered by the Fermi-Dirac statistics. The current is therefore flowing through the few first quasi-bound or resonant tunneling states in the well and becomes very low compared to a standard MOSFET.

When increasing V_G, however, the channel and tunnel barriers are pushed down in energy. When the TBs pass below E_{FS}, the filtering action of the Fermi-Dirac statistics vanish and high non-confined transmission states above the well start to drive a significant amount of thermionic current (Fig. 2.B). This gives RT-FETs two different thresholds. The first, V_{th1} (=0.19V in Fig.3), related to the resonant tunneling (RT)-current, happens like in a standard transistor when the top of the channel barrier (TCB) passes below the source Fermi level E_{FS}. The second, V_{th2} (=0.43V in Fig.1.B), related to the thermionic current above the well, happens when the TBs passes below E_{FS}. For $V_G>=V_{th2}$, an important additional thermionic current will start flowing enabling further improvement of the slope and current ratio with V_G and hence very high on-current (Fig. 1.B and Fig.2.B). The transistor is recovering the on-current level of a MOSFET.

The RT-limited subthreshold regime also drives other interesting specifics characteristics: RT-FETs are intrinsically immune to source-drain tunneling, i.e. diffuse tunneling under the channel barrier (as in a standard MOSFET) is quantum mechanically excluded. RT-FETs therefore appear promising candidate for extending the roadmap below 10nm channel length and boosting the on current with alternative channel materials (Fig.3). Subthreshold slope below the kT/q limits are possible through gate modulation of the RT-states and the resonance condition allowing for new resonant levels to drive the current when increasing V_G (Fig.3.B). Under certain condition is it also possible to create a zone of negative resistance, which is of interest for memory application for example (5-6).

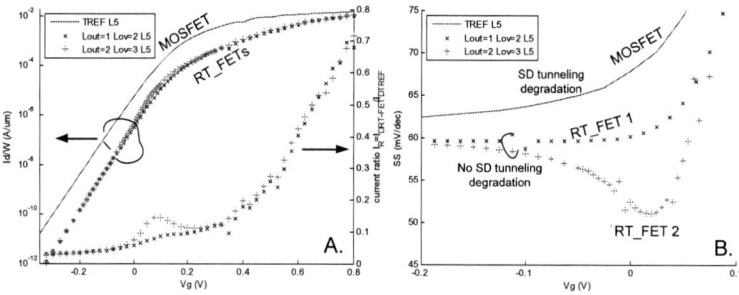

Figure 3: $I_D(V_G)$ and SS(V_G) curves of TREF, and RT-FETs with L_{out}=1nm and L_{ov}=2nm, and with L_{out}=2nm and L_{ov}=3nm. TB_S=0.2eV, TB_D=0.41eV. V_D=0.7V. L=5nm. L_C=2nm. t_{si}=w_{si}=2nm. The current ratio, $I_R(V_G)$ is also shown and further enlightens the interesting properties of RT-FETs, i.e. low leakage, sharp turn-on and high on-current levels. Also on the contrary to the MOSFET, no degradation of the subthreshold slope due to SD tunneling is observed in RT-FETs.

Performance Comparison

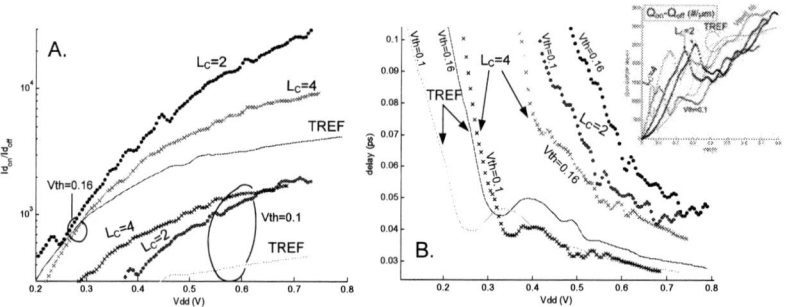

Figure. 4: A) I_{ON}/I_{OFF} ratio , B) delay, and charge $Q_{ON} - Q_{OFF}$ (inset) vs. V_{DD}, of TREF and RT-FETs with L_C=2nm and L_C=4nm for V_{th} set to 0.16 and 0.1V.

In Fig. 1.B, $I_D(V_G)$ curves of RT-FETs with tunnel barriers, with increasing L_C and overlap L_{ov} are shown and compared to those of an identical classical "reference" nanowire MOSFET without barrier or overlap (noted TREF below). We also show the current ratio of the two devices I_R=I_{DRTFET}/I_{DTREF} vs. V_G. We further compare the different devices in term of I_{ON}/I_{OFF} ratio and delay vs. supply voltage V_{DD} (=V_{ON}-V_{OFF}) for 2 different threshold voltages, i.e. 0.16 and 0.1V, changing the gate workfunction to have same V_{th} for all devices. The delay is estimated by $(Q_{ON}-Q_{OFF})/I_{ON}$, Q_{ON} and Q_{OFF} being the total charge in the device at $V_G=V_{ON}$ and V_{OFF} respectively (Fig. 4). Depending on the tunnel barriers width, L_C, the subthreshold current can be carried by resonant tunnelling states (corresponding to sharp peaks in the transmission) in the well (RT-current) and/or free or quasi free states above the well ("thermionic" like current) (both currents flow in case of RT-FET of Fig. 5.A). A transistor already dominated by the thermionic current below threshold, i.e. with L_C too large, will have characteristics very similar to a MOSFET but with a shifted threshold voltage, i.e. $V_{th}=V_{th2}$ (e.g. transistor with L_C=8nm in Fig. 5.B). Its performances are not enhanced and therefore not investigated further here. A transistor dominated by the RT-current below its threshold voltage, V_{th}, can achieve low off current and steep slope region owing to the RT effect. Its actual V_{th} will be equal to V_{th1}. The RT-FET with L_C=2nm has the lowest I_{OFF} and enhanced I_{ON}/I_{OFF} ratio compared to the reference MOSFET for both sets of V_{th}. However for $V_G>=V_{th2}$, an important additional thermionic current will start flowing (e.g. transistor with L_C=2nm, Fig.1.B) and its delay rapidly decreases in this range (Fig.4). Transistors having intermediate L_C (i.e. L_C=4nm in Fig.1.B) will have both thermionic and RT-current flowing in subthreshold (Fig.5.A). They will present a regime between V_{th1} and V_{th2} where their slope and current characteristics are in between subthreshold and above threshold. Their effective threshold voltage, V_{th}, will also be in between V_{th1} and V_{th2}. Although their off-current compared to RT-FET with L_C=2nm is not as low, they feature very sharp turn-on just above threshold which ensures best delay characteristics and comparatively good I_{OFF} performances for very low threshold voltages (i.e. 0.1V and below). This makes them the most suitable devices for ultra low voltage especially at ultra low threshold for high speed application where lower I_{ON}/I_{OFF} ratio can be traded-off for low delay and high I_{ON} (Fig.4). Note also that the capacitance, or charge variation, increase of the RT-FETs due to the overlaps, is more or less compensated by a charge reduction owing to the DoS reduction in the TBs (there are very few electrons in the tunnel barriers) and the quantum well related to RT effect. In addition, in this case, extra peaks in the capacitance and charge curves related to the 0D DOS can be observed for RT-FET with L_C=2 and 4nm (inset of Fig.4.B). For V_{th}=0.1V, the RT-FET with L_C= 4nm has its delay as good as that of a MOSFET from V_{DD} as low as about 0.35V, i.e. before full recovery of the MOSFET current (I_R about 60%). Also its off current is reduced by a factor 5 compared to that of the MOSFET. This enlightens the potentially superior delay-power trade-off for high-end application of the RT-FET with L_C=4nm.

Influence of Scattering

So far we have presented ballistic results. Next we consider the effect of dissipative transport by including phonon elastic and inelastic scattering using a self-consistent Born approximation (9). Our results confirm that what has been experimentally observed for standard nanowires (10) keeps true both for TREF and RT-FETs. In a 10nm long nanowire, scattering does not affect strongly subthreshold characteristics, while intervalley inelastic scattering can degrade the on current significantly, especially for ultra-thin cross-section (below 4x4nm²), further reducing I_{ON}/I_{OFF} ratio. It is therefore very important to reduce inelastic scattering by optimizing the cross-section of the nanowire, using strain engineering, and/or alternative channel materials. In any case, both TREF and RT-FET devices are affected in the same fashion by scattering, so that the improvement of I_{ON}/I_{OFF} ratio and steeper slope due to the change of regime between RT-current to thermionic current above V_{th_2} are still present, implying that scattering should not change fundamentally the interesting properties of RT-FETs discussed above.

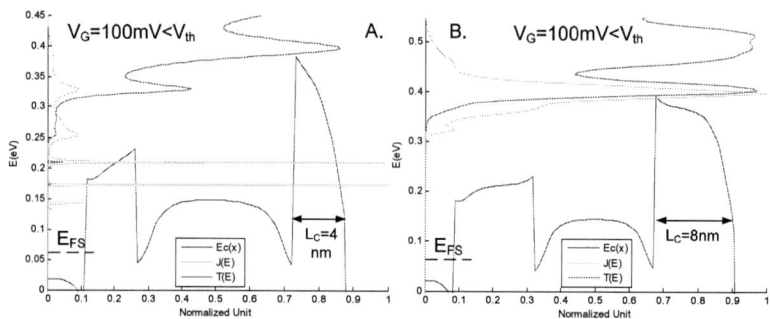

Figure.5: E_C vs. normalized distance, and J(E) (normalized) and T(E) (non normalized) vs. energy below threshold for RT-FET in subthreshold regime. A) with $L_C=L_{ov}=4$nm. Due to the thicker L_C when compared to $L_C=2$nm in Fig.2.A, both thermionic and RT-current are flowing in subthreshold. B) with $L_C=L_{ov}=8$nm. Due to the even thicker L_C, only thermionic current is flowing in subthreshold and no on-regime steep slope region is observed in Fig. 1.B.

Conclusion

Gate-Modulated Resonant-Tunneling Transistors with tunnel barriers width on the order of a few nanometers are promising devices for future generation low power, high speed applications owing to low leakage, immunity to source-drain tunnelling, improved I_{ON}/I_{OFF} ratio in a large supply voltage range and good delay characteristics. A 10nm-long RT-FET with barriers of the order of 4-5nm could operate with a supply voltage as low as 0.5V, a threshold voltage of 0.1V, same delay than a MOSFET but I_{OFF} (leakage power) reduced by a factor 5 and active power reduced by over 30% compared to a standard MOSFET.

Acknowledgments

This material is based upon works supported by FRS-FNRS Belgium.

References

1. Borkar S., *IEEE Micro*, Jul.-Aug. 1999, pp 23-29.
2. Kish, L., *Phys. Lett. A*, *305*, 144 (2002).
3. http://www.itrs.net/
4. Q. Rafhay, et al., Solid-State Electronics, 52, p. 1474 (2008).
5. A. Afzalian , J.-P. Colinge, and D. Flandre, Solid-State Electronics, in press.
6. A. Afzalian, D. Flandre, Proc. of ESSDERC 2010 Conference.
7. A. Afzalian, et al., ECS Transactions, Vol. 19 (4), pp. 229-234 (2009).
8. A. Afzalian et al., J. of Computational Electronics, 8, N° 3-4, pp. 287-306 (2009).
9. S. Jin, Y.J. Park, and H. S. Min, J. Appl. Phys. **99**, 123719 (2006).
10. S.D. Suk et al., IEDM 2007, p. 891.

CHAPTER 13

DEVICE PHYSICS AND TECHNOLOGY

302

ECS Transactions, 35 (5) 303-312 (2011)
10.1149/1.3570810 ©The Electrochemical Society

Numerical Modeling of Noise and Transport in SOI Devices

B. Meinerzhagen[1], A. T. Pham[2], S.-M. Hong[3], C. Jungemann[3]

[1]TU Braunschweig, 38023 Braunschweig, Germany, Email:b.meinerzhagen@tu-bs.de

[2]Physics Modelling and Simulation, IMEC, 3001 Leuven, Belgium

[3]Universität der Bundeswehr München, 85577 Neubiberg, Germany

Abstract

Accurate numerical modeling for SOI devices of technical interest has always been challenging concerning the accuracy of the physical models as well as concerning the convergence properties of the numerical algorithms. For partially depleted devices the most important underlying reason is the necessity of a precise modeling of low impact ionization and diffusion currents in order to get reasonable results for the kink effect and noise. On the other hand in fully depleted devices the DIBL effect at smaller channel lengths can only be sufficiently suppressed using thin Silicon films. Therefore a precise modeling of size quantization based on the Schrödinger Equation is typically needed for fully depleted devices. Today even more challenges exist due to the application of strain, hetero junctions and non standard interface and channel orientations. In this paper it is demonstrated that most of these challenges can be adequately addressed today.

Introduction

Even for the design of advanced double gate (DG) or fully depleted (FD) SOI field effect transistors(FETs) with channel lengths of 30 nm and less the classical Drift-Diffusion (DD) device model is still widely applied though it is well known that for such devices the DD model cannot provide physically accurate solutions [1]. One reason for this is certainly that alternative more accurate models are typically based on the solution of Boltzmann's Transport Equation (BE) and therefore much slower. However, another very important reason is that the BE is typically solved with a Monte Carlo (MC) algorithm. This implies that many of the favorable numerical properties typically available for the DD model and definitely needed for SOI simulations like an easy control of numerical accuracy and simulation time, the availability of true DC solutions and the Jacobian of the complete discrete equation system allowing simultaneous Newton iterations as well as AC and noise analysis are lost if a MC algorithm is used.

The MC CPU-Time accuracy trade off

In order to show that already a closer look on numerical solution accuracy can easily identify important problems that are hard to solve with a MC algorithm, the calcu-

303

lation of the linear drain current response at equilibrium for a DG MOSFET in weak or strong inversion is studied first. To evaluate this response with a high numerical solution accuracy is an easy task for a DD model if the usual solution methods are used. However, for a MC algorithm the situation is completely different. For a self-consistent multi particle solution of BE and Poisson's Equation (PE) with the MC algorithm, where all particles have equal weights, the following equation holds for the CPU-time T_{CPU} necessary to reach a relative numerical error r for the DC drain to source current I_D with a probability of 95% [2]:

$$T_{CPU} = \frac{\alpha_{cost} \, 8 U_T \, Q_{tot}}{r^2 \, V_D \, I_D} \tag{1}$$

U_T is the thermal voltage, Q_{tot} is the total electron charge for the n-channel and the total hole charge for the p-channel case. V_D is the drain to source voltage, the absolute value of which should be smaller than 1 mV in order to suppress nonlinear transport effects in nanoscale FETs. Moreover, α_{cost} is the ratio of the CPU time and the simulated time divided by the number of simulated particles. Thus, α_{cost} depends on the hardware available for the simulation and can be easily determined by running the MC algorithm for a short time. For one 1 GHz CPU $\alpha_{cost} \approx 2.5 \cdot 10^{10}$ holds. First of all and most important this formula shows clearly that $r \sim 1/\sqrt{T_{CPU}}$, which is valid for all MC simulations. Moreover, it becomes clear that T_{CPU} can vary easily by 10 orders of magnitude and more depending on the size of the drain current and the simulation accuracy r required which is not the case for DD simulations, where thanks to the availability of Newton's method CPU-time is only a weak function of the required numerical accuracy. Finally, T_{CPU} becomes extremely large if the linear response must be calculated with high accuracy in the subthreshold region as for example for magnetotransport measurements. The relative magnetoresistive response may be as small as 10^{-3} as shown below so that r should be 10^{-5} in order to calculate the magnetoresistive response with an accuracy of 1%. For such a problem a MC solution is basically impossible. In case single particle or weighted particle MC algorithms are used formula (1) is not exactly valid any more but the basic trends described above remain still the same.

Direct Solution of the BE

The solution methods for the BE that avoid the MC algorithm, which will be addressed as direct BE solution methods throughout the rest of this paper, are typically based on the expansion of the distribution function in spherical harmonics in the 3D k-space. This has a long history and so many scientists have contributed to this art that we can cite only a few pioneering papers [3–6]. The biggest disadvantage of this method is the high memory requirement if a 2D real space is considered. Therefore until recently only low order expansions were possible so that the accuracy of these BE solution methods was still inferior compared to the MC alternative. The situation has completely changed in the last couple of years due to the availability of random access memories of 100 GByte and more, which make higher order expansions with very little residual error possible and allow the direct self-consistent solution of BE and PE (BE-PE system) [7] thanks to the existence of efficient linear equation solvers [8]. Moreover even the multi subband BE in inversion layers can now be solved by a direct

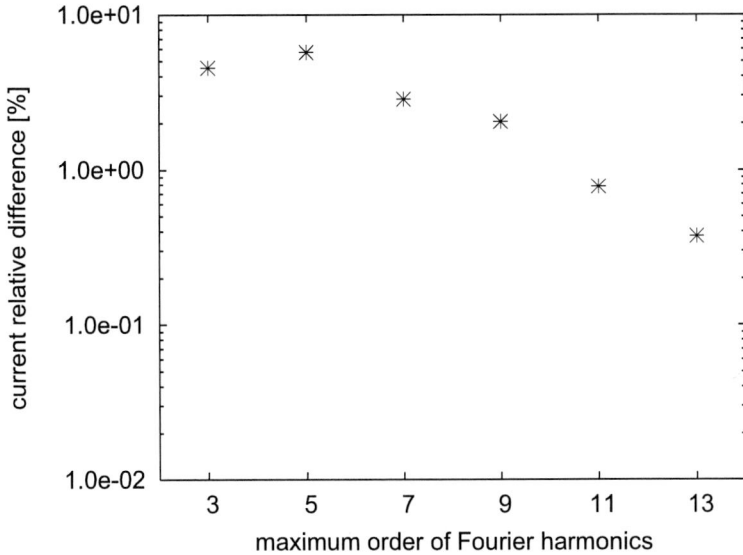

Figure 1: Relative error of the DC drain current evaluated with lower order Fourier expansions in comparison to the current resulting for maximum order 15 for a Ge PMOS double gate transistor with 16 nm channel length.

method based on a Fourier expansion of the distribution function in the 2D k-space. Today already a simulator exists that solves the 1D Schrödinger Equation (SE), PE and the multi subband (MSB) BE (BE-PE-SE system) self-consistently for FET devices without using MC algorithms [9]. In order to show that simulators allowing harmonic expansions of the distribution function of variable order are really necessary even for fundamental unknowns in Fig. 1 the dependence of the drain current in an advanced Ge DG PMOSFET on the maximum Fourier expansion order is shown. The simulator solving the BE-PE-SE system described in [9] was used for this study. Please note that a too low expansion order may introduce an error of nearly 10% even for the drain current. For direct BE solution methods the box integration method can be used for the space discretization, which comparable to the DD model case allows to make the discretized transport model charge conservative [7]. This implies that for the discretized model Kirchhoff's current law holds for the terminal currents as soon as the transport equation is solved with sufficient accuracy. One key features of the classical DD TCAD model is the availability of the Scharfetter-Gummel discretization scheme for the particle current densities which makes numerically stable CPU efficient solutions on sparse space grids possible. Comparable discretization schemes have meanwhile been developed as well for the direct solution of the BE based on the maximum entropy dissipation scheme [10] and the H-transform [5] as described in [7]. More details about the discretization methods of the BE on which the direct solution methods are based can be found in [7,9].

Accuracy and CPU-Time for the direct methods

After discretization of the BE within the k-space based on spherical or Fourier harmonics and in the real space based on the box integration method the discrete BE can be solved without substantial problems even for two space dimensions due to the availability of efficient linear solvers [8] and large random access memories. Even if the BE is nonlinear because the Pauli principle is considered, the Jacobian of the discrete BE is readily available and Newton's method can be used to establish a solution of the BE. A very important solution method for the DD TCAD model is Gummel's nonlinear relaxations scheme. A similar scheme can be applied for the direct solution of the BE-PE or BE-SE-PE systems by solving each equation successively one after the other and considering the coupling of the model equations iteratively by repeating this solution loop until convergence is established. Fig. 2 shows by the dashed line for a typical DG PMOS case the relative drain current error r as a function of the number of nonlinear Gummel type relaxations for the BE-SE-PE system. In the log/linear plot shown in Fig. 2 an upper bound for this error is given by the straight dotted line. Since this dotted line represents a decaying exponential function and each relaxation step consumes exactly the same CPU-time it is clear that for this Gummel relaxation method $r < C_1 \cdot \exp(-C_2 \cdot T_{CPU})$ with positive constants C_1 and C_2 holds for the error r established after a total CPU-time T_{CPU}. This shows that for the direct solution methods the error improves at least exponentially with CPU-time. Therefore a high accuracy can be easily achieved in contrast to the alternative MC solution method where this is often impossible due to the square root dependence on CPU-time. The second very important solution algorithm for the DD TCAD model is the simultaneous Newton algorithm considering the complete coupling between all equations. This solution method has already been demonstrated for the BE-PE system [7]. For the BE-PE-SE system the consideration of the coupling between the equations is more difficult since SE is an eigenvalue problem which must be solved in order to establish the coupling between PE and BE. First order perturbation theory can be used to evaluate this coupling in a linearized manner. Based on this idea an efficient incomplete simultaneous Newton method for the BE-PE-SE was realized and reported in [11]. In Fig. 2 it can be seen by the drawn line that even this incomplete Newton scheme leads to a substantial improvement of convergence speed if it is combined with the Gummel like relaxation scheme.

Accuracy of calibrated DD

One successful idea to improve the accuracy of classical TCAD DD or hydrodynamic (HD) device models was the hierarchical modeling approach. The basis of this method is the extraction of the DD and HD transport parameters under homogeneous material and field conditions using a physically more accurate model based on the BE and the application of these parameters in the TCAD models using a table model approach [2]. Of course such a calibration is at least partly possible as well using special test structures and experimental data. However, for very advanced SOI FETs with channel lengths in the order of about 40 nm or less calibration even on the device level does no longer lead to models with sufficient accuracy. This is demonstrated

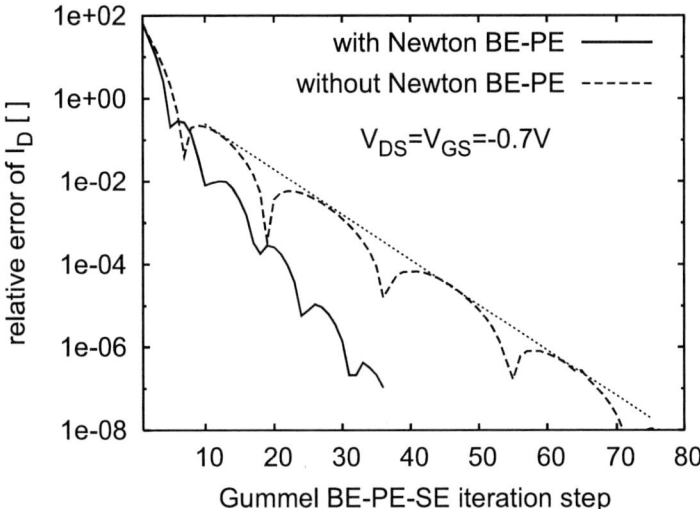

Figure 2: Relative error of DC drain current versus number of relaxation loops for the BE-PE-SE system and a Si Double Gate PMOSFET with 16 nm channel length. The dashed line refers to the Gummel type relaxation scheme and the drawn line refers to the relaxation method where the coupling between PE and the BE is considered by one incomplete simultaneous Newton step as part of the relaxation loop.

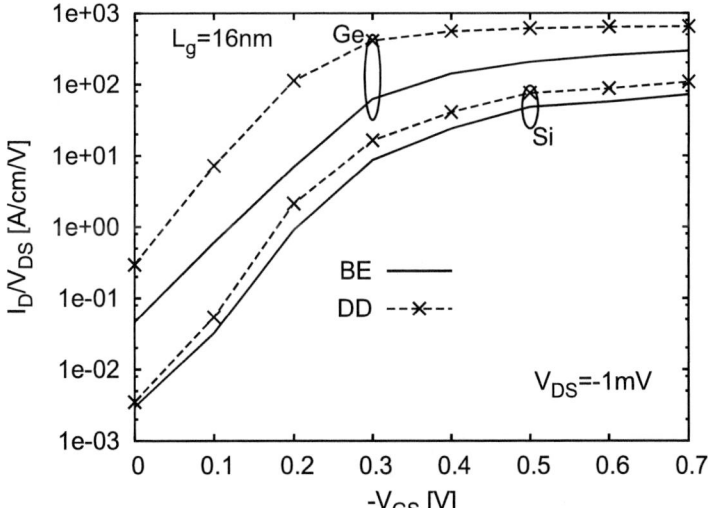

Figure 3: Characteristic of channel self conductance evaluated based on the solution of the BE-PE-SE system and formula (2) for two short channel double gate PMOSFETs. Two cases (unstrained Si ((001) surface/[110] channel) and uniaxially stressed Ge ((110) surface/[110] channel, the stress direction is parallel to the channel direction)) are considered. $V_{DS} = 1$mV.

next for the DD model and and the linear response of ultrashort DG PMOSFETs at equilibrium. For simplicity 2D FETs that are homogeneous in width direction are considered. It can be shown that for the linear response drain self-conductance at equilibrium the following equation is an exact result following from the DD ansatz [1] even in the case of a degenerate particle gas:

$$ g_D^{DD} = qW \left\{ \int \left[N_{\text{inv}}(x) \mu_{\text{eff}}(x) \right]^{-1} dx \right\}^{-1} \qquad (2) $$

The integration is performed along the channel from source to drain. N_{inv} is the inversion charge per unit area and μ_{eff} the effective mobility. This formula can be perfectly calibrated for a given device with respect to the BE-PE-SE system if $\mu_{\text{eff}}(x)$ and $N_{\text{inv}}(x)$ are evaluated locally in the device based on the results of the BE-PE-SE system solver using the Kubo-Greenwood formula [1]. Fig. 3 compares the results of the calibrated formula (2) and the self-conductance at equilibrium resulting from the solution of the BE-PE-SE system for a DG PMOSFET with 16nm channel length and unstrained Si with (001) interface orientation or uniaxially strained Ge with (011) interface orientation as channel material, respectively. It can be seen that in the Si case the difference is typically larger than a factor of two and reaches an order of magnitude in the Ge case. Though the calibration of the DD model is an ideal one on the device level large differences between both approaches are observable for these advanced devices. These differences are due to the DD ansatz itself, which is no longer valid for such devices [1]. Therefore, calibration of TCAD DD and HD models does in general not lead to accurate models for devices with such small dimensions.

Hard SOI simulation problems

Finally, the excellent numerical properties of the direct BE solution methods especially for SOI simulation problems are demonstrated for some examples, for which accurate results could not be provided before these new methods became available. As already pointed out above one such example is the magnetoresistive effect in sub 40 nm DG devices especially if this effect is small. For the direct PE-SE-BE device solver presented in [9] such simulations with high numerical accuracy are no problem as shown in Fig. 4. Please note that for a small magnetic field the relative magnetoresistive response can be well below 0.1% so no MC simulator can be used if any reasonable accuracy is required. The physical accuracy of the PE-SE-BE model based only on the direct non MC solution methods is demonstrated for ultra thin body p-channel FD SOI structures by a comparison with experimental data in Fig. 5 and 6. Figure 5 shows the large influence of Si film thickness on the effective mobility. A comparison is shown between simulation and experiment for Si film thicknesses of infinity (bulk), 28 nm and 3-4 nm, single gate (SG) and symmetrical double gate (DG) structures and otherwise standard interface and channel orientations. Figure 6 shows that more complicated SOI structures involving SiGe hetero junctions (HOI), biaxial strain and nonstandard interface and channel orientations can be handled with high accuracy as well. Another notoriously difficult simulation example even for larger channel lengths and body thicknesses are PD SOI MOSFETs with high

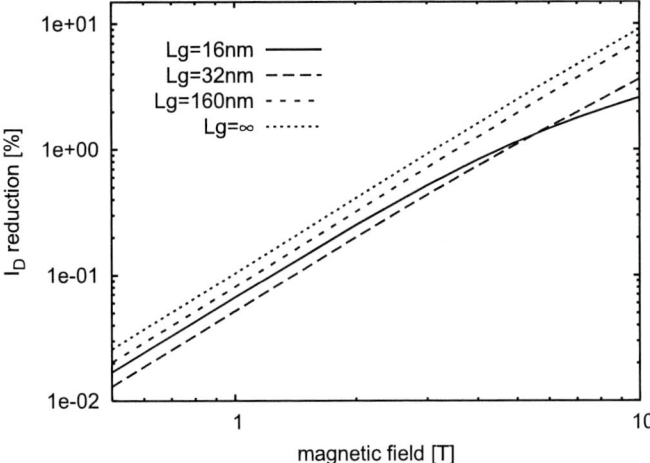

Figure 4: Relative reduction of drain current due to the magnetoresistive effect as a function of magnetic field for biaxially strained Si double gate PMOSFETs with 10nm body thickness. VGS = 0.7V. VDS = 1mV for finite gate length ($L_g \neq \infty$) and 1V/cm lateral electric field for the homogenous channel case ($L_g = \infty$). T = 300K. Reprint from [9].

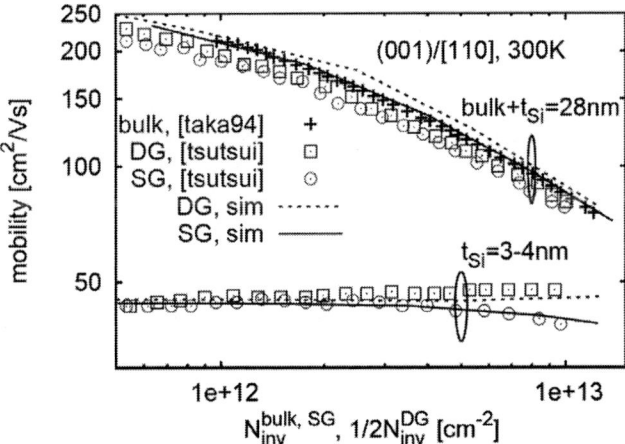

Figure 5: Comparison of simulated and measured effective hole mobilities versus hole inversion charge for different Si film thicknesses. Exp. are from [12], [13].

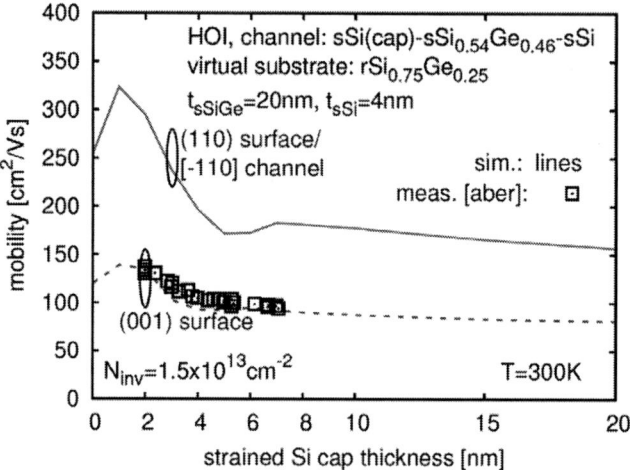

Figure 6: Comparison of simulated and measured effective hole mobility versus Si cap thickness for a biaxially strained HOI structure and different orientations. Experiments are from [14].

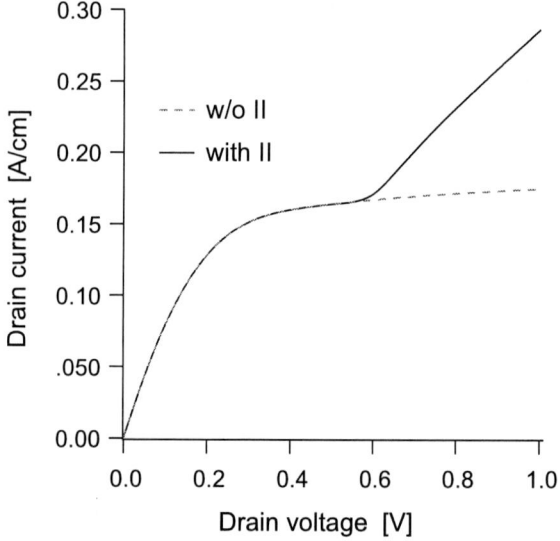

Figure 7: Output characteristics of the SOI MOSFET for $V_{GS} = 1.0$ V with and without impact ionization. Reprint from [7].

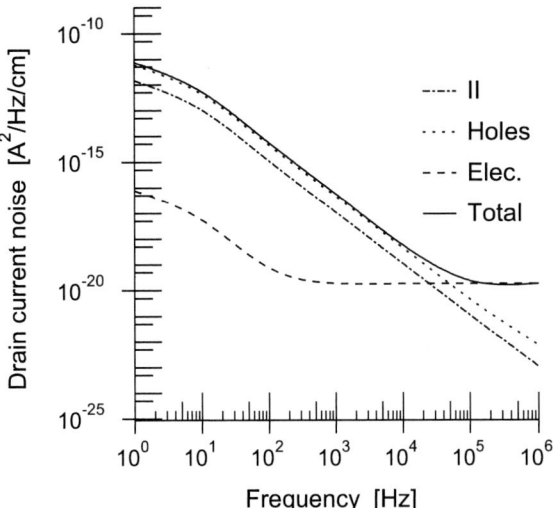

Figure 8: Spectral intensity of the drain current fluctuations for the SOI device and $V_{GS} = 1.0$ V and $V_{DS} = 1.0$ V. Reprint from [7].

substrate doping levels. In order to describe the kink effect in these devices correctly, impact ionization (II) must be modeled with high accuracy. Therefore, the DD model yields incorrect results even if nonlocal II models are used. The situation is better for HD models concerning the modeling of II, but HD models suffer from a non physical diffusion of the channel carriers into the quasi-neutral substrate. This effect was discovered long ago [15] and is of no concern for MOSFETs with a bulk contact. For SOI, however, this effect leads to a non-physical charging of the isolated substrate with minority carriers and can even lead to a non-physical negative differential resistance of the output characteristic [16]. Moreover, these devices cannot be simulated based on the BE and MC algorithms, since due to the charging of the substrate not only very small currents but as well effects with extremely different time constants have to be resolved, which is another situation, where MC algorithms, which typically require time steps smaller than femto seconds, lead to extremely large CPU-times. Finally, even to calculate a DC solution without a good initial solution is a real challenge where many DD and HD solvers fail. Figures 7 and 8 demonstrate, however, that the numerical solution algorithms of the BE-PE solver described in [7] are already so mature that not only a DC solution for a PD SOI NMOSFET with $2 \cdot 10^{17}$ cm^{-3} substrate doping and 500 nm channel length can be calculated but the kink effect with high accuracy and the low frequency drain current noise as well. The output characteristic in Fig. 7 is shown with and without the consideration of II. Please note that there is no negative differential resistance even if II is switched off. Figure 8 shows that PD SOI NMOSFETs show a strong increase of the drain current noise at low frequencies which is due to impact ionization and amplified by the backgate effect which is controlled by the II generated holes in PD SOI NMOSFETs [17]. Please note that due to the full consideration of the BE these noise simulations don't require any additional model parameters but are entirely based on the scattering models of the BE. Moreover noise is evaluated based on the Langevin approach using the CPU-time efficient frequency domain method exploiting the full Jacobian of the linearized BE-PE system for the calculation of appropriate terminal current Green's functions as described in [18].

References

[1] C. Jungemann et.al., *IEEE Trans. Electron Devices*, **52**, 2404 (2005).

[2] C. Jungemann and B. Meinerzhagen, *Hierarchical Device Simulation: The Monte-Carlo Perspective*, p. 173, Springer, Wien New York (2003).

[3] G. A. Baraff, *Phys. Rev.*, **133**, A26 (1964).

[4] N. Goldsman et al., *Solid–State Electron.*, **34**, 389 (1991).

[5] A. Gnudi et al., *Solid–State Electron.*, **36**, 575 (1993).

[6] K. Rahmat et al., *IEEE Trans. Computer–Aided Des.*, **15**, 1181 (1996).

[7] S. M. Hong and C. Jungemann, *J. Compu. Electr.*, **8**, 225 (2009).

[8] Matthias Bollhöfer and Yousef Saad, "ILUPACK — preconditioning software package." release 1.1 available online at www.math.tu-berlin.de/ilupack/ (2004).

[9] A. T. Pham, C. Jungemann and B. Meinerzhagen, *J. Compu. Electr.*, **8**, 242 (2009).

[10] C. Jungemann et al., *SIAM J. Numerical Analysis*, **41**, 64 (2003).

[11] A. T. Pham, C. Jungemann and B. Meinerzhagen, *Proc. SISPAD*, **115** (2009).

[12] S. Takagi et al., *IEEE Trans. Electron Devices*, **41**, 2357 (1994).

[13] G. Tsutsui and T. Hiramoto, *IEEE Trans. Electron Devices*, **53**, 2582 (2006).

[14] I. Aberg et al., *IEEE Trans. Electron Devices*, **53**, 1021 (2006).

[15] I. Bork et al., *NUPAD Tech. Dig.*, **5**, (1994).

[16] B. Polsky et al., *IEEE Trans. Electron Devices*, **52**, 500 (2005).

[17] C. Jungemann et.al., in *Simulation of Semiconductor Processes and Devices 2004*, G. Wachutka, G. Schrag, Editors, p. 235, Springer, Wien New York (2004).

[18] C. Jungemann et.al., *IEEE Trans. Electron Devices*, **54**, 1185 (2007).

Functionalization of Silicon Nanowires for Specific Sensing

V. Passi[a], E. Dubois[b], C. Celle[c], S. Clavaguera[c], J.-P. Simonato[c], J.-P. Raskin[a]

[a] Institute of Information and Communication Technologies, Electronics and Applied Mathematics, Université catholique de Louvain, 1348 Louvain-la-Neuve, Belgium
[b] Institut d'Electronique, de Microelectronique et de Nanotechnology, Silicon Microelectronics Group, Avenue Poincaré, F-59652, Villeneuve d'Ascq Cedex, France
[c] CEA–Grenoble, LITEN/DTNM/LCRE, F-28054 Grenoble Cedex 9, France

> Thanks to their large surface-to-volume ratio, silicon nanowires (Si NWs) are extremely sensitive to all phenomena which could alter their surface potential and charge distribution. Those surface variations lead to a change of the Si NWs equivalent conductance. The use of the output conductance of a silicon nanowire as a compact transducer for direct detection of (bio)chemical molecules or gases has gained immense attention these last years. In this paper, fabrication of silicon nanowires using top-down approach is shown, with simple calculations to determine hole mobility, hole concentration and resistivity. Transfer characteristics and sampling measurements were performed on the nanowires with and without surface functionalization under various ambient conditions indicating the importance of functionalization in order to avoid any environment effect on the transport properties of the nanowires.

Introduction

Owing to different properties when compared to the bulk material, unique applications in all fields, silicon nanowires have now become a subject of immense study. It is well acknowledged that one dimensional (1-D) silicon nanowires are excellent systems to investigate the dependence of electrical transport and chemical properties. Silicon nanowires (Si NWs) are of exceptional interest because their surface can be modified to act as both interconnect (1) (2) and functional units such as immobilizing matrices in fabricating electronic, electrochemical devices (3) (4) (5) (6) with nanoscale dimensions. Due to large surface-to-volume ratio and quasi 1-D characteristics, Si NWs can be used as ultra high sensitivity sensors for DNA detection (7) (8) (9). However, Si NWs themselves have no chemical specificity and in order to detect certain target species the surface of the Si NWs has to be functionalized with molecules that can interact with the specific target species which are to be detected. The sensing mechanism is related to: (i) change in density of silicon surface states, (ii) creation of a net charge in the molecules which act as an effective top-gate, or (iii) charge transfer between silicon and the functionalized molecules (10) (11) (12) (13). Selective functionalization of Si NW array is important since it improves the sensitivity, selectivity and detection limit of the sensor while keeping the surrounding areas inert. In this paper, we report the fabrication of silicon nanowires of various widths and lengths, using top-down approach and graft the surface of the wires using either (i) 3-(4-ethynylbenzyl)-1, 5, 7-trimethyl-3-azabicyclo [3.3.1] nonane-7-methanol (EBTAM), or (ii) OctadecylTrichloroSilane (OTS). Electrical measurements were carried out on the nanowires before and after functionalization with these molecules and reduction of hysteresis are observed on functionalized wires.

Fabrication

Single crystalline Silicon nanowires of various dimensions are fabricated using top-down approach using electron beam lithography and reactive ion etching. The fabrication process is presented in an earlier work (14). In brief, starting with a silicon-on-insulator wafer, hydrogen silsesquioxane (HSQ) - a negative tone electron beam (e-beam) resist - is spin-coated. Followed with electron beam exposure the resist is developed in tetramethylammonium hydroxide solution (TMAH-25%). Figure 1a shows the nanowires after development of HSQ. Using Chlorine chemistry and HSQ as an etch mask, top-silicon is etched to define nanowires along with large pads in silicon. HSQ is removed by immersing the wafer in HF-1% for 30 s as shown in Fig. 1b. Then, PMMA is spin-coated followed by exposure of patterns in order to open windows to deposit metal (Fig. 1c). After the development of the resist and the dip of the wafer in HF-1% in order to remove native oxide, platinum is deposited to a thickness of 20 nm. Excess platinum is removed by lift-off in acetone. After inspecting the wafer with scanning electron microscope (SEM) (Fig. 1d), annealing of the metal is completed. Array of 10 nanowires of width of 25, 50, 75, and 100 nm, with length of 0.3, 0.6, 1.2 and 2.4 μm and spacing of 0.3 μm, and 1 nanoribbon (shown as the inset of Fig. 1d) of width 1 μm, length 10 μm are fabricated on the same wafer, functionalized and measured.

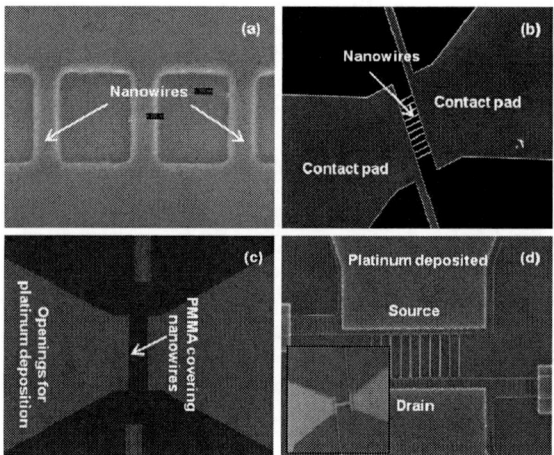

Figure 1. Top-view SEM image of silicon nanowires at various stages of the fabrication. (a) After resist development, (b) after top-silicon etching and HSQ removal, (c) after exposure of PMMA to make windows for metal deposition, and (d) after lift-off of platinum on the source and drain pads. The inset of Fig. 1d shows the fabricated single microwire: width = 1 μm and length = 10 μm.

Electrical Measurements

Electrical characterization of the nanowires before and after functionalization is performed using B1500 Agilent semiconductor device analyzer. Figure 2 shows the transfer characteristics of an array of 10 nanowires of width 25 nm and length 300 nm, measured in ambient air at fixed drain-source voltage (V_{ds}) of 2 V. The backgate bias is varied from +20 V to -20 V in step of 1 V with a hold and delay time of 0.5 s and 0.2 s, respectively. The value of transconductance can be determined by fitting the linear part of the graph. The values of transconductance (G_m) and turn-on threshold voltage (V_{th}) are 1080 nS and 12 V, respectively (extracted from Fig. 2a). A simple calculation to determine the resistivity is shown below.

The capacitance, hole mobility, hole concentration and resistivity are calculated using the equations [1]-[4] (15), respectively, as given below.

$$C = n\, \varepsilon_0\, \varepsilon_{SiO2}\, L_{nw}\, W_{nw}\, /\, T_{box} \qquad [1]$$

$$G_m = \delta I_{ds}\, /\, \delta V_{bg} = \mu_{hole}\, C\, V_{ds}\, /\, L_{nw}^2 \qquad [2]$$

$$n_{hole} = V_{th}\, C\, /\, (\, n\, q\, W_{nw}\, T_{nw}\, L_{nw}\,) \qquad [3]$$

$$\rho = 1\, /\, n_{hole}\, q\, \mu_{hole} \qquad [4]$$

Where C is the capacitance in Farad, n is the number of wires (10 in our case), ε_0 is the permittivity of free space (8.854×10^{-12} F/m), ε_{SiO2} is the dielectric constant of silicon-dioxide, L_{nw} is the length of the nanowire (300 nm), T_{box} is the thickness of buried-oxide (400 nm), W_{nw} is the width of the silicon nanowire, V_{ds} = 2 V, T_{nw} is the thickness of top-silicon/nanowire (50 nm). The values for capacitance, hole mobility, hole concentration and resistivity are 6.47×10^{-18} F, 75.1 cm^2 / V.s, 1.27×10^{17} / cm^3, 0.64 Ω-cm, respectively.

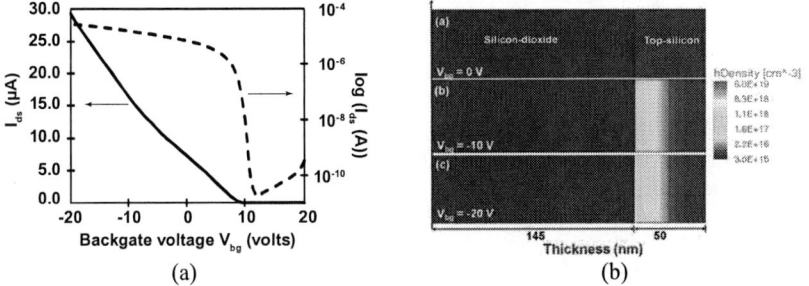

Figure 2. (a) Transfer characteristics (I_{ds}-V_{bg}) of an array of nanowires of width 25 nm and length 300 nm measured in ambient air. Solid line is linear scale on the left hand side and the dotted line is the log scale on the right hand side, respectively. (b) 2-D simulations (without considering the traps at interfaces) showing the hole density in top-silicon of thickness 50 nm for three different backgate bias values for V_{ds} = 0.1 V.

The value of hole mobility obtained from the measurement is much lower than expected bulk mobility since the carrier transport is mainly in the thin surface layer (surface mobility) rather than in the core of the wires. The increase in scattering centers (created due to reactive-ion-etching of the top-silicon to define the nanowires), also accounts to the reduction of carrier transport in the nanowires. Figure 2b shows the hole concentration in the nanowires for various backgate voltages (0 V, -10 V, and -20 V) for fixed V_{ds} = 0.1 V. There is accumulation of holes closer to the buried-oxide and top-silicon interface as V_{bg} is decreased from 0 V to -20 V.

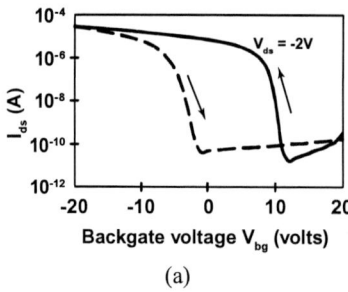 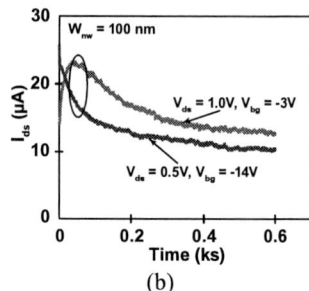

(a)   (b)

Figure 3. (a) Transfer characteristics (I_{ds}-V_{bg}) of an array of nanowires of width 100 nm and length 300 nm measured in ambient air. Solid black line is forward sweep from +20 V to -20 V and the dotted black line is reverse sweep, respectively. (b) Sampling measurements showing the current variation with time under various backgate biases.

Hysteresis is observed as can be seen from Figure 3a, when the measurements are carried out in ambient air on nanowires without any functionalization. Figure 3b represents sampling measurements for a 50 nm-thick top-silicon thickness, (an array of 10 nanowires) 100 nm-wide and 300 nm-long for two different backgate voltages (-3 V and -14 V) in ambient air. Not even after 10 minutes of measurement a steady state is reached. In order to understand the hysteresis effect measurements were performed in dry air and under vacuum (~80 mTorr). Figures 4a-4c show the I_{ds}-V_{bg} measurements on nanowires under ambient air, dry air and vacuum, respectively. Reduction of hysteresis is observed when measurements are performed in dry air and vacuum. The hysteresis of the nanowires depends on the environment; however, this can be missed since it takes relatively long time for appreciable effects to take place (16). By changing the environment from ambient air to dry air, noticeable decrease in hysteresis is observed (Figure 4a versus Figure 4b). By measuring the nanowires under vacuum (~80 mTorr) further reduction in hysteresis is observed but does not vanish completely indicating that the vacuum level is not sufficient. The causes of hysteresis could be the charge trapping by water molecules on the surface of the nanowires when exposed to ambient air. There are two possible types of charge traps, (i) water molecules which are weakly adsorbed on the nanowire surface, which can be easily removed by pumping in vacuum, (ii) silicon-dioxide surface bound water in close proximity to the nanowires. Silicon surface consists of Si-OH silanol groups which are hydrated by water molecules that are hydrogen bonded to the silanol (17) (18) (19). It is well known that hydrophilic silicon surface contains adsorbed water layer in atmospheric conditions and at temperature below 200°C (20), and a monolayer or sub-monolayer of hydrogen-bonded water remains on the silicon-dioxide and cannot be removed by pumping in vacuum (17) (18) (19). This water is known to act as a slow charge traps in conventional MOS devices (21) (22), which leads to the conclusion that surface water is largely responsible for the hysteresis in the nanowires.

In order to avoid any environmental effect on the transport in nanowires (the hysteresis and current variation with time) functionalization of the nanowires with either (i) EBTAM or (ii) OTS was carried out. Grafting of EBTAM was done on hydrogen terminated silicon surface obtained by short HF dip whereas OTS molecules are grafted onto native oxide formed on the nanowire surface (23). Figures 5a-5c show the transfer characteristics of nanowires and nanoribbons after functionalization with OTS. Drastic reduction of hysteresis is observed after grafting with either of the molecules indicating the reduction of adsorbed water molecules on the surface of the nanowires (24). The above results clearly show the impact of surface states/charges which are predominant at the nanoscale. Understanding the effect of surface states/ions on the transport properties at the nanoscale is not straight forward but in order to do so it is evident that surface states must be well controlled to avoid any misleading or incorrect interpretation of the results.

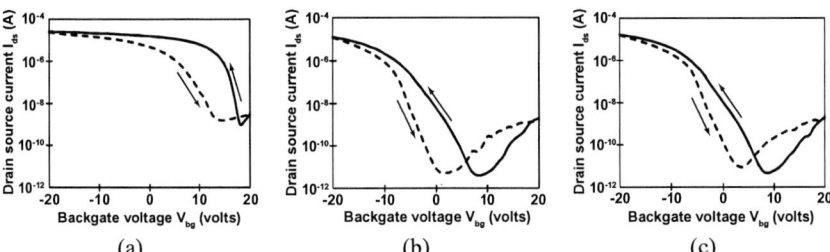

(a)   (b)   (c)

Figure 4. Transfer characteristics of nanowires under various environment conditions: (a) ambient air, (b) dry air and (c) vacuum. Solid lines represent forward (+20 V to -20 V) and dashed lines represent backward sweep (-20 V to +20 V) respectively.

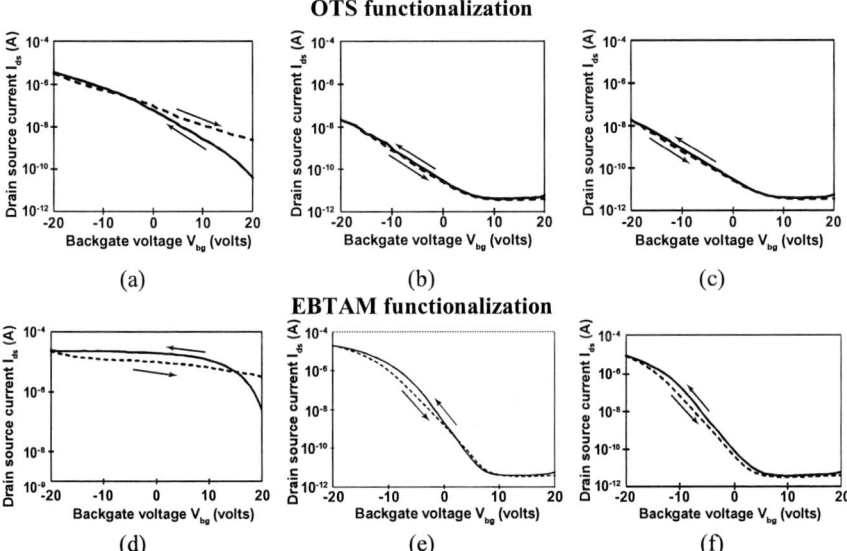

OTS functionalization

(a)   (b)   (c)

EBTAM functionalization

(d)   (e)   (f)

Figure 5. Transfer characteristics of functionalized nanowires under various environment conditions: (a, d) ambient air, (b, e) dry air and (c, f) vacuum.

Conclusion

Single crystalline silicon nanowires were fabricated using top-down approach and transport properties of these nanowires were studied. It was found that the resistivity of the nanowires is lower than the original wafer. The conductivity of the nanowires exhibits very high sensitivity to humidity. Surface modification with EBTAM and OTS make the nanowires immune to the surroundings. These functionalized nanowires can be used as gas sensors. Transfer characteristics demonstrate that surface states dominate the transport properties, and surface passivation can improve the device performance and reliability.

References

1. V. Jousseaume, V. T. Renard, *Interconnect Technology Conference*, pp. 1-3, 2010.
2. A. D. Wissner-Gross, *Institute of Physics Publishing, Nanotechnology*, pp. 4896-4990, 2006.
3. A. K. Wanekaya *et al.*, *Electroanalysis,* 18, 533 (2006).
4. R. E. Chee, J. H. Chua, A. Agarwal, S. M. Wong, G. J. Zhang, in *ICBME/2009, C. T. Lim, J. C. H. Goh, Editors,* 23, p. 838.
5. M.-W. Shao, H. Wang, Y. Fu, J. Hua, D-D-D. Ma, *J. Chem. Sci.*, 121, 323 (2009).
6. L. Mai, Y. Dong, L. Xu, C. Han, *Nano Letters*, 10, 4273 (2010).
7. G.-J. Zhang, G. Zhan, J. H. Chua, R-E. Chee, E. H. Wong, A. Agarwal, K. D. Buddharaju, N. Singh, Z. Gao, N. Balasubramanian, *Nano Letters*, 8, 1066 (2008).
8. C.-H. Lin, C.-H. Hung, C.-Y. Hsiao, H.-C. Lin, F.-H. Ko, Y.-S. Yang, *Biosensors and Bioelecronics*, 24, 3019 (2009).
9. J.-I. Hahm, C.M. Lieber, *Nano Letters*, 4, 51 (2004).
10. D. Cahen, R. Naaman, Z. Vager, *Adv. Func. Mat.*, 15, 1571 (2005).
11. O. Shaya, M. Shaked, A. Doron, A. Cohen, I. Levy, Y. Rosenwaks, *Applied Physics Letters*, 93, 043509 (2008).
12. N. Elfström, R. Juhasz, I. Sychugov, T. Engfeldt, A. E. Karlström, J. Linnros, *Nano Letters*, 7, 2606 (2007).
13. R. Cohen, L. Kronik, A. Shanzer, D. Cahen, A. Liu, Y. Rosenwaks, J. K. Lorenz, A. B. Ellis, *J. of Am. Chem. Soc.*, 121, 10545 (1999).
14. V. Passi, F. Ravaux, E. Dubois, J.-P. Raskin, *The 23rd IEEE Intl. Conf. on Micro Electro Mechanical Systems MEMS 2010*, Hong Kong, pp. 464-467, 2009.
15. G. D. Yuan, Y. B. Zhou, C. S. Guo, W. J. Zhang, Y. B. Tang, Y. Q. Li, Z. H. Chen, Z. B. He, X. J. Zhang, P. F. Wang, I. Bello, R. Q. Zhang, C. S. Lee, S. T. Lee, *ACS Nano*, 4, 3045 (2010).
16. W. Kim, A. Javey, O. Vermesh, Q. Wang, Y. Li, H. Dai, *Nano Letters*, 3, 193 (2003).
17. N. Tas, T. Sonnenberg, H. Jansen, R. Legtenberg, M. Elwenspoek, *J. Micromech. Microeng.*, 6, 385 (1996).
18. R. K. Iler, The chemistry of silica: Solubility, polymerization, colloid and surface properties and biochemistry, Wiley - New York (1979).
19. M. L. Hair, Infrared spectroscopy in surface chemistry, Marcel Dekker - New York (1967).
20. L. T. Zhuravlev, *Colloids and Surface A: Physicochemical and Engineering Aspects*, 173, 1 (2000).
21. J. F. Zhang, B. Eccleston, *IEEE Trans. Electron Devices*, 41, 740 (1994).
22. J. S. Chou, S. C. Lee, *IEEE Trans. Electron Devices*, 43, 599 (1996).
23. D. K. Aswal, S. Lenfant, D. Guerin, J. V. Yakhmi, D. Vuillaume, *Analytica Chimica Acta*, 568, 84 (2006).
24. D. Wang, Y.-L. Chang, Q. Wang, J. Cao, D.-B. Farmer, R.-G. Gordon, H. Dai, *J. AM. Chem. Soc*, 11602 (2004).

Ultra Low Power 3-D Flow Meter in Monolithic SOI Technology

N. André, B. Rue, G. Scheen, L. A. Francis, D. Flandre and J.-P. Raskin

Institute of Information and Communication Technologies, Electronics and Applied
Mathematics, Université catholique de Louvain, Place du Levant, 3
B-1348 Louvain-la-Neuve, Belgium

> Silicon-on-Insulator technology, with unique properties such as
> harsh environment resistance and lower power consumption (1), is
> presented here as a platform for CMOS and MEMS co-integration.
> An original CMOS-compatible process has been developed for the
> design and the co-fabrication of out-of-plane movable cantilevers
> and ring oscillators circuits on the same chip. The measured
> transducer, by deflection of the out-of-plane MEMS component,
> shows until 10% variation of the frequency under different flow rates.

I. Introduction

Co-integration of CMOS circuits and MEMS within a single package or die is the objective of many research groups to improve sensors performance or integration level as well as the cost. To interface extremely small capacitance variations (several hundred femtoFarads depending on the out-of-plane beams deflection) from a capacitive transducer, co-integration of the MEMS transducer with an integrated read-out circuit is mandatory to minimize parasitic elements and thus provide high resolution. A SoC (System-on-Chip) technology, building the sensor and the circuit on the same substrate using thin film Silicon-on-Insulator (SOI) technology wafers and traditional CMOS-compatible layers, followed by a post-process release as in (2)-(3), is presented in this work.

II. SoC description

A. Fabrication

Classically with micro-electro-mechanical system (MEMS), microcantilevers are widely used in atomic force microscopy (AFM), mass sensing, and contact sensing and force measurements. A change in surface tension or surface stress, due to interfacial interactions between the surface and the environment or intermolecular interactions on the surface, is detected electrically either via a resistive or capacitive transducer.

In this section, we introduce a method to fabricate three-dimensional cantilevers using both microfabrication techniques and mechanical stress in multilayered thin films and to co-integrate this mechanical sensing component and the complementary metal-oxide semi-conductor (CMOS) circuit on a same silicon chip. The term three-dimensional (or 3-D) is used here to describe a structure presenting a non-flattened geometry, i.e. an out-of-plane curvature when the reference plane is the silicon substrate surface. Movable

3-D cantilevers offer detection as a result of a stimulus changing their deflection. This change in deflection (due to a flow for example) bends downwards or upwards the cantilevers, respectively, increasing or decreasing their capacitance. For 3-D MEMS based cantilevers, several techniques have been proposed in the literature: projection micro-stereolithography (4), plastic deformation under magnetic field (5), reflow of solder hinges (6), multi-stack silicon-direct wafer bonding (7). In (8), the incorporation of a probe tip under the released structures is even done to reach such a shape. Finally in our work (9), a novel miniaturization technique, based on appropriate use of built-in stresses, is demonstrated to obtain the 3-D MEMS component.

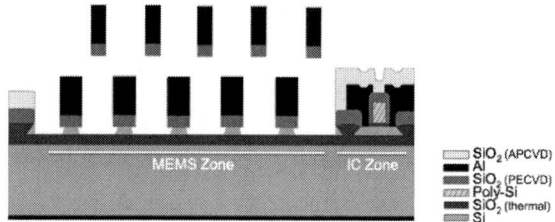

Fig. 1. Schematic cross section of an SOI flow anemometer.

In the purpose of going further towards the embedded microsystem, MOS circuits and MEMS sensors need to be co-integrated. The most critical steps for our chosen process are the thermal annealing and the release of MEMS without degrading CMOS circuits. Figure 1 presents the co-integrated flow anemometer, immediately after the release with an IC-respectful dry SF_6-plasma etching (25 W, 30 min). In that case, co-integration of our microsystem with the circuit requires 3 extra lithographic steps, mainly to protect the MEMS areas when processing only the IC part (implantations, etchings, etc.) and protect the IC part when processing only the MEMS devices (dry etching and release).

B. Ring oscillator

The desired geometry, as a result of the presented process, by the co-integration of a variable 3-D MEMS capacitor and sensed by a 5-stage ring oscillator, leads to a capacitance-to-frequency flow anemometer (Fig. 2). According to the airflow rate, the out-of-plane MEMS components will bend downwards, increasing the variable capacitance, and the oscillating frequency of the simple circuit consisting in these inverters and their associated MEMS capacitors will exhibit a shift.

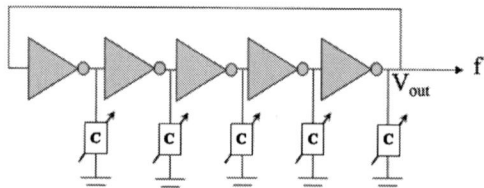

Fig. 2. Electrical assembly schematic – oscillation frequency sensitive to C (C variable with airflow rate).

Moreover, in order to decorrelate the temperature effect from the airflow, a quasi-identical 5-stage ring oscillator without out-of-plane movable beams is measured at the same time.

III. Micromechanical analysis

A. Multimorph bending

The 3-D shape relies precisely on the control of the internal stresses in multilayered structures originating from thermal expansion mismatch between constituting layers as well as on the control of the plastic yielding of a metallic layer (fully described in (9)). The dimensions for the interdigitated capacitors are the following: 200 μm-long, 10 μm-wide and 1.25 μm-thick (composed of 250 nm SiO_2 and 1 μm Al-Si alloy) for out-of-plane cantilevers and identical dimensions for in-plane beams except their width which is equal to 30 μm. Each couple composed of 10 curved beams and 10 flat beams (in-plane) corresponds to a small 250 fF capacitance connected directly to the output node of a CMOS inverter.

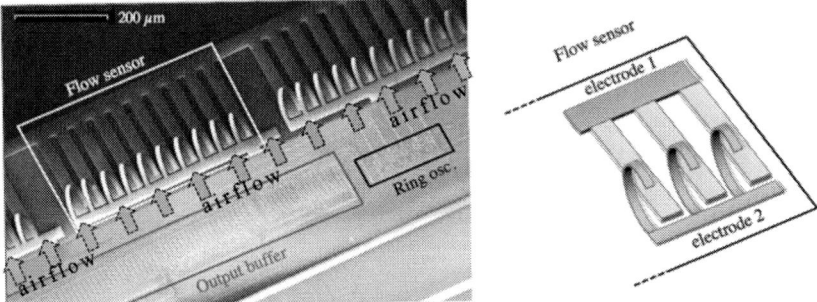

Fig. 3. SEM picture of a flow anemometer in the 1 μm fully depleted (FD) SOI process, with ring oscillator, output buffer and variable MEMS capacitors.

B. Fluid-structure interactions

Following the analysis in Kao et al. (10) of the forces acting on a curved microbeam under an airflow, we can divide the interactions on three main forces caused by:
- the change in momentum of fluid flow, F_{fluid}, with a downwards bending action;
- the difference in pressure between top and bottom of the beam, $F_{Bernoulli}$, with an upwards bending action;
- the elastic restoring force, F_k, with an upwards bending action.

To express F_{fluid} when an airflow velocity v_1 impacts laminarly a circular beam, we consider two fluid velocities components: v_t, tangential to the beam and v_p, perpendicular to the beam (Fig. 4). With a no-friction hypothesis, v_t follows the beam curve and escapes with an angle θ, as v_2. We consider also than v_p is totally dissipated and, as developed (10) for [1] and [2]:

$$F_{fluid} = \sqrt{F^2{}_x + F^2{}_y} = \sqrt{(2\rho r w v_1^2 (2+\cos\theta)\sin^2\frac{\theta}{2})^2 + (-\frac{1}{2}\rho r w v_1^2 \sin^3\theta)^2} \qquad [1]$$

$$F_{bernoulli} = \Delta P.surface = \frac{\rho}{2}wlv_1^2 \qquad [2]$$

$$F_k = -k\Delta z = -\frac{3EI_z}{l^3}\Delta z = -\frac{Ewt^3}{4l^3}\Delta z \qquad [3]$$

with the density ρ, the radius r, the deflection Δz, the Young modulus E, and w, t and l, respectively, the beam width, thickness and length. To obtain [3], we approximate our beam by a one-anchored cantilever deflected by its tip. We consider also v_1 as an average velocity, no boundary effect having been taken into account.

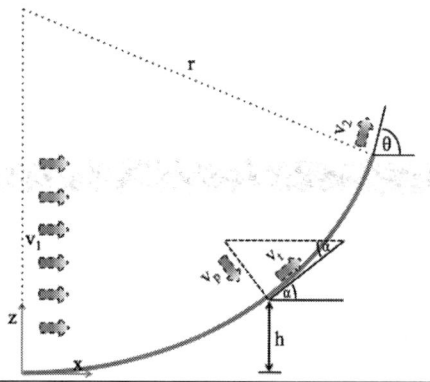

Fig. 4. Airflow force on a circular curved beam.

Searching equilibrium between these 3 forces, for a given velocity, the beam first bends downwards, increasing the restoring force, and thus vibrates into these 2 positions. For high velocities, important deflections are however allowed. Table I provides numerical examples for $\rho_{air}=1$ kg/m^3, $E = 70$ GPa and beam dimensions as described in Section III. F_{fluid} and $F_{Bernoulli}$ follow a velocity square-law while F_k is velocity-independent and equals to zero if no deflection.

TABLE I. Numerical values of forces acting on a curved beam under airflow.

Inlet velocity	F_{fluid} ($\theta=90°$)	F_{fluid} ($\theta=0°$)	$F_{Bernoulli}$	F_k ($\theta=90°$)	F_k ($\theta=0°$)
10 m/s	1.8 10^{-7} N	0	-1 10^{-7} N	0	-3.5 10^{-6} N
50 m/s	5 10^{-6} N	0	-2.5 10^{-6} N	0	-3.5 10^{-6} N
100 m/s	2 10^{-5} N	0	-1 10^{-5} N	0	-3.5 10^{-6} N

IV. Results

The measurement set-up includes a Suss microtech hollow nozzle placed at 100 μm above the circuit through which a constant pressure can be applied (Fig. 5). As a flow is constantly escaping, the movable out-of-plane cantilevers bend correspondingly. A white light interferometer Polytec MSA 500 measures the cantilevers topography under flow through a transparent window and a 10x Mirau objective placed at the top of the nozzle. Beams tips are in average 140 μm-height from the substrate surface when no pressure is applied, and almost totally flat for 2 bars applied pressure.

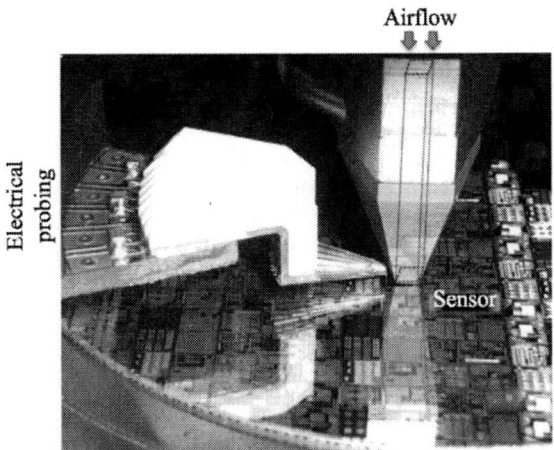

Fig. 6. Measurement set-up for flow stimulus and electrical probing.

Figure 7 describes the frequency variation as a function of the applied pressure inside a Suss*-made pressure nozzle, showing 10% variation between no beam deflection (i.e. no flow or pressure) and maximum beam deflection (movable beams being in the substrate plane). For a range of flow velocities from 0 to 120 m/s (i.e. from atmospheric pressure to applied 2 bars), there was a measurable change in oscillating frequency over a range of 270 kHz to 240 kHz.

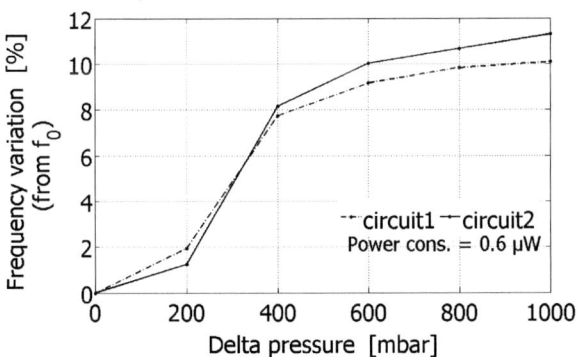

Fig. 7. Frequency variation for different applied pressures, with a Suss*-made pressure nozzle, and different flow anemometers.

Ultra low power sub-microwatt consumption of 0.6 µW for 300 kHz oscillating frequency is reached with this minimalist transducer.

IV. Conclusions

3-D SOI sensors were built and characterized under various airflows illustrating the 3-D stressed-cantilever concept. Such micro-integrated sensors can easily be built incorporating their associated electronics on the same chip. Such minimalist flow transducers offer a list of advantages such as CMOS compatibility, SOI process, simple to build 3-D microstructure, no direct current consumption due to capacitive detection, extremely low power consumption, of the order of 0.1 µW as well as occupy small chip area (1.25 mm^2 including CMOS circuits and MEMS). The presented microsystem can then be seen as a technology for the fabrication of highly integrated low-power MOS circuits, built with co-integrated MEMS sensors.

Acknowledgments

The authors are very grateful to P. Simon and D. Van Vynckt for their support in characterization tests and C. Renaux for the CMOS process.

References

1. D. Flandre, J.-P. Raskin and D. Vanhoenacker, in *Selected Topics in Electronics and Systems*, M.J. Dean and T.A Fjedldly, Editors, pp. 273-362, World Scientific Publishing Co (2002).
2. S. -H. Tseng, Y. -J. Hung, Y. -Z. Juang and M. Lu, *Sensors and Actuators A*, **139**, 187-193 (2007).
3. J. L. Lopez, J. Verd, A. Uranga, G. Murillo, J. Giner, E. Maringo et al., *Proceedings of the Eurosensors XXIII conference, Procedia Chemistry 1*, 1131-1134 (2009).
4. J. -W. Choi, R. Wicker, S. -H. Lee, K. -H. Choi, C. -S. Ha and I. Chung, *J. of Materials Processing Technology*, **209**, 5494-5503 (2009).
5. P. Argyrakis and R. Cheung, *Microelectronic Engineering*, **86**, 2176-2179 (2009).
6. S. A. Wilson, R. Jourdain, Q. Zhang, R. Dorey, C. Bowen, M. Willander et al., *Materials Science and Engineering R*, **56**, 1-129 (2007).
7. N. Miki, X. Zhang, R. Khanna, A. A. Ayon, D. Ward and S. M. Spearing, *Sensors and Actuators A*, **103**, 194-201 (2003).
8. E. Kolesar, M. Ruff, W. Odom, J. Howard, S. Ko, P. Allen et al., *Thin Solid Films*, **398-399**, 566-571 (2001).
9. F. Iker, N. André, T. Pardoen and J. –P. Raskin, *J. of Microelectromechanical Systems*, **15**, no. 6, 1687-1697 (2006).
10. I. Kao, A. Kumar and J. Binder, *IEEE Sensors Journal*, **7**, no. 5, 713-722 (2007).

Performance of Ultra-Low-Power SOI CMOS Diodes Operating at Low Temperatures

M. de Souza[a], B. Rue[b], D. Flandre[b], and M. A. Pavanello[a]

[a] Department of Electrical Engineering, Centro Universitário da FEI,
São Bernardo do Campo, Brazil
[*]e-mail: michelly@fei.edu.br
[b] Microelectronics Laboratory, ICTEAM Institute, UC Louvain,
Louvain-la-Neuve, Belgium

In this work the low temperature performance of ultra-low-power SOI CMOS diodes is presented. Experimental measurements performed in fabricated devices from 148K to 373K show that the temperature lowering can promote a significant leakage current reduction and increase of the forward current. Two-dimensional numerical simulations are used to extend the studied temperature range and analyze the doping concentration influence on the low temperature operation of these diodes.

I. Introduction

A common-way to implement diodes in CMOS technology is to connect the drain and gate of transistors together, as shown in Fig. 1A. However, these diodes present, as major drawbacks, the high leakage current (I_L) and high threshold voltage (V_{TH}) values. The leakage current of these diodes is given by MOS transistor current at zero gate-to-source voltage, which increases both as the device is further scaled and reversely biased, as well as when the diode voltage is lowered by using transistors with low threshold voltage (1).

Aiming to reduce the MOS diode reverse leakage current and threshold voltage at the same time, the so-called ultra-low-power (ULP) diode has been proposed (2). It consists of an arrangement of nMOS and pMOS transistors as shown in Fig. 1B. Some works reported the high-temperature operation of SOI CMOS ULP diodes (1, 3) showing an important reduction in the leakage current, differently to that observed for standard MOS diodes. However, despite the advantages provided by the operation of MOS devices at low temperatures (such as increased carrier mobility, reduced subthreshold slope, and leakage current (4)), no study on the operation of ULP diodes at low temperatures has been presented so far.

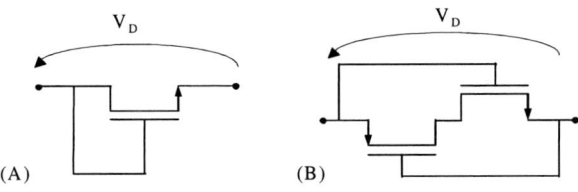

Fig. 1. Standard (A) and ULP (B) CMOS diodes.

This work presents the first evaluation of ultra-low-power diodes performance at low temperatures. Experimental results are shown for devices operating from 373K down to 198K. Two-dimensional numerical simulations were also performed in order to extend the studied temperature (T) range down to 100 K and to evaluate the impact of doping concentration (N_A) in the low temperature performance of ultra-low-power diodes.

II. Experimental Results

The studied ULP diodes have been fabricated according to the fully-depleted SOI process described in (5), which features silicon film thickness (t_{Si}) of 80nm, and gate (t_{oxf}) and buried oxide (t_{oxb}) thickness of 31 nm and 390 nm, respectively. In all cases, transistors present channel length L=2μm and channel width W=100μm. Two ULP diodes, featuring different doping concentrations, $N_{A,P}$, for the pMOS transistors were characterized. Diode 1 presents the standard doping concentration in the pMOS channel, leading to threshold voltage, $V_{TH,P}$ of -0.4V at room temperature (T=300K) and Diode 2, with higher doping level, resulting in $V_{TH,P}$=+0.5V at room temperature. In both cases, nMOS transistors present intrinsic doping concentration (actually a lightly doped p-type region with $N_A=10^{15}$ cm^{-3}) in the channel region, leading to threshold voltage ($V_{TH,N}$) of -0.3V at room temperature.

Fig. 2 plots the absolute diodes current (I_D) as a function of the applied voltage (V_D) for the measured devices, at different temperatures, ranging between 198K and 373K for both measured devices. From these curves one can see that the increase of N_A in p-type transistor leads to higher values of forward and reverse ULP diode currents at any temperature. In addition, the temperature lowering reduces the reverse leakage current for both ULP diodes, independent of the pMOS transistor doping level. This reduction is associated to the nMOS $V_{TH,N}$ increase with temperature roll-off and is more pronounced in the diode featuring lower $V_{TH,P}$.

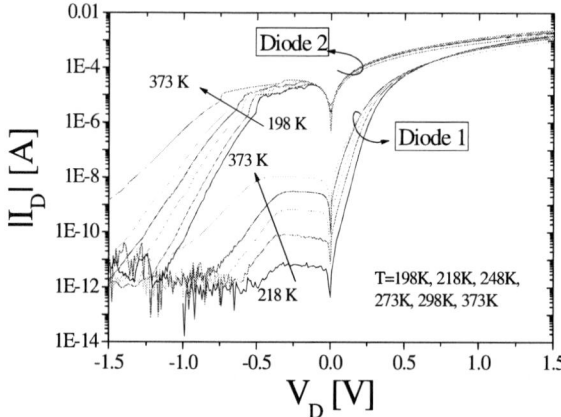

Fig. 2. Experimental I_D versus V_D of two different ULP SOI CMOS diodes for temperatures ranging between 198K and 373K.

III. Two-Dimensional Numerical Simulation Analysis

In order to extend the studied temperature range and further investigate the performance of ULP diodes at low temperatures, Atlas (6) numerical 2D simulations were performed. Ultra-low-power diodes were simulated considering transistors with similar technological parameters as from the experimental samples, namely t_{Si}=80nm, t_{oxf}=31nm and t_{oxb}=390 nm. The channel length was set to 2 μm and the channel width to 1 μm. For all simulated nMOS the channel was kept at the natural wafer doping concentration level, N_A=10^{15} cm^{-3}, and the doping level of pMOS ($N_{A,P}$) has been varied, according to the data reported in (7) leading to different $V_{TH,P}$, as shown in Table I.

TABLE I. P-channel transistors doping concentration and resulting threshold voltage ($V_{TH,P}$) at room temperature.

Doping concentration, $N_{A,P}$ [cm^{-3}]	p-MOS threshold voltage, $V_{TH,P}$ [V]
4.0×10^{16}	-0.42
5.6×10^{16}	-0.10
9.6×10^{16}	0.55

Physical models accounting for mobility dependence on temperature, velocity saturation, and doping concentration, bandgap narrowing, Auger and SRH recombination, doping-dependent lifetime and impact ionization were included in the simulation files. In addition, no optimization of model parameters has been made, which is beyond the scope of this analysis and may affect the quantitative results but does not affect the qualitative trends.

Curves of diode current as a function of the applied voltage were simulated for ULP diodes with different pMOS transistors doping concentration, $N_{A,P}$=4.0×10^{16} cm^{-3}, 5.6×10^{16} cm^{-3}, and 9.6×10^{16} cm^{-3}, and the obtained results are presented in Figure 2A, B and C, respectively. As in the experimental data, the increase of pMOS transistor threshold voltage diminishes the reverse leakage current. In addition, as lower the $V_{TH,P}$ value, larger is the decrease of I_L with temperature reduction.

This reduction is clearly seen in Figure 4, which presents the absolute diode current as a function of the temperature, extracted from the |I_D/W| *versus* V_D curves presented in Figure 2, at different V_D values (-0.5V, -0.25V, -0.1V, and 1V). Contrarily to the behavior presented for standard CMOS diodes (3), the reverse current of ULP diodes decreases as the device is further reversely biased and this decrease becomes even more pronounced as the temperature is lowered. In addition, the forward current increases with temperature reduction as a result of the increased carrier mobility (4). This way, the ratio between the forward and reverse current of ULP diodes is increased both by reducing the temperature and p-channel transistor doping concentration, as can be seen in the right y axis of Figure 4, in which the ratio between the diode current at V_D=1V and V_D=-0.25V is presented.

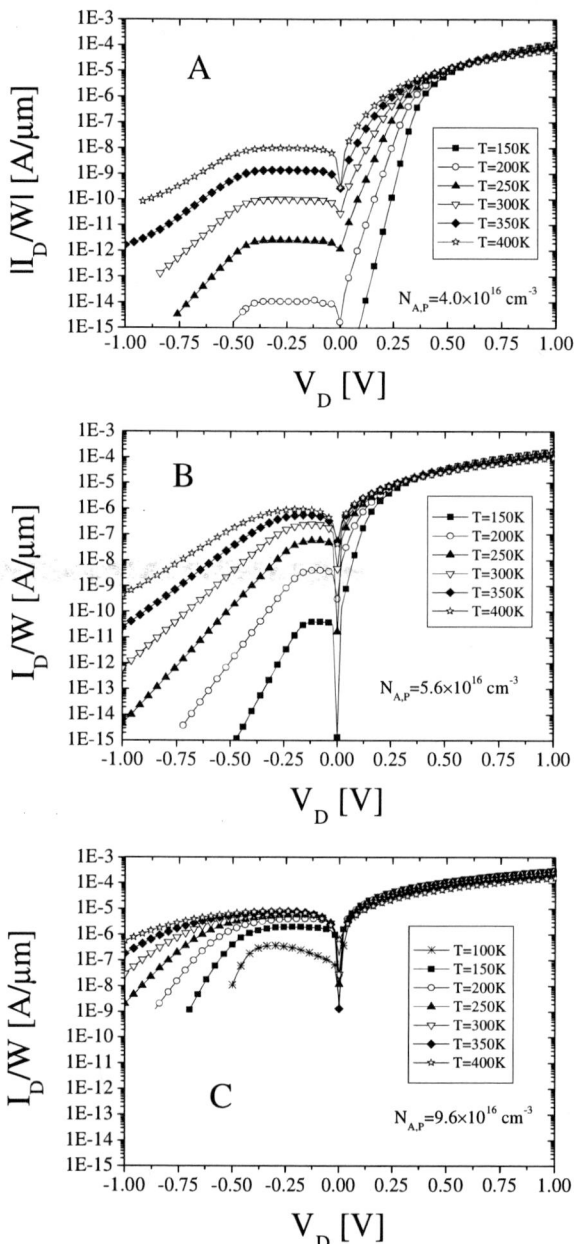

Fig. 3. Simulated $|I_D/W|$ versus V_D of ULP SOI CMOS diodes with different doping concentrations for the pMOS transistor, obtained at temperatures ranging between 100K and 400K.

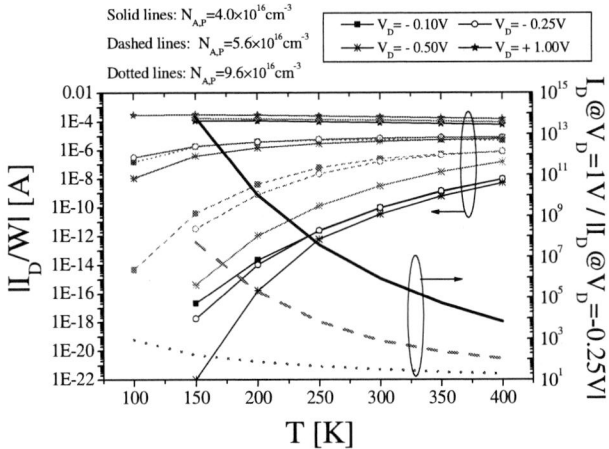

Fig. 4. Simulated I_D/W *versus* T of different ULP SOI diodes at several values of V_D and relation between $|I_D|$ at $V_D=1V$ and -0.25V.

As reported in (2), these ULP diodes may present a region with negative impedance. Some circuits such as memory cells (2) and current mode adders (7) have been presented to take advantage of this region. This region is more noticeable when $V_{TH,N}<0$ and $V_{TH,P}>0$ as can be seen in Figure 5A that presents I_D/W *versus* V_D in linear scale for the simulated diodes at room temperature. Figure 5B presents the same curves for the diode with higher $N_{A,P}$, in which the negative impedance region is larger. According to the curves presented in Figure 5B, the temperature reduction diminishes $|I_D/W|$ and the negative impedance region tends to disappear due to the increase of $V_{TH,N}$ promoted by the temperature roll-off.

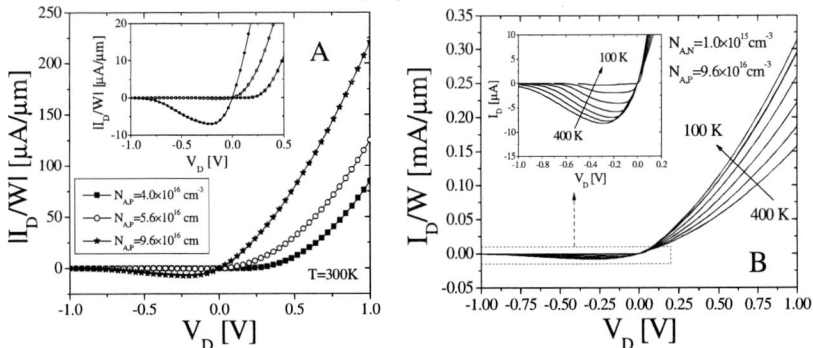

Fig. 5. Simulated I_D *versus* V_D of different ULP SOI diodes at room temperature (A) and at different temperatures (B).

IV. Conclusions

In this work the operation of ultra-low-power SOI CMOS diodes at low temperature has been presented. The analysis has been performed through experimental measurements and two-dimensional numerical simulations. The obtained results showed that the temperature lowering further reduces the reverse leakage currents of ULP SOI CMOS diodes and increases the forward current, improving the ratio between forward and reverse current. The negative impedance region observed in ULP diodes tends to disappear with temperature roll-off, as a result of the increase of nMOS threshold voltage.

Acknowledgments

The authors acknowledge to the Brazilian research-funding agencies CAPES and CNPq for the financial support.

References

1. D. Levacq, C. Liber, V. Dessard, D. Flandre, *Solid-State Electronics*, **48**, 1017 (2004).
2. D. Levacq, C. Liber, V. Dessard, D. Flandre, in *Proceedings of IEEE International SOI Conference*, 19 (2003).
3. B. Rue. D. Levacq, D. Flandre, in *Proceedings of IEEE International SOI Conference*, 65 (2006).
4. E. A. Gutierrez, J. Deen and C. Claeys. *Low Temperature Electronics: Physics, Devices, Circuits and Applications*. Academic Pres (2001).
5. D. Flandre, S. Adriaensen, A. Akheyar, A. Crahay, L. Demeûs, P. Delatte, V. Dessard, B. Iniguez, A. Nève, B. Katschmarskyj, P. Loumaye, J. Laconte, I. Martinez, G. Picun, E. Rauly, C. Renaux, D. Spôte, M. Zitout, M. Dehan, B. Parvais, P. Simon, D. Vanhoenacker, J.-P. Raskin. *Solid-State Electronics*, **45**, 451 (2001).
6. Atlas User's Manual, Silvaco (2010).
7. G. Gosset, B. Rue, D. Flandre, in *2008 IEE International Conference on RFID*, p. 134 – 140.
8. I. Hassoune, A. Drummond, A. Gaudissart, D. Bol, D. Levacq, D. Flandre, J. D. Legat, *Solid-State Electronics*, **49**, 1185 (2005).

Author Index

Afzalian, A.	295	Emam, M.	129
Agopian, P. G.	145	Erfurth, W.	43
Aksamija, Z.	267		
Alati, D. M.	163	Fabian, J.	3
Andre, N.	319	Faynot, O.	247
Aulnette, C.	29, 239	Fenouillet-Beranger, C.	247
		Ferain, I.	63, 73, 283
Bae, Y.	247	Figuet, C.	29
Bawedin, M.	79, 103	Flandre, D.	295, 319, 325
Bhaskar, U. K.	221		
Bonafos, C.	157	Francis, L. A.	319
Bonnin, O.	239	Fukuchi, K.	85
Bourdelle, K. K.	29	Fukuda, H.	227
Cauchy, X.	239	Galeti, M.	253
Celle, C.	313	Gamiz, F.	195
Chang, S.	79, 103	Gimenez, S. P.	163, 259
Checka, N.	179	Girard, C.	239
Cirne, K. H.	259	Groeseneken, G.	15
Claeys, C.	15, 145, 151, 189, 253, 289	Guiot, E.	29
Clavaguera, S.	313	Hähnel, A.	43
Colinge, C.	63	Hayashi, O.	85
Colinge, J.	63, 73, 283	Hayazawa, N.	43
Collaert, N.	151, 253	Heyns, M.	15
Cristoloveanu, S.	79, 93, 103, 157, 195, 247	Hirose, K.	201
		Hong, S.	303
		Houri, S.	221
Damlencourt, J.	157		
Daval, N.	29, 239	Ikeda, H.	201
De Gendt, S.	15	Ino, D.	85
De Lima, J. A.	259	Ionica, I.	157
de Souza, M.	283, 325	Ishida, M.	213
Dehdashti Akhavan, N.	63, 73, 283	Ishikawa, Y.	227
Doria, R. T.	189, 283	Itabashi, S.	227
Drazek, C.	29		
Dubois, E.	313	Jungemann, C.	303
El Hajj Diab, A.	157	Kato, R.	51

Kawabata, N.	39, 51	Petzold, M.	43
Kawata, S.	43	Pham, A.	303
Keast, C. L.	179		
Kedzierski, J.	179	Raskin, J.	129, 169, 221,
Knezevic, I.	267		313, 319
Kobayashi, D.	201	Razavi, P.	63, 73, 283
Kranti, A.	73, 283	Reiche, M.	43
Kranti, A.	63	Roda Neve, C.	169
Kurosawa, M.	39, 51	Rodrigues, M.	151, 253
		Rodriguez, N.	195
Landru, D.	29	Rooyackers, R.	15
Lee, C.	63, 73, 283	Rue, B.	319, 325
Lee, J.	79, 103		
Leonelli, D.	15	Sadoh, T.	39, 51, 55
Letertre, F.	29	Saito, H.	201
		Saracco, E.	157
Makarov, A.	277	Sawada, K.	213
Maleville, C.	239	Scheen, G.	319
Martinez, F.	103	Schwarzenbach, W.	239
Martino, J. A.	145, 151,	Seixas Jr., L. E.	259
	189, 253,	Selberherr, S.	117, 277
	289	Senawiratne, J.	123
Mazure, C.	29	Shinojima, H.	227
Meinerzhagen, B.	303	Silveira, M.	259
Miyao, H.	213	Simoen, E.	145, 151, 189,
Miyao, M.	39, 51, 55		253, 289
Moutanabbir, O.	43	Simonato, J.	313
		Sokolov, L. V.	135
Naumann, F.	43	Sonnenberg, V.	151
Nazarov, A.	73	Stanojevic, Z.	117, 277
Nguyen, B.	29, 239	Sverdlov, V.	117, 277
Nishi, H.	227		
		Takahashi, K.	213
Ohata, A.	247	Takahashi, R.	227
Ohta, Y.	55	Tarun, A.	43
Omura, Y.	85	Toko, K.	55
Osintsev, D.	277	Trevisoli, R. D.	283, 289
		Tsuchizawa, T.	227
Pardoen, T.	221		
Park, J.	39	Usenko, A.	111, 123
Passi, V.	221, 313		
Pavanello, M. A.	189, 283, 289, 325	Vahoenacker-Janvier, D.	129
Perreau, P.	247	Valenza, M.	103

Vandooren, A.	15
Verhulst, A.	15
Vitale, S. A.	179
Wada, K.	227
Watanabe, T.	227
Weinbub, J.	277
Wyatt, P. W.	179
Xiong, W.	79
Yamada, K.	227
Yan, R.	63, 73, 283
Yokoyama, H.	55
Yu, R.	63, 73, 283